GAṆITASĀRAKAUMUDĪ

THE MOONLIGHT OF THE ESSENCE OF MATHEMATICS

by ṬHAKKURA PHERŪ

Edited with Introduction, Translation,
and Mathematical Commentary
by SaKHYa

MANOHAR
2009

SaKHYa stands for
Sreeramula Rajeswara Sarma
Takanori Kusuba
Takao Hayashi
Michio Yano

First published 2009

ISBN 978-81-70304-809-8

Published by
Ajay Kumar Jain *for*
Manohar Publishers & Distributors
4753/23 Ansari Road, Daryaganj
New Delhi 110 002

Printed at
Salasar Imaging Systems
Delhi 110 035

Gaṇitasārakaumudī

Dedicated to the Memory of

Professor David Pingree

Contents

Abbreviations vi
Preface vii

Part I Introduction

Section 1 Ṭhakkura Pherū: Life and Works xi
Section 2 Mathematics of the *Gaṇitasārakaumudī* xviii

Part II Text

Chapter 0 Introduction 3
Chapter 1 Prathamo 'dhyāyaḥ 9
Chapter 2 Dvitīyo 'dhyāyaḥ 17
Chapter 3 Tṛtīyo 'dhyāyaḥ 19
Chapter 4 Caturtho 'dhyāyaḥ 27
Chapter 5 Pañcamo 'dhyāyaḥ 35
Chapter 6 Sūtrānukramaṇikā 39

Part III English Translation

Chapter 1 Twenty-five Fundamental Operations 45
Chapter 2 Eight Classes of Reduction of Fractions 57
Chapter 3 Eight Types of Procedures 61
Chapter 4 Four Special Topics 75
Chapter 5 Quintet of Topics 85
Chapter 6 List of Rules 91

Part IV Mathematical Commentary

Chapter 1 Twenty-five Fundamental Operations 97
Chapter 2 Eight Classes of Reduction of Fractions 125
Chapter 3 Eight Types of Procedures 133
Chapter 4 Four Special Topics 163
Chapter 5 Quintet of Topics 183

Appendices

A Concordance of the *Gaṇitasārakaumudī* and Other Works 195
B The Type Problems 207
C Index to the Numbers in the Text 215
D Glossary-Index to the Text 223

Bibliography 257

Index of Mathematical Terms 267
Index of Things Mentioned in the Text 273
Index of Sanskrit/Prakrit Authors and Titles 277

Abbreviations

AB	*Āryabhaṭīya* by Āryabhaṭa I
AD	*Anuyogadvārasūtra*
AHK	*Apabhraṃśa-Hindī Kośa* by N. Kumar
AP	*Aparājitapṛcchā* by Bhuvanadeva
Apte	*The Practical Sanskrit-English Dictionary* by V. S. Āpaṭe
ASA	Abhayadeva's commentary on the SA
BAB	Bhāskara I's commentary on the *Āryabhaṭīya*
BG	*Bījagaṇita* by Bhāskara II
BKS	*Bṛhatkṣetrasamāsa* by Jinabhadra Gaṇi
BM	*Bakhshālī Manuscript*
BSS	*Brāhmasphuṭasiddhānta* by Brahmagupta
CCM	*Caturacintāmaṇi* by Giridharabhaṭṭa
DP	*Dravyaparīkṣā* by Ṭhakkura Pherū
EI	*Epigraphia Indica*
GK	*Gaṇitakaumudī* by Nārāyaṇa
GM	*Gaṇitamañjarī* by Gaṇeśa II
GP	*Gaṇitapañcaviṃśī* by Śrīdhara
GSK	*Gaṇitasārakaumudī* alias *Gaṇitasāra* by Ṭhakkura Pherū
GSS	*Gaṇitasārasaṃgraha* by Mahāvīra
GT	*Gaṇitatilaka* by Śrīpati
HGA	*Historical Grammar of Apabhraṃśa* by G. V. Tagare
Hobson-Jobson	*Hobson-Jobson: A Glossary of Colloquial Anglo-Indian Words and Phrases, etc.* by H. Yule and A. C. Burnell
JS	*Jyotiṣasāra* by Ṭhakkura Pherū
L	*Līlāvatī* by Bhāskara II
MGED	*The Modern Gujarati-English Dictionary* by B. N. Mehta and B. B. Mehta
MMW	*A Sanskrit-English Dictionary* by M. Monier-Williams
MS	*Mahāsiddhānta* by Āryabhaṭa II
MU	*Mānasollāsa* by Someśvara III
OHED	*The Oxford Hindi-English Dictionary* by R. S. McGregor
PG	*Pāṭīgaṇita* by Śrīdhara
Pischel	*A Grammar of the Prākrit Languages* by R. Pischel
PM	Patan Manuscript
PSM	*Pāiya Sadda Mahaṇṇavo* by Haragovind Das T. Sheth
PV	*Pañcaviṃśatikā*, anonymous
SA	*Sthānāṅgasūtra*
SGS	*Ratnaparīkṣādi-saptagranthasaṃgraha* by A. Nāhaṭā and B. Nāhaṭā
SGT	Siṃhatilaka Sūri's commentary on the GT
SS	*Siddhāntaśekhara* by Śrīpati
TP	*Tiloyapaṇṇatti* by Jadivasaha
Tr	*Triśatikā* by Śrīdhara
UTA	Umāsvāti's commentary on his *Tattvārthādhigamasūtra*
YJ	*Yavanajātaka* by Sphujidhvaja

Preface

Of the few scholars working on Sanskrit texts on astronomy and mathematics today, the largest group is concentrated in Japan in the ancient city of Kyoto. This group has the admirable habit of meeting once a week to study an original text. One of the valuable fruits of this joint study is the *Studies in Indian Mathematics: Series, Pi and Trigonometry* (Tokyo, 1997), which was awarded the Publication Prize by the Mathematical Society of Japan in 2005.

When I spent a semester at Kyoto University as Visiting Professor of Indian Science in 2002, I was invited to join the group, then consisting of Professors Michio Yano, Takao Hayashi and Takanori Kusuba. I had met Professor Yano for the first time in 1976 during the International Symposium on Āryabhaṭa in New Delhi; since then we have been meeting often in India and elsewhere. Professor Kusuba and I were together at Brown University in the academic year 1992–93. I came to know Professor Hayashi in 1986 through his marvelous PhD thesis on the *Bakhshālī Manuscript*.

For our weekly meetings, we chose the *Gaṇitasāra-kaumudī* of Ṭhakkura Pherū who held a high position in the treasury of ᶜAlā' al-Dīn Khaljī at Delhi. This work is not only the first mathematical work to be composed in Middle Indic, but it also extends the range of mathematics beyond the traditional framework. In those six months, we translated the text and drafted a mathematical commentary. It took, however, some time to prepare our work for publication, as all of us were busy with various academic and administrative duties in the intervening years. The work is now complete, thanks to the diligent efforts of Professor Hayashi who acted as the editor of the team, put together our individual inputs and prepared the press copy in LaTeX. It is also he who suggested the name for our group.

The group name SaKHYa emphasizes the warm bond of friendship that developed among us during the past several years through the common interest in the history of science; it has an additional significance as well, for it was generated mathematically by the permutation of the initial letters of our names. Permutation and combination have been important elements in Indian mathematics; they also played a significant role in systems of classification in other *śāstras*. Indeed, they were employed, for the very first time — in the classification of poetic metres — by Piṅgala Nāga in his *Chandaḥsūtra*, where he taught, among others, how to arrange the permutations in a tabular form called *prastāra*. The initial letters of our four names produce a set of permutations (*caturakṣara-prastāra*) consisting of 24 terms.[1] From the fourteenth term of this series is derived the group name SaKHYa.

SaKHYa is now pleased to offer to the historians of mathematics this edition, translation, mathematical commentary, and a detailed glossary-index of Ṭhakkura Pherū's *Gaṇitasāra-kaumudī*, 'The Moonlight of the Essence of Mathematics,' also known as *Gaṇitasāra*, 'The Essence of Mathematics.'

As we complete this joint venture, we recall the memory of those scholars who inspired us in this task in various ways. Sri Bhanwar Lal Nahata of Kolkata

---

[1]KYSH, YKSH / KSYH, SKYH / YSKH, SYKH / KYHS, YKHS / KHYS, HKYS / YHKS, HYKS / KSHY, SKHY / KHSY, HKSY / SHKY, HSKY /YSHK, SYHK / YHSK, HYSK / SHYK, HSYK.

discovered the writings of Ṭhakkura Pherū in a single manuscript copy and published them in collaboration with Sri Agar Chand Nahata. When I brought out Ṭhakkura Pherū's *Rayaṇaparikkhā* with an English translation and commentary in 1984, Bhanwar Lal-ji sent me a warm letter of appreciation, which I greatly cherish, and shared with me his own notes on Pherū's other writings. The link between Bhanwar Lal-ji and me was Sri Hazari Mull Banthia of Kanpur from whom I learnt much about contemporary Jainism.

We also remember fondly our mentor and friend Professor David Pingree; the four of us had the privilege of spending some time with him at Brown University; it is his monumental *Census of the Exact Sciences in Sanskrit* which first introduced us to Ṭhakkura Pherū's scientific writings. We dedicate our book to the memory of Professor David Pingree.

We are highly indebted to Professor Irfan Habib who took a keen interest in our work and very kindly translated for us relevant passages from Indo-Persian sources, especially from the *Dastūr al-Albāb fī ᶜIlm al-Ḥisāb* by Ḥājjī ᶜAbd al-Ḥamīd Muḥarrir Ghaznavī.

Our sincere thanks are due to Mr Ramesh Jain of the Manohar Publishers & Distributors for agreeing to bring out this work; he and his staff deserve all praise for the fine production of the volume. Ṭhakkura Pherū composed the *Gaṇitasāra-kaumudī* in Delhi; we are happy that our study of this work also is being published from Delhi.

Düsseldorf, December 2008 SREERAMULA RAJESWARA SARMA

Part I

Introduction

1 Ṭhakkura Pherū: Life and Works

That the history of Indian mathematics is dominated by Sanskrit texts is understandable because Sanskrit has been the pan-Indian medium of scholarly discourse throughout the centuries. Nevertheless, there exist some texts composed in Middle Indic and in the Dravidian languages, where efforts were made to reach wider strata of society. No systematic survey of such texts has been undertaken so far, nor was any individual text of this genre studied thoroughly.

One such text is the *Gaṇitasārakaumudī*, composed in the first quarter of the fourteenth century in Delhi by Ṭhakkura Pherū.[1] His writings were brought to the notice of the scholarly world by Sri Agar Chand Nahata and Sri Bhanwar Lal Nahata, who made immense contribution to the history of Jain literature and culture.[2] In 1946 they discovered a manuscript containing seven works by Pherū and published them in 1961 under the title *Ṭhakkura-Pherū-viracita-Ratnaparīkṣādi-sapta-grantha-saṃgraha*.[3] Five of these texts are dated. The earliest, *Kharataragaccha-yugapradhāna-catuḥpadikā*, a eulogy of the pontiffs of the Kharatara sect, was composed in VS 1347 (AD 1291). The *Ratnaparīkṣā* on gemmology and the *Jyotiṣasāra* on astronomy and astrology were completed in VS 1372 (AD 1315). On the Vijayadaśamī in the same year (9 September 1315), the *Vāstusāra* on architecture and iconography was completed, while the *Dravyaparīkṣā* on assay and exchange of coins was composed in VS 1375 (AD 1318). No dates are given for the remaining two: *Dhātūtpatti* on metallurgy, and *Gaṇitasārakaumudī* on mathematics.

From the introductory and concluding stanzas of some of these works, we learn that Pherū belonged to a town called Kannāṇā or Kannāṇa-pura. Known to Jain Sanskrit literature as Kanyānyana, this town survives today as Kaliyana (28°33′ N; 76°12′ E) in the Bhiwani district of Haryana state. According to contemporary Jain records, this was an important place of pilgrimage in the thirteenth and fourteenth centuries. Pherū was born in the Śrīmāla caste and was a member of the Kharatara sect of Śvetāmbara Jains. His father was Ṭhakkura Canda (Sanskrit: Candra) and his grandfather is variously referred to as Kalasa (Sanskrit: Kalaśa) or Kāliya and had the title *siṭṭhi* (Sanskrit: *śreṣṭhin*), 'merchant-banker'. Pherū had a son Hemapāla for whom he composed the *Ratnaparīkṣā*.[4]

---

[1] On his life and works, see Sreeramula Rajeswara Sarma, *Ṭhakkura Pherū's Rayaṇa-parikkhā: A Medieval Prakrit Text on Gemmology,* Aligarh, 1984, Introduction, pp. 1–20.

[2] For their contribution, see *Nāhaṭā-Bandhu Abhinandana Grantha*, Varanasi, 1976; *Śrī Bhaṃvaralāla Nāhaṭā Abhinandana Grantha*, Calcutta, 1986; Śāradā Gosvāmī, *Itihāsa Puruṣa Śrī Agaracanda Nāhaṭā: Jīvana aura Sāhitya*, Calcutta, 2000.

[3] Agaracanda and Bhaṃvaralāla Nāhaṭā (ed), *Ṭhakkura-Pherū-viracita-Ratnaparīkṣādi-sapta-grantha-saṃgraha*, Rājasthāna Purātana Granthamālā 60, Jodhpur, 1961; reprinted, 1996 (henceforth SGS).

[4] While the other known members of this family had names derived from Sanskrit, the name 'Pherū' is rather strange. According to Monier-Williams, *Sanskrit-English Dictionary*, *pherū* is a Sanskrit word meaning 'jackal'. William Crooke reports that in Punjab, Pherū is a deity of whirlwinds (*The Popular Religion and Folklore of Northern India*, reprint: Delhi, 1968, vol. 1, p. 81). But this term has been in use also as a personal name among the Hindus, Jains and Sikhs. On the certificate of good conduct issued to Warren Hastings in Varanasi towards the end of the eighteenth century, one of the signatories was Pherū Misir, obviously a Brahmin (cf. Motīcandra, *Kāśī kā Itihāsa*, Bambaī, 1962, pp. 234, 444). A Pherū-malla is mentioned in an inscription which records the consecration of an image of Pārśvanātha on

Pherū's earliest known work is dated AD 1291 and the last chronological reference to him relates to 1323, as will be shown below. It is therefore probable that Pherū was born sometime in the second half of the thirteenth century, perhaps around 1270. Nothing is known about his early life and education. In the *Kharataragaccha-yugapradhāna-catuḥpadikā*, he states that he composed this work at Kannāṇā in 1291 in the presence of Vācanācārya Rājaśekhara. It is likely that he was brought up and educated at Kannāṇā, and Rājaśekhara may have been one of his teachers. His education was wide-ranging. Besides the Jain religious texts, he studied several Sanskrit and Prakrit texts on astronomy, astrology, mathematics and architecture. His writings, moreover, reveal his practical experience in the trade of gems and perfumery articles, and in minting and money exchange. In his *Ratnaparīkṣā*, Pherū states that he was composing the work for his son Hemapāla, who therefore seems to have been embarking on the gem trade. The *Dravyaparīkṣā* was written for his son and brother to initiate them into the profession of money exchangers. Apparently the family was engaged in the trade of luxury goods, and in banking and money exchange.

The Śrīmāla Jains of this period were especially noted for their expertise in minting and banking. In the *Lekhapaddhati*,[5] a collection of model documents from early medieval Gujarat, the coins used in various public and private transactions are often described as *śrī-śrīmālīya-khara-ṭaṃkaśālā-hata-triparīkṣita*, implying that the coins were struck (*hata*) in a mint (*ṭaṃkaśālā*) belonging either to the city of Śrīmāla (modern Bhinmal, 25°0′ N; 72°15′ E, in the state of Rajasthan) or to persons belonging to the Śrīmāla clan and that these coins were tested thrice (*triparīkṣita*) for their metal content, or more precisely for the content of silver or gold. It is not clear what *khara* in this expression denotes. It is possible that it refers to the *kharatara-gaccha* of Śvetāmbara Jains. Then the expression would mean that the coins were produced at a mint maintained by Śrīmāla Jains of the Kharatara sect, to which Pherū also belonged. After minting the coins, these were tested three times to ensure that they had the correct weight and contained the correct amount of gold or silver, which determines the intrinsic value of the coin.[6] The fact that this expression occurs in as many as twelve documents[7] shows that this must have been a standard formula in the early medieval Gujarat to express the genuineness of a particular coin. When the Muslim rule was established in Delhi in the twelfth century, the Sulṭāns did not begin fresh coinage with Arabic legends. Instead, they adapted the fabric of existing Chauhan coinage and added their respective names in

---

4 November 1824 (cf. A. Führer, 'Prabhosâ Inscriptions,' *Epigraphia Indica*, vol. 2, 1894, pp. 240–44). Thirdly, the father of the second Sikh Guru, Angad (AD 1504–52) bore the name Bhāi Pherū Mal. Even today Pherū is used as a personal name, especially in Punjab.

[5] Cf. Ingo Strauch, *Die Lekhapaddhati-Lekhapañcāśikā: Briefe und Urkunden im mittelalterlichen Gujarat*, Text, Übersetzung, Kommentar, Glossar (Sanskrit-Deutsch-Englisch). Monographien zur indischen Archäologie, Kunst und Philologie, Band 16. Berlin: Dietrich Reimer, 2002.

[6] On the method of testing the purity or fineness of gold and silver coins, see Sreeramula Rajeswara Sarma, 'Varṇamālikā System of Determining the Fineness of Gold in Ancient and Medieval India' in *Aruṇa-Bhāratī: Professor A. N. Jani Felicitation Volume*, Baroda, 1983, pp. 369–89.

[7] Cf. Ingo Strauch, *Die Lekhapaddhati-Lekhapañcāśikā*, op. cit., 2.17.5 (pp. 139, 311); 2.32.1 (pp. 171, 372–73); 2.32.2 (pp. 171, 373); 2.33.1 (pp. 173, 379); 2.33.2 (pp. 174, 380); 2.35.1.1 (pp. 177, 384); 2.35.1.2 (pp. 177, 385); 2.35.2.1 (pp. 180, 389); 2.36.1 (pp. 182, 391); 2.36.2 (pp. 182, 392); 2.37 (pp. 183, 394); Z 16 (pp. 208, 440); see also pp. 70–71.

Nāgarī script. Because the minting in the Gujarat-Rajasthan-Delhi region was largely controlled by the Jains, their cooperation was sought by the Sulṭāns for conducting banking and minting operations. Owing to these commercial and monetary reasons, the Jains had good relations at the Delhi court. Even before the establishment of the Delhi Sultanate, contacts existed between the Jains and Muslims on the west coast of India, which area had a long history of maritime relations with the Persian Gulf and the Arabian Peninsula. Thus the Jains came to play the role of mediators between the Islamic and the Sanskritic traditions of learning in the early medieval period.[8] They also were the pioneers in learning the Persian language and propagating it through the medium of Sanskrit.[9]

Ṭhakkura Pherū was one such mediator. Coming from a family of merchant-bankers, Pherū found a ready appointment at the treasury of the Khaljī Sulṭāns of Delhi.[10] It is not known precisely when he entered the services of the Sulṭāns, but it must have been quite some years before 1315, because in this year he completed the *Ratnaparīkṣā*, where he states that he composed this work 'in Saṃvat 1372 during the victorious reign of ᶜAlā' al-Dīn,[11] ... after having seen with his own eyes the vast ocean-like collection of gems in ᶜAlā' al-Dīn's treasury.'[12] Having entered the service of ᶜAlā' al-Dīn Muḥammad Khaljī (r. 1296-1316) sometime before 1315, Pherū continued the service under ᶜAlā' al-Dīn's successors, Shihāb al-Dīn ᶜUmar (r. 1316) and Quṭb al-Dīn Mubārak Shāh (r. 1316–20), and possibly also under Ghiyāth al-Dīn Tughluq (r. 1321–25). In 1318 Pherū occupied a high position in the mint of Quṭb al-Dīn Mubārak Shāh. The *Dravyaparīkṣā*, which was completed in this year, is based on his experience at the Delhi mint.

Pherū is mentioned twice in a contemporary chronicle entitled *Kharataragacchālaṃkāra-yugapradhānācārya-gurvāvalī*,[13] which describes the lives and activities of the pontiffs of the Kharatara sect from the beginning of the eleventh century up to 1336. The account up to 1248 was written by Jinapāla Upādhyāya; the rest must have been periodically added by the clerks of the pontiffs concerned. This chronicle narrates that in the spring of 1318 Ṭhakkura Acala Siṃha secured a decree (*farmān*) from Sulṭān Quṭb al-Dīn Mubārak, granting permission to organize a pilgrimage of the Jain community. Along with other prominent Jains of Delhi, Pherū joined the pilgrim group, who visited, among other holy places, Pherū's native place Kanyānayana and worshipped Vardha-

[8] Cf. Sreeramula Rajeswara Sarma, 'From Yāvanī to Saṃskṛtam: Sanskrit Writings inspired by Persian Works,' *Studies in the History of Indian Thought*, Kyoto, 14 (November 2002), pp. 71–88.

[9] Cf. Sreeramula Rajeswara Sarma, 'Sanskrit Manuals for Learning Persian' in: Azarmi Dukht Safavi (ed), *Adab Shenasi*, Aligarh, 1996, pp. 1–12.

[10] The contemporary Jain chronicles mention the names of several prominent Jain citizens of various cities in Northern India. Some of these persons had the title 'Ṭhakkura'. A close examination of such names shows that they were all residents of Delhi. We are inclined to believe that the title 'Ṭhakkura' in this case refers to their association with the court of the Sulṭāns. In Pherū's family, the title 'Ṭhakkura' was enjoyed by both Pherū and his father but not by his grandfather. This seems to indicate that like Pherū his father had also been employed at the Delhi court.

[11] *Ratnaparīkṣā* 132: *teṇiha rayaṇaparikkhā vihiyā niya-taṇaya-hemapālakae/ kara-muṇi-guṇa-sasi-varise (1372) allāvadī-vijayarajjammi//* (The last quarter is metrically irregular.)

[12] Ibid, 4: *allāvadīṇa-kalikāla-cakkavaṭṭissa kosamajjhatthaṃ/ rayaṇāyaru vva rayaṇuccayaṃ ca niyadiṭṭhie daṭṭhuṃ//*

[13] Printed in: Jina Vijaya Muni (ed), *Kharataragaccha-bṛhadgurvāvalī*, Bombay, 1956 (Singhi Jain Series 42), pp. 1–88.

mānasvāmin there.[14] In 1323 Pherū joined another pilgrimage to Śatruñjaya in Gujarat. This was organized by a wealthy resident of Delhi named Rayapati, who was also of Śrīmāla caste, under a decree from Sulṭān Ghiyāth al-Dīn Tughluq.[15] It is not known whether Pherū occupied any official position at this date, but his very mention by name among the prominent Jains of Delhi suggests that he may have continued his services under Ghiyāth al-Dīn Tughluq as well.

Leaving aside the *Kharataragaccha-yugapradhāna-catuḥpadikā*, which is a small work of piety, the other six texts composed by Pherū deal with diverse scientific and technical subjects, a knowledge of which was apparently required by successful merchant-bankers of those times. It is to Pherū's credit that he composed these scientific texts, not in scholarly Sanskrit, but in popular Apabhraṃśa so that these were accessible to a wider range of people.[16] Thus Pherū became a mediator in several respects: mediator between Sanskrit and Islamic traditions of learning, mediator between the elite Sanskrit and popular Apabhraṃśa, and also mediator between *śāstra* and commerce.

Of the six scientific texts, the *Ratnaparīkṣā*,[17] in 132 *gāthās*, was composed in 1315. It is the first Apabhraṃśa work on gemmology. Apart from the study of important works by the earlier writers, Pherū also had the advantage of practical experience of handling gems in ᶜAlā' al-Dīn's treasury. Pherū generally follows the framework of his Sanskrit models, but adds interesting details from the contemporary gem trade, in particular a very detailed tariff of prices, which is valuable for the economic history of the period.[18]

The *Jyotiṣasāra*,[19] also composed in 1315, consists of 242 *gāthās*. It is noteworthy that in one table (p. 19) oblique ascensions are given for the latitudes of Delhi and Āsī (i.e. modern Hansi in Haryana). At the beginning of the work, Pherū mentions that he consulted the writings of Haribhadra, Naracandra, Padmaprabha Sūri, Jaüna (Sanskrit: Yavana), Varāhamihira, Lalla, Parāśara and Garga.

The *Vāstusāra*[20] on architecture and iconography was completed on 19 September 1315. It contains 205 *gāthās* and is divided into three chapters; these are *Gṛhalakṣaṇa-prakaraṇa* on residential architecture, *Bimbaparīkṣā-prakaraṇa* on iconography and *Prāsādavidhi-prakaraṇa* on temple architecture. V. S. Agra-

---

[14]Ibid, pp. 66–68; esp. p. 66: *ṭhakkurarāja-acala-suśrāvakeṇa pratāpākrāntabhūtalapātaśahi-śrī-kutabadīna-suratrāṇa-phuramāṇaṃ niṣkāsya ... prārambhite ... tīrthayātrotsave ...//*

[15]Ibid, pp. 72–77, esp. pp. 72, 74.

[16]But the titles of the texts are in Sanskrit, and the sub-headings and the colophons are composed in a kind of hybrid Sanskrit prose. An analysis of Pherū's language, though desirable, has to await another occation. Here it would have distracted attention from the mathematical content of the work.

[17]SGS, pt. 1, pp. 1–16; Agar Chand Nahata & Bhanwar Lal Nahata (ed & tr into Hindi), *Ratnaparīkṣā*, Calcutta, 1963–64; Sreeramula Rajeswara Sarma (ed & tr), *Ṭhakkura Pherū's Rayaṇaparikkhā: A Medieval Prakrit Text on Gemmology*, with an introduction, Sanskrit *chāyā*, translation into English and commentary, Aligarh, 1984.

[18]See Sreeramula Rajeswara Sarma (ed &tr), *Ṭhakkura Pherū's Rayaṇaparikkhā*, pp. 73–78.

[19]SGS, pt. 2, pp. 1–40; Jyotirvid Śrī Sītārāma Eḍvoketa, *Śrī-Ṭhakkura-Pherū-kṛta Prākṛta Gāthā baddha Jyotiṣasāraḥ, 'Sūrya' Hindi ṭīkā-sahita,* Varanasi, 1978.

[20]Pandit Bhagawan Das Jain, *Parama-Jaina Candrāṅgaja Ṭhakkura Pherū viracita Vāstusāra-prakaraṇa*, text with a translation into Gujarati, Jaina Vividha Granthamālā, Puṣpa 4, Atmananda Sabha Bhavan, Jaipur, 1939; SGS, pt. 2, pp. 75–103; R. P. Kulkarni (tr), *Vastusara Prakaranam: A Prakrit Treatise of Vastushastra by Thakkura Pheru*, Sanskrit-Sanskriti-Samshodhika, Pune, 1987.

wala believes that this work 'must have served as a practical handbook for architects of Jain temples in the early Sultanate period.'[21]

While these three works were generally written within the traditional framework of the *śāstra* concerned, the other three contain much original material.

The *Dravyaparīkṣā*,[22] consisting of 149 *gāthās*, was written in 1318 in the reign of Quṭb al-Dīn Mubārak. Pherū states that he wrote this work on the basis of his direct experience of various types of coins while he was employed in the Delhi mint.[23] The term *dravya-parīkṣā* means the examination of the metal content in the coins. Since there was no official rate of exchange at that time for different currencies, the official or private money exchangers priced a coin on the basis of its metal content. For this purpose the coin had to be assayed either by rubbing it on the touchstone or by melting some samples. Pherū terms this type of money exchange *nāṇavaṭṭa* (Sanskrit: *nāṇaka-vartana*).

The *Dravyaparīkṣā* can be divided into two parts. The first part (*vv.* 1–50) deals mainly with the techniques of refining gold and silver and of determining their fineness,[24] and thus provides the necessary technical background for currency exchange. The second part (*vv.* 51–149) can be termed a coin catalogue and is most valuable for the monetary history of the period. Here are described name (*nāma*), provenance (*ṭhāma*), weight (*tullu*), metal content (*davvo*), and exchange value in terms of the Khaljī currency (*mullu*), of some 260 types of coins issued by various kingdoms of north India in the thirteenth and early fourteenth centuries. This data is given in verses and also in convenient tables. The metal content of each coin type is expressed as follows. In the case of gold and silver coins, the degree of fineness is given in a scale of 12 for gold and in a scale of 20 for silver. For coins made of alloy, the weight of each metal per 100 specimens is listed.

The most interesting and comprehensive list is naturally of the coinage issued by Quṭb al-Dīn Mubārak. This Sulṭān abandoned the prevailing moulds of Chauhan coinage and introduced a completely new fabric in which he issued, as Pherū reports, as many as sixty-three coin types during the brief span of his reign from 1316 to 1318. Apart from the number, the quality of coinage is said to be far superior to that of his predecessors. Nelson Wright observes: 'The coinage of Qutbuddin Mubarak stands out for its boldness of design and variety of its inscriptions. ... There is perhaps no finer coin in the whole

---

[21] V. S. Agrawala, 'A Note on Medieval Temple Architecture,' *The Journal of the United Provinces Historical Society*, XVI.1 (July 1943), p. 112.

[22] V. S. Agrawala published the text under the title 'Ṭhakkura Pherū viracitā Prākṛtabhāṣābaddhā Dravyaparīkṣā' in *Indian Numismatic Chronicle*, vol. IV, pt. 1 (1964–65), pp. 75–94, and an English translation of *gāthās* 51–149 under the title 'A Unique Treatise on Medieval Indian Coins' in *Ghulam Yazdani Commemoration Volume*, Hyderabad, 1966, pp. 81–101; the same was reprinted as 'Dravyaparīkṣā of Ṭhakkura Pherū in *Indian Numismatic Chronicle*, vol. VII (1969), pp. 100–14. The full text was published in SGS, pt. 1, pp. 17–44. Again Bhanwar Lal Nahata published the text with his own Hindi translation: *Ṭhakkura Pherū viracitā Dravyaparīkṣā aur Dhātūtpatti*, Vaishali, 1976. See also John S. Deyell, *Living Without Silver: The Monetary History of Early Medieval India*, Oxford University Press, New Delhi, 1990, where additional bibliography is given. However, an annotated translation in English of the whole text is a desideratum.

[23] *Dravyaparīkṣā* 2: *je nāṇā muddāiṃ siriḍhilliya taṃkasāla kajjaṭhie/ aṇubhūya karivi ...//*

[24] Cf. Sreeramula Rajeswara Sarma, '*Varṇamālikā* System of Determining the Fineness of Gold in Ancient and Medieval India,' op. cit.

pre-Mughal series than the broad square gold tankah of high relief struck at Qutbabad Fort.'[25] In this radical process of reforms, Pherū must have played a significant role. Important for posterity is the fact that he left an excellent guide to Quṭb al-Dīn Mubārak's coinage in his *Dravyaparīkṣā*. The uniqueness of this text cannot be overemphasized; there has been no such text before or after in India.

Pherū's undated work *Dhātūtpatti* (literally, 'origin of minerals'), consisting of 57 *gāthās*, contains no invocation at the beginning, nor a concluding verse at the end. Perhaps it is part of a larger work which is no longer available. In the form it has come down to us, the *Dhātūtpatti* contains three unconnected sections: (1) mythical origin of minerals; (2) techniques of preparing or extracting brass, copper, lead, tin, bronze, mercury, vermillion and red lead; (3) properties and provenance of perfumery articles like camphor, aloe wood, sandalwood, musk and saffron. The middle section on the extraction of metals, though brief, is valuable for our understanding of medieval metal technology.[26] Likewise, the last section throws interesting light on the trade of perfumery articles.

The *Gaṇitasārakaumudī*, also known as *Gaṇitasāra*,[27] is not dated. This treatise subdivides the silver *ṭaṃka* into 50 *drammas* (1.4a). But according to the *Dravyaparīkṣā* (*vv.* 134–36; 144–46 and the corresponding tables), the silver *ṭaṃka* issued by ᶜAlā' al-Dīn Muḥammad was equal to 60 *drammas* and this rate was continued under Quṭb al-Dīn Mubārak as well. Therefore, the *Gaṇitasārakaumudī* must have been written before 1318 and possibly during the earlier part of ᶜAlā' al-Dīn's rule.

Like Pherū's other works, the *Gaṇitasārakaumudī* is also composed in Apabhraṃśa and, to a large extent, in *Gāhā* (or *Gāthā*) metre.[28] Besides *Gāhā*, Pherū also employs occasionally *Aḍillā*, *Uggāhā*, *Chappaya*, *Dohā*, *Raḍḍā*, *Rolā*, and two unidentified metres.[29]

The *Gaṇitasārakaumudī* is not only the first full-fledged mathematical text composed in Apabhraṃśa, but it also extends the range of mathematics beyond

---

[25] H. Nelson Wright, *The Coinage and Metrology of the Sultanṣ of Delhi*, reprinted, New Delhi, 1974, pp. 107–08.

[26] Cf. R. K. Dube, 'Copper Production Process as described in an early Fourteenth Century Prakrit Text composed by Ṭhakkura Pherū,' *Indian Journal of History of Science*, 41 (2006) 297–312; idem, 'The Extraction of Lead from its Ores by the Iron-Reduction Process: A Historical Perspective,' *The Journal of the Minerals, Metals & Materials Society*, 58.10 (October 2006) 18–23.

[27] It is published under this title in SGS, pt. 2, pp. 41–74.

[28] *Gāhā* is the Prakrit counterpart of the Sanskrit *Āryā*. Like the *Āryā*, the *Gāhā* contains 12, 18, 12 and 15 morae (*mātrās*) respectively in its four feet. According to the *Prākṛta-Paiṅgala* (1.54–62), the two halves of the verse should contain 7 *mātrā-gaṇas* each, followed by a final *guru*. In the first half, the sixth *gaṇa* should be either *ja-gaṇa* (*laghu*, *guru*, *laghu*) or a *na-gaṇa* followed by a *laghu* (i.e. 4 *laghus*). In the second half, the sixth *gaṇa* should consist of just one *laghu*. Furthermore, the odd-numbered (i.e. the first, third, fifth and seventh) *gaṇas* should never be *ja-gaṇas*. The *Prākṛta-Paiṅgala* (1.62) also lays down how to recite a verse in *Gāhā* metre:

> *paḍhamaṃ vī haṃsapaāṃ bīe sīhassa vikkamaṃ jāā/*
> *tīe gaāvaraluliam ahivaraluliaṃ caütthae gāhā//*
> 'The first foot ⟨moves⟩ like the ⟨slow⟩ gait of the swan, the second like the ⟨awesome⟩ stride of the lion, the third ⟨like⟩ the swinging walk of the majestic elephant and the fourth ⟨like⟩ the undulating crawl of a large serpent.'

[29] In our edited text the verses composed in metres other than the *Gāhā* are marked #, and the name of the metre is given in footnotes.

the traditional framework of the earlier Sanskrit texts, and includes diverse topics from the daily life where numbers play a role. Pherū states at the outset (1.2) that he borrowed some material from the past teachers (*puvv-āyaria*), some from his own direct experience and some from his contemporaries. The past teachers or the previous writers on mathematics who influenced Pherū are mainly Śrīdhara, the author of the *Pāṭīgaṇita* and the *Triśatikā*, and Mahāvīra who composed the *Gaṇitasārasaṃgraha*. Of these two, the influence of the former's *Pāṭīgaṇita* and *Triśatikā* is more striking.[30] Indeed some of Pherū's Apabhraṃśa verses look like phonetic adaptation from Śrīdhara's Sanskrit.[31]

The first three chapters of the GSK are well structured like the mathematical texts produced by his predecessors. What Pherū had learnt from his own experience and from that of his contemporaries is included as supplementary material in the fourth and fifth chapters (and also at the end of the third chapter). Pherū may have gathered this material from diverse sources of floating or oral literature, or from contemporary Indo-Persian sources. Because of this reason, there are occasional repetitions, and a certain looseness of structure in the fourth and fifth chapters. Pherū's aim is not to merely compose just one more neutral text on mathematics, but to produce a practical manual which is useful for all numerate professionals like bankers, traders, accountants and masons.[32] The value of the *Gaṇitasārakaumudī* lies, to a large extent, in this supplementary material, which offers us a glimpse into the life of the Delhi-Haryana-Rajasthan region in the early fourteenth century as no other mathematical work does.

The supplementary material includes mechanical shortcuts in commercial arithmetic (4.6–12; 5.18–21), mathematical riddles (notably 4.3, 47, 49, 50, 51, 63), rules for converting dates from the Vikrama era to the Hijrī era and vice versa (4.17) and classification and construction of magic squares (4.38–45).

The innovations in architecture which were being introduced by the Muslim rulers in this period find an echo in the section of solid geometry (3.69-86) where Pherū lays down rules for calculating the volumes of domes (*gommaṭa,* from Persian *gumbad*), square and circular towers with spiral stairways in the middle (*pāyaseva*), minarets with fluted columns (*munāraya,* from Persian *mīnār*), arches (*tāka,* from Persian *tāq*), bridges erected on supporting arches (*pulabaṃdha,* from Persian *pul* and Sanskrit *bandha*), and so on. The mathematical relevance of these rules lies in the fact that the chief mason or the merchant supplying the building material will be able to calculate the number of bricks or stones needed for these constructions, as is evident also from the heading of the section, viz. 'computation of bricks' (*iṭṭānāṃ gaṇanā*). These references to the arch and dome are also historically significant, because the true arch and the true dome were employed successfully for the first time in the ᶜAlā'ī

[30] See Appendix A where the parallels between the GSK and other mathematical works are tabulated.

[31] Compare, for example, *Gaṇitasārakaumudī* 1.31:

> *kheva-samaṃ khaṃ joe rāsi avigayaṃ kha-joḍaṇe hīṇe/*
> *sunna-guṇan-āi sunnaṃ sunna-guṇe sunna sunneṇa//*

with *Pāṭīgaṇita* 21:

> *kṣepasamaṃ khaṃ yoge rāśir avikṛtaḥ khayojanāpagame/*
> *khasya guṇanādike khaṃ saṅguṇane khena ca kham eva//*

[32] He pursues a similar aim in his other works also.

Darwāza, the gateway erected in 1311 by ᶜAlā' al-Dīn Khaljī as part of his extension plans to the Quwwāt al-Islām mosque, which contains the famous Qutub Minar. Equally significant is Pherū's definition of the minaret (*munāraya*): 'The minarets are like the circular spiral stairways with regard to everything in the middle. But this is the difference: the walls are like half triangle and half circle' (3.80). The meaning of the last cryptic sentence is this: in the horizontal cross-section of the minaret, the outer circumference consists of alternate triangles and semicircles. It should be borne in mind that about a hundred years before this time, Quṭb al-Dīn Aybak had built the famous Qutub Minar, and Pherū's employer ᶜAlā' al-Dīn himself began to build—but did not complete—another minaret twice as high. Now, the walls of the lower storey of the Qutub Minar consist of alternately angular and circular columns, the second storey of circular columns and the third storey of angular columns.

The fifth chapter contains an interesting section (5.1–13) which enumerates the average yield per *bīghā* of several kinds of grains and pulses, the proportions of different products derived from sugarcane juice, and the amount of ghee that can be obtained from milk. This valuable data has naturally attracted the attention of economic historians.[33]

Thus Ṭhakkura Pherū's *Gaṇitasārakaumudī* throws valuable light on the development and popularization of mathematics in northern India in the early fourteenth century and also on the economic conditions of that period. However, it is not always easy to fully comprehend Pherū's language. The frequent elision of consonants, avoidance of case endings for metrical or other reasons, employment of many vernacular terms which are not attested elsewhere, phonetic variations of Persian terms, and similar features give rise to great ambiguity. Where parallel procedures or problems occur in other Sanskrit mathematical texts, their help was taken in interpreting Pherū's terse language. But the supplementary material included in this work is unique in the sense that such material occurs for the first time in a mathematical work and consequently no parallels exist in Sanskrit mathematical texts. Here Pherū's style of presentation causes difficulties in understanding the procedures involved. This is particularly so in the case of the section on the construction of various kinds of tents (*khīma*, from Arabic *khaima*, 4.18–37) and the section entitled 'the yield of regional tax' (*deśakaraphala*, 5.14–17). Our translation of these sections cannot but be tentative.

2 Mathematics of the *Gaṇitasārakaumudī*

2.1 Structure of the Work

Our revised text of the *Gaṇitasārakaumudī* [GSK] given in Part II is based on the Nahatas' edition, which in turn is based on a single manuscript discovered by them in 1946 (see §0.7 in Part II). The text in the manuscript consists of five parts, whose titles given in the colophon of each part are as follows.

[33]See the Mathematical Commentary (Part IV) on this section, pp. 183–88.

1. ... *pamcavimśatiparikarmmasūtra .../ prathamo 'dhyāyaḥ* ('chapter one: rules for twenty-five fundamental operations').

2. ... *aṣṭau bhāgajātayaḥ .../ dvitīyo 'dhyāyaḥ* ('chapter two: eight classes of ⟨reduction of⟩ fractions').

3. ... *aṣṭau vyavahārāṇi .../ tṛtīyo 'dhyāyaḥ* ('chapter three: eight types of procedures').

4. ... *deśādhikārādyāḥ catvāri adhikārāṇi* ('four topics beginning with the topic on region').

5. ... *uddesapaṃcagam* ('quintet of topics').

At the end of the first section (on weights and measures) of Chapter 1 (GSK 1.15), Pherū says that there are 45 'gateways' (*dāra*) to mathematics, namely, 25 fundamental operations (*parikamma*), 8 classes (*jāti*) of reduction of fractions, 8 kinds of procedures (*vivahāra*), and 4 special topics (*ahigāra*). These 45 'gateways,' in fact, comprise the first four parts of the present *Gaṇitasāra-kaumudī* as shown above. There is, therefore, no doubt that those four parts were genuine components of the original *Gaṇitasāra-kaumudī* written by Pherū.

As it is not included in the '45 gateways' to mathematics, the fifth and the last part was presumably added as a supplement. This part is not so well structured as the preceding parts. The last section (GSK 5.26–23) deals with masonry but this is not included in the 'five topics' mentioned in the first verse of this part (GSK 5.1). Moreover, this part repeats several verses of the foregoing parts: 5.3b = 1.4b, 5.22a = 3.45a, 5.22b ≈ 3.43b, 5.25b = 3.65b, 5.26–27 = 3.67–68. This loose structure may raise a question about the genuineness of this part. But it should be borne in mind that the last verse of this part explicitly mentions Pherū as the author. Moreover, the manuscript used for the Nahatas' edition of the seven works of Pherū was copied by a person named Purisaḍa in 1347, that is to say, within Pherū's life time and soon thereafter. This Purisaḍa copied the GSK in less than onc month (see §0.7 of Part II). Hence there would not have been enough time either for the GSK to 'grow' from the original state or for Purisaḍa to add a new chapter. The reason for the looseness of structure is perhaps to be sought for in Pherū's innovative attempt to includc in his book not only thc traditional, common topics but also the raw material he had learnt from his own experience and from that of his contemporaries (see §1 of Part I).

The Nahatas' text of the GSK is accompanied by a list of the contents (note that a similar list is appended also to the text of the *Jyotiṣasāra* in Purisaḍa's manuscript). It is written in a sort of mixed Sanskrit just like the short prose sentences in the work (such as introductory phrases to verses) and is placed immediately after the colophon of the scribe at the end of Part 5. This list seems to have been made carelessly by someone other than Pherū. This is suggested by the facts (1) that it (in 6.37) calls the eighth 'class' (*jāti*) for reduction of fractions *staṃbhoddesa* (S. *stambha-uddeśa*, 'pillar-problem') but this nomenclature is of Śrīdhara and not of Pherū who calls it *staṃbhaṃsaka-jāti* (S. *stambha-aṃśaka-jāti*, 'pillar-part class'), (2) that it makes a wrong grouping of verses at two places (6.46 and 6.78), and (3) that it (in 6.61) lists 'pentagonal figures' (*paṃcakoṇa-kṣetra*) as a topic of the procedure for plane figures (*kṣetra-vyavahāra*) but any figure of that category is not treated in the GSK.

It should be noted that only the first three parts are called 'chapter' (*adhyāya*) in the colophons cited above. Generally speaking, not all surviving colophons of Indian manuscripts may retain their original readings. In the case of the GSK, too, it is not certain whether or not those colophons correctly preserve the words of the original author himself. But whoever be the author of the colophons, he seems to have regarded these three parts as belonging to the same category in contrast to the remaining two parts.

It should also be pointed out in this connection that the list of contents mentioned above covers the first three chapters only, which points to a conclusion similar to the above: the author of the list may have made a distinction between the first three parts and the rest. On the other hand, it is stated at the head of the list that the total number of verses is 311, which is the total of all the five parts. Therefore, the original list may have included all the five parts.

Whatever the case may be, it is true that the first three parts of the GSK can be grouped together from the historical viewpoint: they treat the traditional topics of ordinary *pāṭī* works usually divided in two categories, namely, 'fundamental operations' (*parikarmāṇi*) and 'procedures' (*vyavahārāḥ*). The reductions of fractions treated in the second part of the GSK under the title of 'eight classes' are usually included in the fundamental operations either as regular members (as in most *pāṭī* works) or as supplements (as in the *Gaṇitasārasaṃgraha* [GSS] and in the *Līlāvatī* [L]). See the table in §2.5 of Part I.

In the fourth part Pherū deals with topics useful in daily life in the first two sections under the titles of 'region' (*deśa*) and 'cloth' (*vastra*), and recreational topics in the remaining two sections under the titles of 'magic squares' (*yantra*) and 'miscellaneous' (*prakīrṇaka*). Many of these topics were new to mathematical works in India. Particularly important from the historical view point is the treatment of magic squares as no mathematical works prior to the GSK are known to have discussed them although the history of Indian magic squares goes back at latest to the sixth century (see Hayashi 1987).

The fifth part is more clearly designed for practical use; it is presumably meant to be a supplement to the foregoing parts, especially to the first two sections of the fourth part.

Hereafter we use the word 'chapter' for all the five parts of the GSK.

2.2 Characteristic Features

2.2.1 Influence of the *Triśatikā* and *Pāṭīgaṇita*

The first three chapters of the GSK, which treat the topics of ordinary *pāṭī* works comprising the two major parts, 'fundamental operations' and 'procedures,' have been written under the strong influence of the *Triśatikā* [Tr] and partly of the *Pāṭīgaṇita* [PG], both by Śrīdhara (ca. 750). This is but natural since the Tr was one of the most popular textbooks on *pāṭī*. It was the best known *pāṭī* work until Bhāskara's *Līlāvatī* assumed this position in the mid-twelfth century and even then it enjoyed a relatively wide circulation as its manuscripts surviving all over India testify.

As the table in §2.5 of Part I shows, the 25 'fundamental operations' in Chapter 1, excepting the eleven-quantity operation (21st) and the purchase and sale (23rd), and the 8 'classes' of reduction of fractions in Chapter 2 correspond to the 31 'fundamental operations' of the Tr. The 8 kinds of mathematical 'procedures' in Chapter 3 naturally correspond to those of the Tr. The correspondence between the two works will be shown in detail in the Mathematical Commentary under each rule and example as well as in Appendix A. Here we point out some important cases.

The most remarkable resemblance between the GSK and the Tr (and the PG) is found in the first two 'fundamental operations,' namely, the sum (*saṃkalita*) and the difference (*vyavakalita*) of integers. Most of the *pāṭī* works listed in the table of §2.5 regard the ordinary sum and difference of integers as the first two 'fundamental operations,' but the GSK and the Tr regard the sum of a finite series of natural numbers as the 'sum' and the difference between two of them as the 'difference.' The only other known text that adopts the same style is the anonymous *Pañcaviṃśatikā* [PV], which too was written under the influence of the Tr. The *Gaṇitasārasaṃgraha* extends the series for the 'sum' and the 'difference' to arithmetical and geometrical progressions.

Pherū puts the rules for zero (GSK 1.31) in between the multiplication and the division of integers. Among major writers on mathematics, the only other person who gives the same location to them is Śrīdhara (Tr 8 = PG 21). Cf. §2.5 below.

Pherū's algorithm for the area of a circle segment (GSK 3.46) is

$$A = \sqrt{\left(\frac{a+h}{2} \cdot h\right)^2 \cdot 10 \cdot \frac{1}{9}}.$$

Exactly the same algorithm is found in the Tr in almost the same expression. Compare the following.

jīvāśaraikyadalahataśarasya vargaṃ daśāhataṃ navabhiḥ/
vibhajed avāptamūlaṃ prajāyate kārmukasya phalam// Tr 47 //

jīvāsarapiṃḍaddhaṃ saraguṇiyaṃ vagga dahaguṇaṃ kāuṃ/
navabhāe jaṃ laddhaṃ tassa pae havaï dhaṇuhaphalaṃ// GSK 3.46 //

The only other known work that prescribes the same algorithm is, again, the PV.

jyāśaraikyadalaṃ bāṇanighnaṃ tadvargadigghateḥ/
navabhaktasya mūlaṃ taddhanurbāṇaphalaṃ smṛtam// PV A24 //

For various algorithms for the area of a circle segment given in Indian texts, see Table 3.1 in Part IV.

There are a number of agreements like this in the details of the algorithms between the first three chapters of the GSK and the Tr, and also agreement in the numerical data of examples between the two works.

2.2.2 Influence of the *Gaṇitasārasaṃgraha*

Some of the rules of the GSK seem to be based on, or generalization of, the rules of Mahāvīra's *Gaṇitasārasaṃgraha* [GSS].

Pherū (GSK 3.15–25) treats in detail the 'procedures for gold' (*suvarṇa-vyavahāra*). Most of the rules given by him are repetitions of those of the Tr, but the last one is a generalization of a rule of the GSS (see under GSK 3.24 in Part IV).

For the three lengths, the chord (a), the arrow (h) and the arc (b), of a segment of a circle, the GSK uses a crude relationship, $a + h = b$, in addition to the Jain traditional one, $a^2 + 6h^2 = b^2$. The only other work that uses these two relationships is the GSS (see under GSK 3.48–49 in Part IV).

Pherū (GSK 4.61) gives a rule for obtaining a product of multiplication consisting of the same digit repeated nine times, viz.

$$12345679 \times a \times 9 = aaaaaaaaa,$$

where a is a positive integer smaller than ten. This is obtained by multiplying by a both sides of

$$12345679 \times 9 = 111111111,$$

which is given in GSS 2.10 as an example for multiplication.

The formula for the volume of a sphere given in GSK 5.25 is equivalent to one of the two formulas given in the GSS, although their algorithms differ from each other. See under GSK 3.65 in Part IV.

2.2.3 Proportionate distribution

One of powerful mathematical tools, the proportionate distribution (*prakṣepa-karaṇa*, lit. 'investment procedure') or the so-called 'partnership' is employed in six rules of the GSK, namely, in the rule for the part-part class of reduction of fractions (2.3), in the separation of the capital and the interest (3.1), in the distribution of the profit among investors (3.7), in the purchase of several kinds of commodities in proportion (3.11), in the calculation of the weights of several gold pieces melted into a single piece (3.24), and in the distribution of the profit in *jīvalaïs* among merchants (4.2).

2.2.4 Weights and measures

Pherū provides us with important information about the weights and measures of fourteenth century north-western India.

Pherū (GSK 1.3–4b) records an interesting system of currency, in which the unit *damma* (S. *dramma*) is divided five times by vigesimal sub-units beginning with *visova* (S. *viṃśopaka*, 'one-twentieth part'). The *visova* is also used for weight, length, and area.

Pherū (GSK 1.5–6) defines two kinds of *aṃgula* ('digit'), a linear measure: one (called *karaṃgula* or 'a hand digit') is equal to 6 *javas* (S. *yavas*) and the other (called *pavvaṃgula* or 'a joint digit') to 8 *javas*. 24 *karaṃgulas* make 1 *kara* and 24 *pavvaṃgulas* (also called *kaṃviyaṃgula*) make 1 *hattha* (also called *kaṃviya* and *gaja*) but *kara* and *hattha* (both mean 'a hand') are often used side by side synonymously.

Pherū (GSK 1.7) gives definitions of three kinds of *hattha*, linear, plane, and solid, according to the dimension. The three dimensions are clearly recognized already in the *Aṇuogaddāra*, which gives definitions of linear, plane, and solid *aṅgula*, but its nomenclature differs from Pherū's.

For expressing the purity of gold, Pherū (GSK 1.10b) uses the 12-*varṇa* system instead of the ordinary 16-*varṇa* system. It agrees with his weight system for gold, in which 12 *māsayas* (S. *māṣakas*) make one *tolaya* (S. *tolaka*).

2.2.5 Specific gravities

Pherū (GSK 3.67–68 = 5.26–27, GSK 5.28) provides specific gravities of 7 kinds of stones and of 6 kinds of other substances such as clay, grains, etc.

2.2.6 List of the names of decimal places

Pherū (GSK 1.12–14) gives a list of 25 terms for decimal numbers/places, but is apparently aware of longer lists. The list is based on traditional Jain lists such as those recorded in the *Aṇuogaddāra* and in the GSS but introduces the new core term, *nīla*, for the numbers/places greater than 10^{16}. The first 19 terms of Pherū's list have survived in the traditional Hindi list of 19 terms with a minor change in positions.

2.2.7 Sum and difference

As mentioned above, Pherū (GSK 1.16–26), following Śrīdhara, regards the sum of a finite series of natural numbers as the first item called 'the sum' (S. *saṃkalita*) among the 25 fundamental operations and the difference between two of them as the second called 'the difference' (S. *vyavakalita*). Pherū, like Śrīdhara, does not treat the ordinary sum and difference of integers probably because they are too elementary for the supposed readers of the GSK.

The third rule for the sum of a natural series contains a superfluous quantity (say x) called 'the ⟨number of the⟩ letters in question' (*paṇhakkhara*):

$$S(n) = \frac{(nx + x) \cdot n}{2x}.$$

The same formula occurs also in other Jain works, both mathematical and non-mathematical, where x is called 'an optional quantity,' 'the ⟨number of the⟩ words,' etc., but the role of x in the formula is open to question.

2.2.8 Verification of multiplication

Pherū (GSK 1.30) provides two methods for the verification of the result of multiplication. The first one occurs also in Āryabhaṭa II's *Mahāsiddhānta* [MS] and the second one (the so-called 'verification by nine') also in Nārāyaṇa Paṇḍita's *Gaṇitakaumudī* [GK] but we rarely meet versified rules of those methods in other Indian mathematical works.

2.2.9 Eight operations with zero

Like Śrīdhara, Pherū (GSK 1.31) puts his rules for zero in between the rules for multiplication and division. This is because those rules are necessary for the 'fundamental operations' with notational places, among which the multiplication occupies the first place in the GSK as well as in the Tr (and the PG), where the sum and the difference of integers are replaced by those of natural series.

2.2.10 Reductions of fractions

As preliminaries to the eight fundamental operations of fractions, Pherū (GSK 1.47–49) prescribes two kinds of reduction of fractions, namely, 'reduction to the same colour' (*savaṃnaṇa*, S. *savarṇana*; *kalāsavaṃṇana* in the introduction to GSK 2.1) or homogenization and 'reduction to the same denominator' (*sadis-accheya*, S. *sadṛśa-ccheda*). It seems that, in Pherū's terminology, the former is concerned with the reduction of a 'composite' fraction to a 'simple' fraction while the latter with the reduction of two fractions having 'different denominators' to two fractions having the 'same denominator' which are equivalent to the original two fractions.

Pherū deals with the former type of reductions at length in Chapter 2, giving rules for seven 'classes' (*jāti*) of them, but in Chapter 1 (GSK 1.47a) he prescribes only for the two kinds usually called 'part-addition class' (*bhāga-anubandha-jāti*) and 'part-subtraction class' (*bhāga-apavāha-jāti*). The rules are

$$\begin{bmatrix} n \\ b \\ a \end{bmatrix} \rightarrow \begin{bmatrix} na+b \\ a \end{bmatrix}, \quad \begin{bmatrix} n \\ -b \\ a \end{bmatrix} \rightarrow \begin{bmatrix} na-b \\ a \end{bmatrix}.$$

In modern notation, these rules can be expressed as

$$n \pm \frac{b}{a} = \frac{na \pm b}{a},$$

but it should be noted that, in the above scheme, not only the right-hand side but also the left-hand side of each arrow are regarded as a single number just as $n\frac{b}{a}$ is called 'a mixed number' in English. The first seven 'classes' of Chapter 2 deal with the reductions of seven kinds of 'mixed' or 'composite' numbers (fractions) to a simple one. The reason that Pherū treated the above two kinds immediately before the eight fundamental operations of fractions separately from those seven kinds of homogenization seems to be that these two kinds of 'mixed fractions' frequently occur even at the elementary levels of mathematical education.

The anonymous commentator on the PG (39–40 and exs. 18–21) calls these operations 'other's part addition class' (*para-bhāga-anubandha-jāti*) and 'other's part subtraction class' (*para-bhāga-apavāha-jāti*), respectively, in contrast to 'one's own part addition class' and 'one's own part subtraction class,' for which see 'part-addition/subtraction classes' below.

The latter type, the reduction to the same denominator, is necessary when and only when the sum or the difference of two fractions is required and hence follows the present location of the rule.

2.2.11 Inappropriate problem

The problem of GSK 1.80 given as an example for the eleven-quantity operation is in fact an example for the three-quantity operation.

2.2.12 Purchase and sale

Pherū (GSK 1.86–89) gives four formulas and two examples for the topic of purchase and sale, which involves the buying rate, the selling rate, the capital, the net profit, and the total gain. The Tr and the PG do not treat this kind of problems. The GSS has only one formula. The BM, the MS, and an anonymous Sanskrit mathematical anthology (HJJM 10727, see Hayashi 2006, 202–04) give more detailed accounts of this topic, although the relationship of the GSK with those works is yet to be investigated.

2.2.13 Sale of living things

Like other authors of mathematical textbooks, Pherū (GSK 1.92–93) gives a rule for the sale of living things. Most authors give examples of sale of human beings (esp. women) but Pherū refrains from doing so. Pherū's only example, which asks the price of five camels nine years old when three camels ten years old cost 108 *ṭaṃkas*, is a Prakrit adaptation of the second example of the PG (ex. 51). Compare the following.

> *daśavarṣoṣṭratritayaṃ purāṇaśatam aṣṭasaṃyutaṃ labhate/*
> *tat kiṃ navavarṣoṣṭrāḥ pañca labhante samācakṣva//* PG (51) //

> *dasavarisā tiya karahā ṭaṃkā saü aṭṭhaähiya pāvaṃti/*
> *tā navavarisā karahā kaï mullaṃ havaï paṃcāṇa//* GSK 1.93 //

The preceding (i.e., the first) example of the PG (ex. 50), which is identical with the only example of the Tr (ex. 56), is concerned with the price of two women twenty years old. This fact implies that Pherū was consulting the PG (or otherwise a more comprehensive work that includes the contents of the PG) in addition to the Tr, which he follows in most cases of the traditional *pāṭī* problems, and that he preferred the example of camels in the PG (or in that comprehensive work) to the example of women common to the two works of Śrīdhara.

2.2.14 Part-part class

The 'part-part class' of reduction of fractions deals with a composite fraction that is expressed in words as '*b* parts among *a* parts of unity' and in digits presumably as $\begin{bmatrix} 1 \\ a \\ b \end{bmatrix}$, which is reduced to the simple fraction, $\begin{bmatrix} b \\ a \end{bmatrix}$ or $\frac{b}{a}$. Pherū's rule for the part-part class (GSK 2.3) includes the proportionate distribution. This is peculiar to him.

2.2.15 Part-addition/subtraction classes

The 'part-addition' and 'part-subtraction' classes of the reduction of fractions in Chapter 2 (GSK 2.6–9) treat composite fractions consisting of the sum or the difference of a fraction and its own part. The anonymous commentator on the PG (39–40 and exs. 18–21) calls these 'one's own part addition class' (*sva-bhāga-anubandha-jāti*) and 'one's own part subtraction class' (*sva-bhāga-apavāha-jāti*), respectively, in contrast to 'other's part addition class' and 'other's part subtraction class,' for which see 'reductions of fractions' above.

2.2.16 Pillar-part class

The eighth and the last 'class' in Chapter 2 is the pillar-part class (*thaṃbhaṃsakajāti*, S. *stambha-aṃśaka-jāti*), which deals with linear equations of the type:

$$x - \frac{b_1}{a_1}x - \frac{b_2}{a_2}x - \cdots - \frac{b_n}{a_n}x = p.$$

Why did Pherū include this algebraic problem (or its solution) in the chapter devoted to the homogenization of mixed fractions? The answer is probably given by Pherū's algorithm (GSK 2.14) for solving this equation, which can be expressed in modern notation as follows.

$$1 - \frac{b_1}{a_1} - \frac{b_2}{a_2} - \cdots - \frac{b_n}{a_n}$$

$$\rightarrow \quad 1 - \left(\frac{b'_1}{a} + \frac{b'_2}{a} + \cdots + \frac{b'_n}{a}\right), \quad \text{where} \quad \frac{b'_i}{a} = \frac{b_i}{a_i}$$

$$\rightarrow \quad 1 - \frac{b}{a}, \quad \text{where} \quad b = \sum b'_i \quad \rightarrow \quad \frac{a-b}{a} \quad \rightarrow \quad x = p \div \frac{a-b}{a}.$$

Pherū seems to have regarded the process for obtaining the divisor of p or $\frac{a-b}{a}$ as a kind of 'homogenization of a mixed fraction.'

Most authors of mathematical works categorize several types of linear and quadratic equations including the above as *jāti*. Thus the word *jāti* ('birth, family, caste, class') is used by them side by side for the two different categories, the mixed fractions and the algebraic equations, each comprising several 'classes.' Śrīdhara uses the word *uddeśa* ('examples' or 'problems') instead of *jāti*. Except in the translations, we shall use the expressions 'type' for the *jāti* or *uddeśa* of equations, and 'class' for the *jāti* of mixed fractions. For the works that treat the 'type' problems, see under the 'type problems' in the table of §2.5. For a classification of the 'type problems', see Appendix B.

2.2.17 Area of a quadrilateral

Pherū gives two formulas for the area of a quadrilateral: the first (GSK 3.36) is the extended Heron's formula and the other (GSK 3.39) is

$$A = \frac{a+b}{2} \cdot h,$$

where a, b, and h denote the base, the top, and the perpendicular. Most authors of mathematical works naturally prescribe the latter formula for trapeziums only but Pherū applies it to 'all quadrilaterals' (*sayalāṇa caüraṃsāṇaṃ*), whose 'perpendicular' appears to be assumed from the 'middle' (*majjha*) of the top. In fact, his only example (GSK 3.40) for this formula is not a trapezium, while Śrīdhara's examples of quadrilaterals for the same formula are all trapeziums (see below).

Pherū's strange manner of application of the formula was caused presumably by Śrīdhara's verse in the Tr which presents the same formula as follows:

> 'In other quadrilaterals (*caturasreṣv anyeṣu*), half the sum of the base and the face is multiplied by the perpendicular.' (Tr 42b)

In the former half of the same stanza (Tr 42a), Śrīdhara treats the areas of squares and oblongs. Therefore, it may well have made the reader of the Tr think that the first phrase of the latter half (Tr 42b), 'in other quadrilaterals,' means 'in quadrilaterals other than squares and oblongs' or 'in all quadrilaterals except squares and oblongs,' which virtually means 'in all quadrilaterals' since the above formula applies to squares and oblongs as well. But, of course, that is not what Śrīdhara intended. If one reads the text carefully, it is clear that the phrase in question is meant to be a sub-condition of another, major condition taken for granted, namely, 'in equi-perpendicular quadrilaterals' (*samalambaka-caturaśra*) or trapeziums including squares and oblongs. In other words, Śrīdhara in the Tr prescribed the above formula for trapeziums other than squares and oblongs. This is confirmed by his three examples for that formula, namely, Tr exs. 77, 78, and 79, which all give examples of trapeziums.

In his PG, Śrīdhara explicitly mentions the objects of the formula.

> 'In the case of equi-perpendicular quadrilaterals (*samalambaka-caturaśra*) and trilaterals (*tryaśra*), the area (*gaṇita*, lit. 'calculated one') is produced ⟨when⟩ half the sum of the base and the top is multiplied by the mid-perpendicular (*madhyama-lamba*).' (PG 115)

Our conjecture that Śrīdhara's notion of 'equi-perpendicular quadrilaterals' included squares and oblongs also is confirmed by his examples that follow the above PG stanza: they are ex. 122 (*sama-caturaśra* or an equilateral quadrilateral, i.e., a square), ex. 123 (*āyata-caturaśra* or an oblong), ex. 124 (*tribhuja* or a ⟨scalene⟩ trilateral), ex. 125 (*sama-tribhuja* or an equilateral trilateral), ex. 126 (*dvisama-tribhuja* or an equi-bilateral trilateral), ex. 127 (*dvisama-caturbhuja* or an equi-bilateral quadrilateral), ex. 128 (*trisama-caturbhuja* or an equi-trilateral quadrilateral), and ex. 129–130 (*viṣama-caturbhuja* or an inequilateral quadrilateral). Of course, the quadrilaterals of the last four examples are *samalambaka* or 'equi-perpendicular.' For 'inequi-perpendicular' quadrilaterals, Śrīdhara reserved the extended Heron's formula, which he stated as follows.

> 'Half the sum of the sides is four times decreased by the sides. The square root of their product is the area (*gaṇita*) in the case of equilateral inequi-perpendicular ⟨quadrilaterals⟩ and of inequilateral inequi-perpendicular ⟨quadrilaterals⟩.' (PG 117)

2.2.18 Ratio of the circumference to the diameter

The approximations to the ratio of the circumference to the diameter of a circle used in the GSK are the traditional Jain values, $\sqrt{10} : 1$ and $19 : 6$ (see GSK 3.43–49 and 3.58), which are employed also in the Tr as well as in the *Tiloyapaṇṇattī*, a Prakrit Digambara Jain work on cosmology. See under GSK 3.45 in Part IV.

2.2.19 Three lengths of a segment of a circle

For the area of a circle segment Pherū (GSK 3.46) adopts Śrīdhara's formula (see §2.2.1 above) but for its three lengths (the chord, the arrow or the height, and the bow or the arc) he (GSK 3.48–49) seems to follow Mahāvīra, adopting the two relationships of different origins used in the GSS, namely, $a^2 + 6h^2 = b^2$ and $a + h = b$. Mahāvīra uses these relationships in different situations, that is, the former in accurate (*sūkṣma*) calculations of circle segments (*cāpa-kṣetra*) and the latter in crude (*sthūla*) calculations of elongated circles (*āyata-vṛtta*). See Table 3.1 in Part IV. Pherū, on the other hand, uses the former for the arc (b) and arrow (h) and the latter for the chord (a) within the same context. Even then the most naturally expected formula (or algorithm) for the chord would be $a = b - h$ but Pherū gives the following odd algorithm equivalent to it.

$$a = \sqrt{\left(b - \frac{b+h}{2}\right)^2 \cdot 4}.$$

Here we are faced with two problems: (1) Why did Pherū not use one and the same relationship for all the three lengths? (2) Why did he rewrite the simple formula (algorithm), $a = b - h$, to obtain the more complicated one?

2.2.20 Volume and surface of a sphere

Two sets of formulas (algorithms) for the volume and the surface of a sphere occur in the present GSK, one in the traditional topic 'procedure for excavations' of Chapter 3 and the other in the fifth chapter.

$$V = \frac{d^3}{2}\left(1 + \frac{1}{9}\right), \quad S = \frac{C}{4} \cdot C \cdot \left(1 + \frac{1}{9}\right) \quad \cdots \quad \text{GSK 3.65}$$

$$V = d^3 \cdot \left(1 - \frac{1}{4}\right) \cdot \left(1 - \frac{1}{4}\right), \quad S = \frac{C}{4} \cdot C \cdot \left(1 + \frac{1}{9}\right) \quad \cdots \quad \text{GSK 5.25}$$

The formula for the surface (S) is common to the two sets. As it is not found in any other Indian text so far, it is probably Pherū's own. So also is the first formula for the volume (V), but the second one is possibly a rewrite of one of Mahāvīra's formulas:

$$V = \left(\frac{d}{2}\right)^3 \cdot \frac{1}{2} \cdot 9 \quad \cdots \quad \text{GSS 8.28b}$$

For a conjecture about the origin of the surface formula see under GSK 3.65 in Part IV.

2.2.21 Brick walls

Usually, the traditional topic of the 'procedure for piling' deals only with brick walls shaped like a simple rectangular solid but Pherū (GSK 3.69–86) treats brick walls of nine kinds of constructions in details: a dome, square and circular spiral staircases, a minaret, an arch, a staircase, a bridge, a cylindrical well, and a stepwell. He devotes most of the 18 verses of this section to the calculations of the net volumes of those constructions to be filled with bricks. It is only in one example (GSK 3.73) that he briefly touches upon the number of bricks necessary for a construction, which is an important item that constitutes the traditional topic of 'procedure for piling.'

2.2.22 A wage table for sawing

Pherū (GSK 3.89–92) records a wage table for timber sawing as 'told by carpenters.' The table divides the thickness of timber into 7 ranges (the highest range is 17/20 to 1 *gaja*) and gives the total length of cutting lines that yields one *koḍī* of wages within each range. Within each range of the thickness of timber, the wages must be proportional to the total length of cutting lines.

2.2.23 Noon shadows and the time of day

Pherū (GSK 3.104) gives a table of noon shadows and a formula for calculating the time of day from the shadow of a gnomon. This is a common item in the traditional topic of 'procedure for shadows,' but the text seems to be corrupt and we are unable to restore the table and the formula correctly. The coefficient, 7, in the formula is peculiar to the GSK and the anonymous *Pañcaviṃśatikā* [PV]. This coefficient, however, does not occur in another formula for the same purpose given by Pherū in his *Jyotiṣasāra* [JS]. See under GSK 3.104 in Part IV.

2.2.24 Four special topics

The fourth chapter of the GSK, entitled 'four (special) topics,' may be characterized as a supplement to the traditional *pāṭī* mathematics treated in the first three chapters. It comprises four sections, each devoted to one 'topic' (*ahigāra*). The first and the second sections treat the practical mathematics useful for accountancy and for sewing, respectively. Several words of non-Indian origins occur in these two sections. The remaining two sections are devoted to recreational mathematics.

In the first section (GSK 4.1–17) called 'region' (*desa*), Pherū treats the proportionate distribution of *jīvalaïs*, percentage of *upakkhaï* in *caṭṭī*, easy division of *dammas* by 20, 100, 1000, 10000, and 100000, easy calculations of wages from the annual wages (*mukkātaï*) and from the monthly wages (*maseliya*), meeting of two runners, and the conversion of the Vikrama and Hijrī dates.

In the second section (GSK 4.18–37) called 'cloth' (*vastra*), Pherū deals with numerical problems concerning sewing. The mathematical point of this section seems to lie in the areas of cloth in various shapes used for constructions such as tents, but the details are not clear to us.

In the third section (GSK 4.38–45) called 'magic squares' (*jaṃta*, lit. a device or a figure), Pherū classifies magic squares and gives their construction methods. Magic squares in India have a long history prior to Pherū. An irregular magic square of order four was utilized by Varāhamihira in his book on divination, *Bṛhatsaṃhitā* (ca. AD 550), in order to instruct how to blend ingredients of perfume and the magic square of order three was recommended as a talisman for an easy birth in several Sanskrit medical works since the ninth century. But, as far as we know, Pherū is the first in India who theorized on magic squares: he gave the classification of magic squares into three categories (odd, even, and evenly-odd) and prescribed general construction methods for the first two of them. Soon after Pherū, in the same century, Nārāyaṇa Paṇḍita included a fully developed mathematical theory of magic squares in his *pāṭī* work, *Gaṇitakaumudī* (AD 1356). In these and other mathematical works, magic squares are treated purely as mathematical exercises without any religious or magical overtones.

In the fourth and the last section called 'miscellaneous' (*prakīrṇaka*), Pherū treats amusing problems such as offering of flowers at a temple, equal distribution of fruit, cloth, etc., rule of inverse operations, mind reading ('think of a number' type), restoration of erased digits, etc., most of which are not found in the extant Indian mathematical works prior to him. The rule of inverse operations is stated in most mathematical works since the *Āryabhaṭīya* (AD 499) as a convenient tool for solving algebraic equations (see the table in §2.5 below), but the reason Pherū included it in this section seems to be its applicability to amusing games just like the mind reading that follows it.

2.2.25 Quintet of topics (*uddesa*)

The fifth and the last chapter of the GSK, entitled 'quintet of topics,' is probably intended to be a kind of supplement to the earlier chapters, especially to the first two sections of the fourth chapter. The five topics treated are grain, sugarcane juice and oil (both are obtained by extraction), regional tax, price, and measurement (GSK 5.1). The mathematical point of this supplement lies in numerical data on the 'yield' (*nippatti*) or the 'fruit' (*phala*) of the things in those five categories, such as agricultural produce, products from sugarcane juice, etc. This chapter is important for the history of agriculture and industry in India.

The fifth section of the fifth chapter (GSK 5.22–25), on measurement, contains several mathematical formulas including those for calculating the areas of the square and the equilateral triangle inscribed in a circle whose area is given and those of the circles inscribed in a square and an equilateral triangle whose areas are given (GSK 5.23–24). This topic is not found in the extant mathematical works prior to the GSK.

The fifth section on measurement is followed by another, extra section that is concerned with masonry. This topic is not included in the 'quintet of topics' listed in the first verse of the fifth chapter.

It is remarkable that this chapter has no mathematical example set in verse. Certainly GSK 5.22 and 25 in the section on measurement, which prescribe formulas for a circle and a sphere, respectively, are accompanied by one example each, but these examples begin with the 'setting-down' (*nyāsa*) of the numerical data without a verbal statement of the problem itself and the answer immediately follows it in the former case and the solution process is worked out in the latter. These prose passages following GSK 5. 22 and 25 apparently presuppose the example given in GSK 3.66.

2.2.26 Vigesimal place-value notation

In the section on measurement of the fifth chapter (GSK 5.22–25), a sort of vigesimal place-value notation is employed. Its characteristic feature is that each notational place can include a multiple of quarter, where a quarter is expressed by a vertical stroke (|). For example, the number, $(12\frac{1}{2})/20$, which is expressed in GSK 5.24 as 'twelve and a half *visovas*' (*visova* or *visuva* means 'twentieth'), is expressed in the accompanying figure as

0||2||

where '0||' belonging to the units' place means $\left(0+\frac{1}{4}+\frac{1}{4}\right)$ unit or a half unit or 10 twentieths, and '2||' belonging to the twentieths' place means $\left(2+\frac{1}{4}+\frac{1}{4}\right)$ twentieths. Thus the notation represents $12\frac{1}{2}$ twentieths. Similarly, we have the following instances (a pair of angular brackets indicates restoration).

0|⟨2⟩||	for	$(7\frac{1}{2})/20$
0||2S	for	12/19 or 12/20 'with difference'
0|||1	for	$(1-1/5)$ or 16/20
4|||	for	$4\frac{3}{4}$
28⟨||⟩	for	$28\frac{1}{2}$
90⟨|⟩	for	$90\frac{1}{4}$
100SS6	for	$100\frac{5}{18}$ or a little less than $100+6/20$
12⟨1||⟩	for	$121\frac{1}{2}$

The sign 'S' in the second row seems to indicate 'with difference' or 'and slightly more,' while the sign 'SS' in the second row from the bottom seems to be used for separating the units' place from the twentieths'.

2.3 Mathematical Terminology

The mathematical terms in the GSK are quite normal in the sense that for most of them we can point out their Sanskrit counterparts that are commonly used in Sanskrit mathematical works. The following is a list of Apabhraṃśa mathematical terms used in the GSK. The list is divided and subdivided according to the mathematical categories and the word order in each subdivision is determined semantically. The Sanskrit counterparts are put in parentheses. Needless to say, the correspondence between the English headings and the Apabhraṃśa terms is not one-to-one. For more information about each Apabhraṃśa (or Sanskrit) term and for the notation see the Glossary-Index (Appendix D).

2.3.1 Number and digit

Number in general, *saṃkha/saṃkhā* (*saṃkhyā*), *saṃkhyā* (Ts.), *aṃka* (*aṅka*) by extension (see 'digit' below).
Quantity in general, *rāsi/rāsī* (*rāśi*).
Positive (number/quantity), *ādāya* (Ts.), *dhaṇa* (*dhana*).
Negative (number/quantity), *riṇa* (*ṛṇa*), *vaya* (*vyaya*), *hīṇa* (*hīna*).
Zero, *kha* (Ts.), *sunna* (*śūnya*).

Notational place in the decimal place-value system, *paya* (*pada*).
Digit (a numerical figure), *aṃka* (*aṅka*).
Integer, *accheya* (*accheda*), *rū* (*rūpa*), *rūva* (*rūpa*).
Even number, *sama* (Ts.).
Odd number, *viṣama* (Ts.), *visama* (*viṣama*).
Evenly-odd number (a number which can be halved only once), *samavisama* (*samaviṣama*).
Fraction, *aṃsa* (*aṃśa*), *kalā* (Ts.), *bhāga* (Ts.), *bhinna* (Ts.).
Numerator, *aṃsa* (*aṃśa*).
Denominator, *cheya* (*cheda*), *hara* (Ts.).
Homogenization of a composite fraction to a simple one, *kalāsavaṃnaṇa* (*kalāsavarṇana*), *kalāsavarṇana* (Ts.), or simply *savaṃnaṇa* (*savarṇana*), otherwise *bhāgajāti* (Ts.). Pherū divides it in 8 *jāī* (*jāti*) or 'classes.'
Sign for *pāya* (*pāda*) or one-quarter, | (vertical stroke). See the prose sentences after GSK 5.22, 24, and 25.
Sign for a negative numerator (*hīṇaṃsa*), *biṃdu* (*bindu*) or a dot (·).
Sign indicating 'and slightly more,' S (the sign resembles the avagraha). See the figure at GSK 5.24.

2.3.2 Arithmetical operations

Computation Computation in general is called *gaṇaṇa* (*gaṇanā*), *gaṇanā* (Ts.), *gaṇita* (Ts.), and *gaṇiya* (*gaṇita*). Algorithmic or procedural mathematics in general is called *pāṭī* (Ts.), *pāḍī* (*pāṭī*), *vivahāra* (*vyavahāra*), and *vyavahāra* (Ts.). Each individual algorithm, that is, a series of mathematical operations designed for a specific type of problems is called *karaṇa* (Ts.), and a rule for it *karaṇa-sūtra* (Ts.).

Addition

Addition, summation, *anubaṃdha* (*anubandha*), *joḍa*/*joḍaṇa* (*yojana*), *saṃkalita* (Ts.), *saṃkaliya* (*saṅkalita*), *samāsaṇa* (*samāsana*).
To add, $\sqrt{}$*joḍa* ($\sqrt{}$*yuj*).
To increase, $\sqrt{}$*vaḍḍha* ($\sqrt{}$*vṛdh*).
Additive term, increase, *ahiya* (*adhika*), *uttara* (Ts.), *kheva* (*kṣepa*), *vuḍḍhi* (*vṛddhi*).
Increased by, added to, -*ahiya* (*adhika*), -*jutta* (*yukta*), -*juya* (*yuta*), -*miśrita* (Ts.), *sa*- (Ts.), -*saṃjutta* (*saṃyukta*), *saha*- (Ts.), -*sahiya* (*sahita*).
Sum, *akka* (*aikya*), *ikka* (*aikya*), *ikkaṭṭhā* (*ekatra*), *egattha* (*ekatra*), *aikya* (Ts.), *gaṇiya* (*gaṇita*), *juī* (*yuti*), *juya* (*yuta*), *juva* (*yuga*), *joya* (*yoga*), *piṃḍa* (*piṇḍa*), *missa* (*miśra*), *samāsa* (Ts.), *saṃkaliya* (*saṅkalita*).

Subtraction

Subtraction, *apavāha*, (Ts.), *vimakaliya* (*vyavakalita*).
To subtract, $\sqrt{}$*soha* (*śodhaya*).
To decrease, $\sqrt{}$*chijja* ($\sqrt{}$*kṣi*).
Subtracted, subtrahend, *ūṇa* (*ūna*), *gata* (Ts.), *gaya* (*gata*), *vimakaliya* (*vyavakalita*) *sohiya* (*śodhita*).

Less by, decreased by, *-uṇa* (*ūna*), *-una* (*ūna*), *-ūṇa* (*ūna*), *-rahiya* (*rahita*), *-vihīṇa* (*vihīna*), *-hīṇa* (*hīna*).
Difference, *aṃtara* (*antara*), *visesa* (*viśeṣa*).
Remainder, *sesa/sessa* (*śeṣa*).

Multiplication

Multiplication, *guṇaṇa* (*guṇana*), *guṇayāra* (*guṇa-kāra*), *guṇākāra* (*guṇa-kāra*).
To multiply, √*guṇa* (√*guṇ*), √*tāḍa* (√*taḍ*), √*haṇa* (√*han*).
Multiplied, to be multiplied, *guṇita* (Ts.), *guṇiya* (*guṇita*), *guṇiyavva* (*guṇitavya*), *tāḍiya* (*tāḍita*), *saṃguṇiya* (*saṃguṇita*), *haya* (*hata*).
Multiplier, *uṇa* (*guṇa*), *guṇa* (Ts.), *guṇaka* (Ts.), *guṇayāra* (*guṇa-kāra*), *guṇarāsī* (*guṇa-rāśi*).
Multiplicand, *gunnarāsi* (*guṇya-rāśi*).
Product, *guṇiyarāsi* (*guṇita-rāśi*), *vaha* (*vadha*), *phala* (Ts.).
Method, *kavāḍasaṃdhī* (*kapāṭa-sandhi*, a multiplication method).

Division The verb √*pāḍa* is used for division several times in the GSK but its Sanskrit counterpart, *pāṭaya* (causative stem of √paṭ, to split), is rarely used in that sense.

Division, *bhāga* (Ts.), *bhāgahara* (Ts.), *bhāgāhara* (*bhāga-hara*), *bhāya* (*bhāga*), *bhāva* (*bhāga*).
To divide, √*pāḍa* (*pāṭaya*) with or without *bhāya* (*bhāga*, in acc. or inst.), √*bheya* (√*bhid*), √*vibhaya* (*vi*-√*bhaj*), √*hara* (√*hṛ*) with or without *bhāya* (*bhāga*, in acc.).
Divided, to be divided, *bhatta* (*bhakta*), *vihatta/vihattha* (*vihṛta*), *hariya* (*hṛta*), *hariyavva* (*hartavya*).
Divisor, *cheya/cheyaṇa* (*cheda/chedana*), *bhāya* (*bhāga*), *hara* (Ts.).
Dividend, *saṃharaṇīya* (Ts.), *haraṇīya* (Ts.), *hāraṇīya* (Ts.).
Quotient, *phala* (Ts.), *laddha* (*labdha*).
Remainder, *sesa/sessa* (*śeṣa*).

Square, square root, cube, cube root

Square, *vagga* (*varga*), *varga/vargga* (Ts.).
Square root, *paya* (*pada*), *mūla* (Ts.), *vaggamūla* (*varga-mūla*), *vargamūla* (Ts.).
Cube, *ghaṇa* (*ghana*), *ghana* (Ts.).
Cube root, *ghanamūla* (Ts.), *mūla* (Ts.).

2.3.3 Proportion

Rules of three, of five, of seven, of nine, of eleven, *tirāsiyaga* (*trairāśika*), *trairāsika* (*trairāśika*), *paṃcarāśika/paṃcarāsika* (*pañca-rāśika*), *saptarāśika/-saptarāsika* (*sapta-rāśika*), *navarāśika/navarāsika* (*nava-rāśika*), *ekādaśa-rāśaya/ekādasarāsika* (*ekādaśa-rāśika*).
Inverse rule of three, *vitthatiyarāsī* (*vyasta-trika-rāśi*), *vyastatrairāśika* (Ts.).
Investment (for proportionate distribution of the profit), *pakkheva* (*prakṣepa*), *prakṣepaka* (Ts.).
Purchase in proportion, *samavisamakaya* (*sama-viṣama-kraya*).

2.3.4 Mathematical series

The mathematical series treated by Pherū are natural series, arithmetical progressions, square series, and cubic series.

Series, *seḍhī* (*śreḍhī*).
First term, *āi/āī* (*ādi*), *ādi* (Ts.).
Last term, *aṃta* (*anta*).
Common difference, *uttara* (Ts.), *caya* (Ts.), *vuḍḍhi* (*vṛddhi*).
Number of terms, *gaccha* (Ts.), *paya* (*pada*), *mūla* (Ts.).
Sum of a series, *gaṇiya* (*gaṇita*), *dhaṇa* (*dhana*), *saṃkalita* (*saṅkalita*), *saṃkaliya* (*saṅkalita*), *savvadhaṇa* (*sarva-dhana*).
Difference between two natural series, *vimakaliya* (*vyavakalita*).
Square series (which consists of the squares of natural numbers), *vagga* (*varga*), *varga/vargga* (Ts.).
Cubic series (which consists of the cubes of natural numbers), *ghaṇa* (*ghana*).

2.3.5 Geometry

Terms for the right-angled triangle and the cubic figure do not occur in the GSK itself although these figures are treated; in the List of Rules attached immediately after the scribal colophon, the former is called *trikoṇa-vikaṭa* ('beautiful triangle'). Its common designation is *jātya-tryasra* ('well-born trilateral'). The vertical side of a right-angled triangle, usually called *koṭi*, is called *laṃba* ('the perpendicular') in the GSK. The length of an arc (*dhaṇu/dhaṇuha*) is called *piṃḍa*, which usually means the 'thickness' of a solid figure.

Dimensions (three kinds of *hattha*)

Linear (one-dimensional), *sara* (*śara*).
Plane (two-dimensional), *paḍa* (*paṭa*).
Solid (three-dimensional), *ghaṇa* (*ghana*).

Measurement

Size, *parimāṇa* (Ts.).
Length, *āyāma* (Ts.), *dīha* (*dīrgha*), parimāṇa (Ts.).
Length of an arc, *piṃḍa* (*piṇḍa*). Cf. 'thickness' below.
Width, *taya* (*tata*)?, *pihula* (*pṛthula*), *vikkhaṃbha* (*viṣkambha*), *vitthara* (*vistara*), *vitthāra* (*vistāra*).
Height, *uṃḍḍatta* (*aunnatya*), *uccatta* (*uccatva*), *udaya* (Ts.).
Height of a spherical stone (= diameter of the stone), *udaya* (Ts.).
Depth, *veha* (*vedha*).
Thickness, *piṃḍa* (*piṇḍa*).
Area, *khitta* (*kṣetra*), *khetta* (*kṣetra*), *phala* (Ts.), *bhūsaṃkhā* (*bhū-saṃkhyā*), *māṇa/māṇaya* (*māna/mānaka*), *vitthāra* (*vistāra*).
Volume, *pāhaṇasaṃkhā* (*pāṣāṇa-saṃkhyā*) or simply *pāṣāṇa* (Ts.) or *pāhaṇa/-pāhāṇa* (*pāṣāṇa*), *pūra* (Ts.), *pūraṃtara* (*pūra-antara*), *phala* (Ts.), *sela* (*śaila*).
Capacity, *khattaphala* (*khāta-phala*), *phala* (Ts.).

Basic concepts concerning figures

Line, *reha* (*rekhā*), *līha* (*lekhā*), *leha* (*lekhā*).
Plane figure, *kṣetra* (Ts.), *khitta* (*kṣetra*).
Side of a plane figure, *aṃsa* (*asra or aśra*), *asa* (*asra or aśra*), *bhuyā* (*bhujā*), *bhuva* (*bhuja*).
Top side of a plane figure, *muha* (*mukha*), *muhabhuva* (*mukha-bhuja*).
Bottom side of a plane figure, *dharā* (Ts.), *bhū* (Ts.), *bhūmi* (Ts.), *bhūmibhuva* (*bhūmi-bhuja*).
Top (imaginary) plane of a pit, *muha* (*mukha*).
Bottom plane of a pit, *tala* (Ts.).
Perpendicular line, *laṃba/laṃbaya* (*lamba/lambaka*).
Angle, corner, *koṇa* (Ts.).
To show (graphically through a figure), *darśana* (Ts.).

Triangles

Triangle/trilateral, *taṃsa* (*try-asra*), *tikoṇa/tikkoṇa/tikkoṇaya* (*tri-koṇa/tri-koṇaka*), *tibhuya* (*tri-bhuja*), *tibhuva* (*tri-bhuja*).
Scalene triangle, *visamatikoṇa* (*viṣama-tri-koṇa*).
Right-angled triangle, treated but not named. Usually called *jātyatryaśra* in Sanskrit.
Base of a right-angled triangle, *bhuva* (*bhuja*).
Upright of a right-angled triangle, *laṃba/laṃbaya* (*lamba/lambaka*).
Hypotenuse of a right-angled triangle, *kanna* (*karṇa*).
Segments of the base of a trilateral (divided by the perpendicular), *ahavā* (*avadhā*).

Quadrilaterals

Quadrilateral, *caübbhuya* (*catur-bhuja*), *caübbhuva/caübhuva* (*catur-bhuja*).
Oblong, *dīhacaürasa/dīhacaüraṃsa* (*dīrgha-catur-aśra*), or simply *dīha* (*dīrgha*).
Square, *caüraṃsa/caürasa* (*catur-asra*).
Diagonal, *kanna* (*karṇa*).

Circle

Circle, *daüra* (from Ar-Pe. *daur*), *vaṭṭa* (*vṛtta*), *vaṭṭakhitta* (*vṛtta-kṣetra*), *vaṭṭaḍa* (*vartula*), *vṛtta* (Ts.).
Diameter, *vikkhaṃbha* (*viṣkambha*), *vitthara* (*vistara*), *vitthāra* (*vistāra*).
Circumference, *parihi/parihī* (*paridhi*).
Segment of a circle, *dhaṇu* (*dhanus*), *dhaṇuha* (*dhanuṣa*).
Chord, *jīvā* (Ts.).
Arc, *dhaṇu* (*dhanus*), *dhaṇuha* (*dhanuṣa*).
Arrow (the height of a segment of a circle), *sara* (*śara*).

Irregular plane figures

Young moon figure (crescent-like figure), *bāliṃda* (*bāla-indu*).
Moon figure (bow-like figure), *caṃda* (*candra*).

Full moon figure, *paripunnacaṃda* (*paripūrṇa-candra*).
Thunderbolt figure (made of two trapeziums), *kulisa* (*kuliśa*).
Elephant's tusk figure, *gayadaṃta* (*gaja-danta*)
Drum figure, *muruja* (*muraja*).
Barleycorn figure, *java* (*yava*).
Cart-wheel-circle figure, annulus, *sagaḍacakkavaṭṭa* (*śakaṭa-cakra-vṛtta*).

Solids

Rectangular solid, *caüraṃsa* (*catur-asra*).
Cube, treated but not named. Usually called *ghana* in Sanskrit.
Sphere, *golaka* (Ts.), *gola/golaya* (Ts./*golaka*).
Pits or excavations of any shape, *khatta* (*khāta*), *khāta* (Ts.).
Walls, *bhitti/bhittī* (Ts.). Pherū treats the volumes of nine kinds of 'walls' such as a dome, etc. See GSK 3.69–86.

2.3.6 Bhūtasaṃkhyā

The following *bhūtasaṃkhyās* are used in the GSK. The references are to the verse numbers.

0: kha 1.31, 4.59.
1: sasi 4.42, 45; sasihara 4.38.
2: kara 4.38, 42, 59; loyaṇa 4.45.
3: aggi 4.38; guṇa 4.42; haranayaṇa 4.45.
4: juga 4.38, 45; juya 4.38, 42; yuga 4.38, 42, 45.
5: iṃdiya 4.38, 45; sara 4.39, 42.
6: rasa 4.17, 38, 42, 45.
7: uvahi 4.38; muṇi 3.104, 4.17, 39, 42, 45; vāīya 5.5.
8: vasu 4.38, 42.
9: gaha 4.45; naṃda 4.17; nihāṇa 4.17; nihi 4.38, 42.
10: disi 4.38, 42.
11: īsara 4.38.
12: diṇayara 4.38; ravi 3.103, 4.40, 42.
14: maṇu 4.40, 42.
15: tihi 4.38, 42.
16: kala 4.42; niva 4.37.
20: nakha 4.40.
24: jiṇa 3.61, 4.40.
28: rikkha 4.40.
32: daṃta 4.39.
33: sura 4.39.

The Sanskrit counterparts of these are commonly used in mathematical and astronomical works written in Sanskrit, with the only exception of *uvahi* (S. *udadhi*, sea) used for seven. The words meaning 'sea' such as *arṇava*, *udadhi*, *samudra*, *sāgara*, etc. are, as a *bhūtasaṃkhyā*, usually used for the number four according to the notion of 'the four oceans.' The GSK provides a very rare usage based on the notion of 'the seven seas.'

2.4 Contents of the Work

	Verses
Chapter 1: Twenty-five fundamental operations	
1 Weights and measures	
Introduction	1–2
Monetary units	3–4a
Area measures	4b
Linear measures	5–6
Three kinds of *hattha*	7
Volume measures	8
Weight measures	9–10a
Purity of gold	10b
Time units	11
Names of twenty-five decimal places	12–14
Forty-five gateways to mathematics	15
2 Eight fundamental operations of integers	
Sum	16–20
Difference	21–26
Multiplication	27–29
Verification of multiplication	30
Eight operations with zero	31–32
Division	33
Square	34–36
Square root	37–38
Cube	39–42
Cube root	43–45
3 Eight fundamental operations of fractions	
Expression of fractions	46
Reduction of fractions	47–49
Sum	50
Difference	51–52
Multiplication	53–54
Division	55–56
Square	57–58
Square root	59
Cube	60–61
Cube root	62
4 Proportion	
Three-quantity operation	63–71
Five- to eleven-quantity operations	72–80
Inverse three-quantity operation	81–85
Purchase and sale	86–89
Barter	90–91
Selling of living things	92–93
Chapter 2: Eight classes of reduction of fractions	
1 Part class	1
2 Multi-part class	2
3 Part-part class	3–5

	Verses
4 Part-addition class	6–7
5 Part-subtraction class	8–9
6 Part-mother class	10–11
7 Chain-reduction class	12–13
8 Pillar-part class	14–17
Chapter 3: Eight types of procedure	
1 Procedure for mixture	
Capital, interest, time	1–6
Proportionate distribution	7–14
Procedures for gold	15–25
2 Procedure for series	
Arithmetical progressions	26–30
Sum of the sums of natural series	31–32
Square series and cubic series	33
Sum of the natural series with two additional terms	34
3 Procedure for plane figures	
Areas of a square and an oblong	35
Areas of a trilateral and a quadrilateral	36–40
Perpendicular and base segments of a trilateral	41
Hypotenuse of a right-angled triangle (Pythagorean Theorem)	42
Circle	43–45
Segments of a circle	46–49
Irregular figures	50–53
4 Procedure for excavations	
Arithmetical mean of lengths	54
Capacity of a rectangular pit	55–57
Capacity of a cylindrical well	58–59
Capacities of various pits with uniform cross-sections	60
Volume of a rectangular stone	61–62
Volumes of various stones with uniform cross-sections	63–64
Volume and surface of a spherical stone	65–66
Specific gravities of stones	67–68
5 Procedure for piling	
List of nine kinds of walls	69
Rectangular walls	70–73
Domes	74–76
Square and circular walls of spiral stairs	77–79
Minarets	80
Arches, staircases, bridge structures	81–84
Cylindrical and other wells	85–86
6 Procedure for sawing	
Number of planks cut out of a piece of timber	87–88
Wages for sawing	89–93
Three notes on sawing	94
Factor of hardness of trees	95
7 Procedure for mounds of grain	
Volume of grain heaped up on the ground	96–99
Volumes of mounds piled against walls	100

	Verses
8 Procedure for shadows	
Height of a pillar, etc.	101–102
Construction of a gnomon	103
Noon shadows and the time of day	104
Chapter 4: Four special topics	
1 Region	
Introduction	1
Proportionate distribution of money	2–3
Percentage	4–5
Easy division of *dammas* by 10, 100, 1000, etc.	6–9
Easy division of *dammas*, etc. by 20	10
Wages for elapsed days	11
Daily wages	12
Meeting of two runners	13–16
Conversion of Vikrama and Hijrī dates	17
2 Cloth	18–37
3 Magic squares	
Classification of magic squares and exs. of order 4	38
Magic squares of the evenly-odd order: an example	39–40
Magic squares of the even order	41–42
Magic squares of the odd order	43–45
4 Miscellaneous	
Offering of flowers at a temple	46–47
Equal distribution of the plucked mangos	48–49
Equal distribution of *varisolaga*	50–51
Equal divisions of a square piece of cloth	52–53
Meeting of two runners	54–55
Rules of inversion	56–57
Mind reading	58–59
Restoration of erased digits	60
Number consisting of the same digits	61
Equal divisions of cows (a common multiple)	62–63
Distribution of cow's milk	64
Chapter 5: Quintet of topics	
1 Yield of grains	
Introduction	1
Irrigated regions	2
Area measures	3
Agricultural produce	4–8
Tax on agricultural produce	9
2 Yield of sugarcane juice and oil	
Quantities of the products from sugarcane juice	10–12
Quantities of the products from sesamum, etc.	13
3 Yield of regional tax	
Tax on cattle and household goods and exemption	14–15
Exceptions and variable nature of the example	16–17
4 Yield of price	
Commodities per *damma* and per *ṭaṃka*	18–19

	Verses
Prices per *maṇa* and per *sera*	20
Prices per score and per article	21
5 Measurement	
Circumference and area of a circle	22
Areas of a square and a circum/in-scribed circle	23
Areas of a triangle and an in/circum-scribed circle	24
Volume and surface area of a sphere	25
6 Masonry	
Specific gravities of stones, clay, grains, oil, ghee, salt	26–28
Quantities of the materials for brick walls	29–30
Quantities of the materials for plastering	31–32
Concluding remark	33

2.5 Fundamental Operations

We give below a comparative table showing the extent of the concept of 'fundamental operations' (*parikarmāṇi*) in various *pāṭī* works. The serial numbers in the table indicate those topics which are included in the 'fundamental operations' in that order in each work. The letter 'r' indicates a rule outside the 'fundamental operations' and 'e' an example. For additional details see the notes following the table. For the 'type problems' see Appendix B. The English names of those 'types' are based on their Sanskrit names but not always literal translations.

Works	BSS	PG	Tr	GSS	MS	SS	GT	SGT	L	GSK	GK
Number of *parikarmāṇi*	20	29	(31)	(16)	(39)	20	(20)	(36)	(24)	25	(24)
Arithmetical Operations											
Integers											
addition	(1)	1*	1*	7*	1	(1)	1	1	1	1*	1
Subtraction	(2)	2*	2*	8*	2	(2)	2	2	2	2*	2
Multiplication	(3)	3	3	1	3	3	3	3	3	3	3
Division	(4)	4	4	2	4	4	4	4	4	4	4
Square	(5)	5	5	3	5	5	5	5	5	5	5
Square root	(6)	6	6	4	7	7	6	6	6	6	6
Cube	7	7	7	5	6	6	7	7	7	7	7
Cube root	8	8	8	6	8	8	8	8	8	8	8
Fractions											
addition	1	9	9	15*	21	1	1	9	9	9	9
Subtraction	2	10	10	16*	22	2	2	10	10	10	10
Multiplication	3	11	11	9	23	3	3	11	11	11	11
Division	4	12	12	10	24	5	4	12	12	12	12
Square	5	13	13	11	25	4	5	13	13	13	13
Square root	6	14	14	12	26	(6)	6	14	14	14	14
Cube	(7)	15	15	13	27	(7)	7	15	15	15	15
Cube root	(8)	16	16	14	28	(8)	8	16	16	16	16

(continued)

Works	BSS	PG	Tr	GSS	MS	SS	GT	SGT	L	GSK	GK
Zero											
addition	r1	r1	r1	r3	9	r1	r1		17	r1	17
Subtraction	r2	r2	r2	r4	10	r2	r2		18	r2	18
Multiplication	r3	r3	r3	r1	11	r3	r3		19	r3	19
Division	r4	r4	r4	r2	12	r4	r4		20	(r4)	20
Square	r5			(r5)	13		r5		21	(r5)	21
Square root	r6			(r6)	(14)				22	(r6)	22
Cube				(r7)	(15)		r6		23	(r7)	23
Cube root				(r8)	(16)				24	(r8)	24
Class fractions											
Part class	9	17	17	r9	20	14	9	17	r1	r9	r1
Multi-part class	10	18	18	r10	19	11	10	18	r2	r10	r2
Part-part class	11	19	20	r11	30	(12)	(11)			r11	
Part-add class	12	20	19	r12	17	9	12	19	r3	r12	r3
Part-sub class	13	21	(21)	r13	18	10	13	20	r4	r13	r4
Part-mother class		22	22	r14	31					r14	
Chain-reduction class			23		29	13	14	21		r15	
Type problems											
Original-part type		r5	24	r15			r7	22	e1	r16	r5
Remainder-part type		r5		r16	32		r8	23	e2		r5
Sum-part type					33						
Part-difference type		r6					r9	24	e3		r5
Root-remainder type		r7					r10	25			r7
Remainder-root type				r18							
Root(∓) type									r5		
Part-root type		r8		r17			r11	26			
Part-root(∓) type									r6		r6
Part-partial-root(∓) type									e4		r9
Two-visible type		r9		r19			r12	27			r8
Visible-part type				r24			r13	28			r12
Partial-root type				r20			r14	29			
Partial-product type				r21							r11
Partial-square type				r22			r15	30			r10
Root-sum type				r23							
Inverse operation	r7	r10			34	15	15	31	r7	r17	r13
Proportion											
Three-qt. operation	14	23	25	r25	35	16	16	32	r8	17	r14
Inv. three-qt. operation	15	24	26	r26	36	17	17	33	r9	22	r15
Five-qt. operation	16	25	27	r27	37	18	18	34	r10	18	r15
Seven-qt. operation	17	26	28	r27	37				r10	19	r15
Nine-qt. operation	18	27	29	r27	37				r10	20	r15
Eleven-qt. operation	19								r10	21	r16
Barter	20	28	30		38	19	19	35	r11	24	r17
Living things		29	31	r28		20	20	36	e5	25	r18
Purchase and sale				r29	39					23	r19

Notes on the above table

BSS (*Brāhmasphuṭasiddhānta* by Brahmagupta, AD 628):

20: The first verse of Chapter 12 named *gaṇita* or 'mathematics,' which treats almost the same topics as later *pāṭī* works, states:

> 'One who knows the twenty *parikarmāṇi* (fundamental operations) beginning with addition severally and the eight *vyavahārāḥ* (procedures) ending with shadow is the calculator.' (BSS 12.1)

9–13: These five rules are called 'the first' to 'the fifth' classes in the BSS.

(1)–(6), (7)–(8): These operations are not treated in the BSS but must be counted among the twenty *parikarmāṇi*. Special rules for the multiplication, division, square, and square root are appended at the end of the same chapter (BSS 12.55–65).

r1–r7: These rules are given in Chapter 18 (on algebra).

PG (*Pāṭīgaṇita* by Śrīdhara, 8th century AD):

29: The five verses that follow the first, introductory verse read:

> 'Addition, subtraction, multiplication, division, square, its root, cube and cube root, and then the same// for fractions, six kinds of reduction of fractions, that is to say, part, multi-part, part-part, and then// ones called partial addition and partial subtraction, and part-mother, the three-quantity operation (rule of three), the same reversed, five-, seven-, and nine-quantity operations,// commodity vs. commodity (barter) and the sale of living things: these are the twenty-nine fundamental operations (*parikarmāṇi*). There shall be nine procedures (*vyavahārāḥ*), that is to say,// mixture in the first place and then series, field, and then excavation and piling, sawing and mounds, shadow, and then the truth of zero.' (PG 2–6)

1*–2*: These are the 'sum' of a natural series and the 'difference' between two of them, respectively.

23: After the three-quantity operation a rule for the 'going forward and backward' problem is inserted.

r1–r4: These rules for zero are given after the multiplication of integers. More comprehensive rules for zero presumably have been given in the lost, additional 'procedure' entitled 'the truth of zero.'

r5–r10: These rules for the 'type problems' and for the inverse operation, together with examples, are given at the end of the 'procedure for mixture.'

Tr (*Triśatikā* by Śrīdhara, 8th century AD):

(31): The number is not explicitly stated but presumably greater than that of the PG by two because the chain reduction (23rd) and the first of the type problems (24th), which occurs under the name 'pillar problem' (*stambhoddeśaka*) together with seven examples, seem to be included in the *parikarmāṇi*.

1*–2*: See 1*–2* under PG.

(21): The published text of the Tr does not have a verse for the part-subtraction class but the original Tr must have had one because the author Śrīdhara says, 'These are the six ways of homogenization of fractions' (*iti ṣaṭ-prakārakalāsavarṇanam*), at the end of his prose comment on the part-mother class. The part-addition and the part-subtraction classes are almost always coupled together in Indian arithmetical works including the PG, Śrīdhara's other work on *pāṭī*.

23: The Tr does not include the chain reduction (*vallīsavarṇana*) in the category of 'the class' (*jāti*) or 'the homogenization of fractions' (*kalāsavarṇana*).

r1–r4: See r1–r4 under PG.

GSS (*Gaṇitasārasaṃgraha* by Mahāvīra, ca. AD 850):

(16): The three verses that follow the weights and measures in Chapter 1 enumerate the names of *parikarmāṇi*:

> 'Now the names of *parikarmāṇi*:
>
> The first one is multiplication; it shall also be the product. The second is called division and the third the square. The fourth is called the square root and the fifth the cube. Then the sixth is the cube root and the seventh is said to be the accumulation (*citi*, i.e., summation); it is also called the sum (*saṅkalita*). Then the eighth is the difference (*vyutkalita*); it is also called the remainder. The same eight ⟨operations exist⟩ for fractions also.' (GSS 1.46–48)

The eight *parikarmāṇi* for integers are treated in Chapter 2 entitled 'Procedure for *parikarmāṇi*' and those for fractions in Chapter 3 entitled 'Procedure for the Homogenization of Fractions.'

7*–8*: The 7th is the 'sum' of an arithmetical progression and that of a geometrical progression and the 8th is the 'difference' between two of those sums.

15*–16*: The 'addition' is the 'sum' of an arithmetical progression whose first term, common difference, and number of terms are fractions, or that of a geometrical progression whose first term and common ratio are fractions. The 'subtraction' is the 'difference' between two of them.

r1–r4 and (r5)–(r8): These rules for zero are given in Chapter 1 entitled 'Technical Terms.' (r5)–(r8) are implicit. The verse that follows the above three verses reads:

> 'A number multiplied by zero is zero. It is unchangeable when divided, increased, or decreased ⟨by zero⟩. The product, etc. of zero is zero. In the addition ⟨of a number to zero⟩, zero has the form of the additive.' (GSS 1.49)

r9–r14: These 'six classes' (*ṣaḍjāti*) of fractions are treated after 16* (subtraction of fractions) in Chapter 3.

r15–r24: These rules for the 'type problems,' with examples, are given in Chapter 4 entitled 'Miscellaneous ⟨Problems⟩.'

r25–r28: These rules for proportion, with examples, are given in Chapter 5 entitled 'Three-Quantity Operation.' r23 (the inverse three-quantity operation) is followed by a rule for the 'going forward and backward' problem.

r29: This rule for purchase and sale, with an example, is given in Chapter 6 entitled 'Procedures for Mixture.'

MS (*Mahāsiddhānta* by Āryabhaṭa II, ca. AD 950):

(39): The MS devotes one chapter (Chapter 15) for *pāṭī* and in it treats almost the same topics that other *pāṭī* works deal with nearly in the same order. But there is no indication that it divides the contents of that chapter into the two categories, *parikarmāṇi* and *vyavahārāḥ*. In fact, the word *parikarman* does not occur in the MS while the word *vyavahāra* does. The number, 39, is that of all the topics (except a few, for which see 32–33 and 39 below) treated before the *vyavahārāḥ*.

(14)–(16): These are implicit. The rules for zero run as follows:

> 'A number is unchangeable when increased by zero. The same is the case with the subtraction ⟨of zero from a number⟩. In the case of multiplication, division, squaring, etc. of zero, ⟨the result is⟩ zero.' (MS 15.10b–11a)

32–33: The remainder-part type is coupled with the sum-part type, which is followed by the 'rule of concurrence.'

39: The purchase and sale is followed by a rule for the 'going forward and backward' problem. A more elaborate treatment of the purchase and sale is found in the section on the 'procedure for mixture.'

SS (*Siddhāntaśekhara* by Śrīpati, ca. AD 1040):

20: The first verse of Chapter 13 (*vyakta-gaṇita* or 'mathematics of visible quantities') reads:

> 'One who knows these twenty fundamental operations (*parikarmāṇi*) and the procedures whose first and eighth are mixture and shadow, respectively, knows the mathematics of visible quantities and enjoys the state of being a leading calculator in the society of those who are well versed in calculation.' (SS 13.1)

(1)–(2), (6)–(8): These are not treated but must be counted.

(12): The 'part-part class' is the most appropriate candidate for the missing item among the 20 *parikarmāṇi*, although it is strange that that topic happens to be omitted both in the SS and the GT (cf. (11) of the GT).

r1–r4: These rules for zero are given in Chapter 14 (*avyakta-gaṇita* or 'mathematics of invisible quantities').

GT (*Gaṇitatilaka* by Śrīpati, ca. AD 1040):

(20): The number is not explicitly stated in the text but seems to be the same as that of the SS by the same author.

(11): The 'part-part class' is the most appropriate candidate for the missing item among the 20 *parikarmāṇi* (cf. (12) of the SS).

r1–r6: These rules for zero are given in a verse located after the eight operations of fractions but seem not to be counted among the 20 *parikarmāṇi*.

r7–r15: These rules for the 'type problems,' together with examples, are located after the 'chain-reduction class.'

SGT (Siṃhatilaka's commentary on the *Gaṇitatilaka*, 14th century AD):

(36): At the end of his commentary on an example for the inverse operations, Siṃhatilaka says:

> '... Thus the inverse problem is completed. Up to now thirty-one *parikarmāṇi* have been completed.' (SGT 94, p. 68)
>
> 'Now the thirty-second, the three-quantity operation, is undertaken. ...' (SGT 95, p. 68)

This statement implies that the last topic of the proportion, 'living things,' is the 'thirty-sixth' *parikarman*.

L (*Līlāvatī* by Bhāskara II, AD 1150):

(24): The chapter for the arithemtical operations consists of three sections, which conclude with the statements, 'These are the eight *parikarmāṇi* ⟨on integers⟩,' 'These are the eight *parikarmāṇi* on fractions,' and 'These are the eight *parikarmāṇi* on zero,' respectively.

r1–r4: These rules for the 'class fractions' are placed immediately before the 9th parikarman as preliminaries to the 8 arithmetical operations of fractions.

r5–r11 and e1–e5: These (rules and examples) are included in the chapter called *prakīrṇaka* (miscellaneous). e1–e3 are given as examples for the optional quantity operation; r5–r6 are called the multiplier operation; and e5 is given as an example for the inverse three-quantity operation.

GSK (*Gaṇitasārakaumudī* by Ṭhakkura Pherū, ca. AD 1315):

25: At the end of the section for the weights and measures in Chapter 1, Pherū says that there are twenty-five *parikamma*. See GSK 1.15.

1*–2*: See 1*–2* under PG.

r1–r3, (r4)–(r8): These rules for zero are given after the multiplication of integers. (r4)–(r8) are implicit.

r9–r16: These eight topics are treated in Chapter 2 entitled 'Eight Classes of Reduction of Fractions.'

r17: This rule for inverse operations is given in Section 4 ('Miscellaneous') of Chapter 4, 'Four Special Topics.'

GK (*Gaṇitakaumudī* by Nārāyaṇa Paṇḍita, AD 1356):

(24): The first section of the chapter on the arithmetical operations is closed with the statement, 'These are the eight *parikarmāṇi* on integers;' the second section begins with the statement, 'Now the eight *parikarmāṇi* on fractions;' and the third, 'Now a versified rule for the ⟨eight⟩ *parikarmāṇi* (pl.) on zero.' Cf. (24) of L.

r1–r4: These rules for the 'class fractions' are placed immediately before the 9th parikarman, that is, the addition of fractions. Cf. r1–r4 of L.

r5–r18: These are treated in Chapter 2, 'Miscellaneous ⟨Problems⟩.' r5 is a rule for the optional quantity operation (regula falsi).

r19: This topic is dealt with at the beginning of the 'Procedure for Mixture.'

2.6 Patan Manuscript

Twenty-one verses of the GSK are cited and/or paraphrased with solutions to the examples in prose Sanskrit in a manuscript from Patan which we shall call Patan Manuscript [PM]. It is a small undated manuscript numbered 8894 of the library of Shree Hemacandracarya Jain Jnan Mandir, Patan, North Gujarat. The manuscript consists of two parts, ff. 1–4*r* and f. 4*v*, which we call A and B. Both parts contain anonymous anthologies. Part A consists of three sections. Section 1 contains fourteen stanzas taken exclusively from the GSK together with comments or solutions in prose Sanskrit, and four prose passages which paraphrase four verses of the GSK. Pherū is mentioned by name at the end of the section: *pherūkṛtagaṇitasāre// cha//.* Section 2 consists of citations from, and paraphrases of, the verses of the PG, GSS, GSK, L, and some unknown work(s), and section 3 consists of twelve stanzas cited exclusively from Śrīdhara's works (PG and Tr). Part B consists of citations from the Tr and some unknown work(s). The PM has been edited and discussed by Hayashi [1995a, 464–484].

The verses of the GSK cited and/or paraphrased in the PM are as follows.

GSK	PM	GSK	PM	GSK	PM
2.003	A33	4.015	A5	4.051	A12*
2.004	A34	4.016	A6	4.052	A13
2.005	A40	4.046	A7	4.053	A14⋆
3.101	A1	4.047	A8*	4.058	A18
3.102	A2	4.048	A9	4.059	A17
4.013	A3	4.049	A10	4.062	A15
4.014	A4	4.050	A11	4.063	A16*

The numbering of the PM passages follows Hayashi's edition. An asterisk (*) indicates that the PM passage renders the verse of the GSK into prose without citing it. A star (⋆) indicates that the PM passage gives only the solution.

We shall give variants from the verses cited in the PM in the footnotes to the Text (Part II), and cite the paraphrases and solutions of the examples in the Mathematical Commentary (Part IV).

Part II

Text

Chapter 0

Introduction

This is a revised edition of the *Gaṇitasārakaumudī* of Ṭhakkura Pherū. Since no manuscript of the work was available, we based our text entirely on the Nahatas' edition published in AD 1961. Wherever we felt it necessary to emend the text for mathematical, grammatical or metrical reasons, we took the liberty of rewriting the text, retaining the original readings in footnotes. We did not 'normalize' or 'correct' irregular Sanskrit forms found in the prose parts, which are written in a sort of mixed Sanskrit. We have added silently the chapter headings, *prathamo 'dhyāyaḥ*, etc.

This edition provides a kind of *pada* (word) text in which words are separated from each other by means of a space or a hyphen (-) or an equal sign (=). The hiatus between successive members of a compound is indicated by a circumflex accent (ˆ). This is useful for the analysis of the vocabulary, for indexing, etc.

0.1 Verse Numbers

In the first three chapters we adopted the verse numbers of the Nahatas' edition. In the fourth chapter of the Nahatas' edition the verses of the fourth and the last section are numbered separately from those of the first three sections but, for convenience, we used a single set of serial numbers for the entire chapter. We regarded the last chapter entitled *uddeśapañcaka* as the 'fifth' although it does not have a chapter number in the Nahatas' edition. For the list of rules given after the colophon of the scribe, we thought it expedient to give each line of the list a serial number together with the chapter number six. See Table 0.1.

0.2 Symbols for Word Separation (-, =)

In this and the next sections, we use the expression, 'C $\leftarrow$ B $<$ A' (or 'A $>$ B $\rightarrow$ C'), where the symbol '$<$' (or '$>$') indicates a word union (sandhi) with or without a euphonic change and '$\leftarrow$' (or '$\rightarrow$') a re-decomposition of the united words in the present edition. That is, B is an expression in the Nahatas' edition, while C is the expression of B in our revised edition and A shows its original elements.

Table 0.1: Verse Numbers

Nahatas' edition		Revised edition	
Chapters	Verses	Chapters	Verses
1	1–93	1	001–093
2	1–17	2	001–017
3	1–104	3	001–104
4, Sec. 1–3	1–45	4	001–045
4, Sec. 4	1–19	4	046–064
Uddeśapañcaka	1–33	5	001–033

(1) '-' indicates a union of two successive members of a compound.

Exs. ti-jaya-nāhaṃ (1.001a) ← tijayanāhaṃ < ti + jaya + nāhaṃ.
ikk-ikka (3.010b, etc.) ← ikkikka < ikka + ikka.
t-aṃsa (2.009b, etc.) ← taṃsa < ti + aṃsa.
s-ega (1.022a, etc.) ← sega < sa + ega.

(2) '=' indicates a word-union outside compounds.

Exs. tah=aṭṭha (1.015a) ← tahaṭṭha < taha/tahā + aṭṭha.
tah=eva (1.003a) ← taheva < taha/tahā + eva.
satta=uvari (3.010a) ← sattovari < satta + uvari.

0.3 Behaviour of the Initial and the Final Vowels

When a word ending in a vowel (x) is followed by another word beginning with a vowel (y), either within a compound or outside the compounds, there are three possible forms of combination in the *Gaṇitasārakaumudī*.

(1) x + y > y (elision of the last vowel).
(2) x + y > xy (hiatus).
(3) x + y > z (a new vowel).

Ex. ṭhāṇa + ahiya (in a compound) >
(1) ṭhāṇahiya → ṭhāṇ-ahiya (1.034b);
(2) ṭhāṇa ahiya → ṭhāṇa^ahiya (this example does not occur in the GSK);
(3) ṭhāṇāhiya → ṭhāṇa-ahiya (1.039b).

Out of these three cases, case (1) occurs most frequently and case (3) rarely. Besides the above example, ṭhāṇāhiya, we have the following for case (3). We have included 'a + ā > ā' in case (1).

a + a > ā
bhāga + aṇubaṃdha > bhāgāṇubaṃdha → bhāga-aṇubaṃdha (2.006b).
bhāga + apavāhaṃ > bhāgāpavāhaṃ → bhāga-apavāhaṃ (2.008b).
navaṃsa + ahiu > navaṃsāhiu → navaṃsa-ahiu (2.017b).
nīla + aruṇa > nīlāruṇa → nīla-aruṇa (4.037a).
hīṇa + ahiya > hīṇāhiya → hīṇa-ahiya (5.011a, 5.017a).

ā + a > ā
tā + aṇukkamihi > tāṇukkamihi → tā=aṇukkamihi (4.038e, 4.062a).
varisolā + aṇukkamihi > varisolāṇukkamihi → varisolā=aṇukkamihi (4.051b).

a + u > o
daha + uttara > dahottara → daha-uttara (3.008b).
daṃta + uvama > daṃtovama → daṃta-uvama (3.051a).
bāliṃda + uvama > bāliṃdovama → bāliṃda-uvama (3.052a).
ghaṇa + uvama > ghaṇovama → ghaṇa-uvama (3.055b).
tala + uvara > talovara → tala-uvara (4.022a).
viviriya + uddesago > viviriyoddesago → viviriya-uddesago (4.057b).
satta + uvari > sattovari → satta=uvari (3.010a).

0.4 Symbols for the Hiatus (ˆ) and the Avagraha (')

(1) The sign ˆ indicates a union of the last and the initial vowels of two successive members of a compound by which no euphonic change occurs (hiatus).

Ex. dasaˆavva (1.012b).

(2) The sign ' indicates an elision (avagraha) of the initial a occuring after a final e or o of the preceding word. Indian pandits (also the Nahatas) insert an avagraha sign also where an initial a/ā is assimilated with a final a/ā. In the Nahatas' text the avagraha sign (S) occurs several times both in the Apabhraṃśa verses and in the ardha-Sanskritic prose parts but most of the former cases can be expressed more properly without it by means of the above symbols.

Ex. taha 'ṭṭha in the Nahatas' text should be tah=aṭṭha (1.015a).

0.5 How To Restore the Original Metrical Text

The original metrical text can be restored by removing the three separation marks according to the following rules.

(1) '-' and '=': Sandhi rules should be applied when these symbols are situated in between two vowels, that is, 'vowel'-'vowel' and 'vowel'='vowel' (see §0.2 and §0.3 above); otherwise just delete them.

(2) 'ˆ': Just delete it. In Roman characters, aˆi and aˆu should be replaced by aï and aü, respectively, in order that they may not be confused with the diphthongs, ai and au, respectively.

0.6 Other Symbols and Abbreviations Used in Our Edition

x] y T: T reads y for x.
N: the Nahatas' edition or the editors (the Nahatas).

N/Ms: the manuscript used for, and mentioned in, N.
PM: Patan Manuscript.
#: metres other than *Gāhā* (or *Āryā*).

0.7 The Manuscript and Its Date

Around 1946, Agar Chand Nahata and Bhanwar Lal Nahata discovered a manuscript containing seven works by Ṭhakkura Pherū in the Śrīmaṇi Jīvan Jain Library in Calcutta. In the catalogue of the library, this manuscript was entered as *Sārā Kaumudī Gaṇita Jyotiṣa.* During the next fifteen years until it was finally published in 1961 under the title *Ṭhakkurapherū-viracita-ratnaparīkṣādi-saptaganthasaṃgraha,* the Nahatas showed the manuscript (or its photocopy) to several scholars like Dashrath Sharma, V. S. Agrwala, Moti Chandra and Jinavijaya Muni. For four of Pherū's works, viz. *Ratnaparīkṣā, Vāstusāra, Jyotiṣasāra,* and *Dhātūtpatti,* there exist one or two other manuscript copies, but no other manuscripts are known of the *Dravyaparīkṣā* and of the *Gaṇita-sārakaumudī.* Nor is there another manuscript copy of the collected works. Therefore this manuscript discovered in Calcutta in 1946 is very valuable. Unfortunately the present location of this unique manuscript is not known. Some details about the manuscript are provided by the Nahatas in their edition.[1]

This paper manuscript, consisting of 60 folios, was copied for his own use by Purisaḍa, son of Sāhu Bhāvadeva, in 1347. The leaves measure 26.7 × 9.7cm, while the written portion occupies an area of 21.6 × 8.4. The sequence of the seven texts copied is as follows: *Jyotiṣasāra* (ff. 1–18), *Dravyaparīkṣā* (ff. 19–27*r*), *Vāstusāra* (ff. 28-35), *Ratnaparīkṣā* (ff. 36–41*r*), *Dhātūtpatti* (ff. 41*v*–43*r*), *Kharataragaccha-yugapradhāna-catuḥpadikā* (ff. 43*v*–44), and *Gaṇitasāra-kaumudī* (ff. 45–60). There are colophons at three places.

1. At the conclusion of the *Dhātūtpatti* (f. 43*r*)

   śrīvikramāditye saṃvat 1403 varṣe phāguṇa śu. 8 candravāsare mṛgasiranakṣatre likhitam/ sā. bhāvadevāṅgaja purisaḍa/ ātmavācana-paṭhanārthe subham astu/

2. At the conclusion of the *Kharataragaccha-yugapradhāna-catuḥpadikā* (f. 44*v*)

   saṃ. 1403 phā. śu. 8 li⟨khitam⟩/

3. At the end of the *Gaṇitasārakaumudī* (f. 60*v*)

   likhitaṃ caitra sudi 5 saṃvat 1404/

Two dates are mentioned in these colophons. According to the Pancanga program by Michio Yano and Makoto Fushimi,[2] the first date, *saṃvat 1403 varṣa phāguṇa śu. 8 candravāsara mṛgasiranakṣatra,* translates to Monday, 19 February 1347. The second date, *caitra sudi 5 saṃvat 1404,* is somewhat problematic in the absence of the weekday, but we can say that the date is most probably Saturday 17 March 1347.

[1] Cf. SGS, Introduction, pp. 3–4, 8; see also V. S. Agrawala, 'Dhātūtpatti', *The Journal of the Uttar Pradesh Historical Society,* 24-25 (1951-52), p. 321.

[2] The program is available at http://www.cc.kyoto-su.ac.jp/~yanom/pancanga/index.html

According to the tables of Sewell and Dikshit [1896] and of Pillai [1982], 13 March 1347 = 1 Caitra. Therefore,

17 March 1347 = 5 Caitra Śaka 1270 current (1269 expired)
= 5 Caitra Vikrama 1405 current (1404 expied).

According to Schram [1908],

Julian Day 2213120 = 0 Caitra Śaka 1269 expired
= 0 Caitra Vikrama 1404 expired.

According to the Pancanga program,

Saturday, 17 March 1347 = Julian Day 2213125
= 4 Caitra Śaka 1269 expired
= 4 Caitra Vikrama 1404 expired.

Since the fraction of *tithi* at the local sunrise in Delhi is 0.901, and in this *pakṣa* the second day was repeated (as *adhidina*), we can admit the possibility that the day was called Caitra 5. In fact, if we run the same program and use the <Try> menu, we get the result,

5 Caitra Vikrama 1404 = Saturday, 17 March 1347.

We do not know when Bhāvadeva's son Purisaḍa commenced the copying of the manuscript. But he completed the *Dhātūtpatti* (i.e. up to f. 43*r*) on Monday, 19 February 1347. On the same day, he also copied the short text *Yugapradhāna Catuṣpadikā* (just three pages from f. 43*v* to f. 44*v*). Again we do not know whether he commenced the copying of the *Gaṇitasārakaumudī* on 19 February itself or later. Whichever the case may be, he completed the work (16 folios from f. 45 to f. 60) in less than a month on 17 March 1347. The last chronological reference to Pherū as shown earlier pertains to the year 1323, when he participated in a pilgrimage to Śatruñjaya in Gujarat. Thus this manuscript was copied twenty-four years later in 1347; this can be either within Pherū's life time itself or shortly thereafter.

The Nahatas' edition reproduces in facsimile four pages from the manuscript: 1. a page from the middle of the *Dravyaparīkṣā*, 2. the last page of the same work (f. 27*r*), 3. the first page of the *Vāstusāra* (f. 28*r*), and 4. the last page of the *Dhātūtpatti* (f. 43*r*).

Chapter 1

Prathamo 'dhyāyaḥ

1.001a namiūṇa ti-jaya-nāhaṃ lacch-īsa-gir-īsa-sayala-deva-jjaṃ/
1.001b lehāṇa gaṇaṇa-pāḍī puvv-āyariehi jaha-vuttā//
1.002a tatto vi kiṃci gahiyaṃ kiṃci vi aṇubhūya kiṃci suṇiūṇaṃ/[1]
1.002b taṃ sayala-loya-heū pherū pabhaṇei caṃda-suo//
1.003a paḍikāiṇi taha kāiṇi paḍivissaṃsā tah=eva vissaṃsā/
1.003b jāva ya hoṃti visovā vīsa^uṇa-kameṇa nāyavvā//[2]
1.004a vīsi visoihi dammo dammihi paṃcāsi ṭaṃkao ikko/
1.004b vīsa kama dīha-vitthari aha kaṃvī saṭṭhi vīgahao//
1.005a pavv-aṃguli caüvīsihi battīsa kar-aṃgulī ya vinneyā/
1.005b aṭṭhi javi tiriya-gehiṃ pavv-aṃgulu ikku jāṇeha//[3]
1.006a caüvīs-aṃgula hattho paṃḍiya cahuṁ hatthi havaï ḍaṃḍu igo/[4]
1.006b bihu sahasi daṃḍi koso cahuṁ kosihi joyaṇo ikko//
1.007a iya bhaṇiyaṃ sara-hatthaṃ vikkhaṃbh=āyāma-guṇiya paḍa-hatthaṃ/
1.007b vitthārahu udaya guṇaṃ taṃ ghaṇa^atthaṃ viyāṇāhi//[5]
1.008a cahuṁ kara-puḍehi pāī cahuṁ pāī egu māṇao bhaṇio/
1.008b cahuṁ māṇehi vi seī solasa seī bhave pattho//
1.009a chahiṁ guṃji māsao hui tehi vi cahu ṭaṃku ṭaṃki dasahi palo/
1.009b chahi palihi ikku sero serihi cālīsi ikku maṇo//
1.010a javi solasehi māsaü tehi vi cahu ṭaṃku tolao ti^uṇo/
1.010b solahi javehi vannī bārahi vannī mahā-kaṇao//
1.011a saṭṭhi-pali ega ghaḍiyā ghaḍiyā-saṭṭhīhi egu diṇu-rayaṇī/
1.011b diṇi-rayaṇi tīsi māso bārahi māsaṃmi varisu igo//
1.012a egaṃ daha saya sahasaṃ dasahasa lakkhaṃ tah=eva dasa-lakkhaṃ/
1.012b koḍiṃ taha dasa-koḍī avvaṃ dasa^avva jāṇeha//
1.013a khavvaṃ taha dasa-khavvaṃ saṃkhaṃ dasa-saṃkha paüma dasa-paümaṃ/
1.013b nīlaṃ taha dasa-nīlaṃ nīla-sayaṃ nīla-sahasaṃ ca//
1.014a dasa-sahasa-nīla taha puṇa nīlaṃ lakkho vi nīla-dasa-lakkhaṃ/
1.014b taha koḍi-nīla iccāi saṃkha^aṃkāiṁ nāmāiṃ//[6]

---

[1]tatto vi N] tatto va N/Ms.
[2]jāva ya] jāvaya N.
[3]gehiṃ] gehaṃ N.
[4]cahuṁ] cahuṃ N.
[5]atthaṃ, error for hatthaṃ?
[6]aṃkāiṁ] aṃkāiṃ N.

// iti 25 gaṇita-aṅka//

1.015a parikammaṃ paṇavīsaṃ tah=aṭṭha jāī ya aṭṭha vivahārā/[7]
1.015b ahigārā cār=eyaṃ paṇayālīsāiṁ dārāiṃ//[8]

1. atha saṃkalitam=āha/[9]

1.016a icchā egi juy=addhe icchā-guṇiyaṃ havei saṃkaliyaṃ/
1.016b ....//[10]

1.017a sama-diṇa-dalam=agga-guṇaṃ visamaṃ aggima-daleṇa saṃguṇiyaṃ/[11]
1.017b jaṃ hui taṃ saṃkaliyaṃ na saṃsayaṃ ittha nāyavvaṃ//
1.018a icchā paṇh-akkharihiṃ guṇijjaï paṇhu meli puṇu iccha haṇijjaï/[12]
1.018b bi^uṇihiṃ paṇhihiṃ bhāu harijjaï laddh-aṃkihi saṃkaliu kahijjaï//#[13]
1.019a egg-āī jāva dasaṃ saṃkaliyaṃ pihagu dasa-guṇāṇaṃ ca/
1.019b eg-uttara-vuḍḍhi-kame bhaṇeha eyāṇa mūla puṇo//
1.020a saṃkaliy=aṭṭha-guṇ=igi juya tassa payaṃ egi hīṇa addheṇa/
1.020b aha bi^uṇa-vagga-mūle sesa-samaṃ saṃkaliya-mūlaṃ//

2. aha vyavakalitam=āha/

1.021a jaha saṃkaliya-paeṇaṃ ikk-ikkaṃ egay-āi vaḍḍhei/[14]
1.021b taha vimakalie chijjaï ikk-ikkaṃ mūla-rāsīo//[15]
1.022a s-egaṃ vimakaliya-payaṃ saṃkaliya-payaṃ ca kīrae sahiyaṃ/
1.022b dunha pay-aṃtari guṇiyaṃ dalī-kayaṃ vimakaliya-sesaṃ//
1.023a saṃkaliya-sahassāo das-āi dasa-das-ahiyassa saṃkaliyaṃ/[16]
1.023b sohevi bhaṇasu paṃḍiya jaṃ hui vimakaliya-ses-aṃkaṃ//[17]
1.024a vimakaliya-sesa sohivi saṃkaliya-dhaṇāu sessa bi^uṇa-kayaṃ/[18]
1.024b tassa payaṃ sesa-samaṃ taṃ hui vimakaliya-mūla-payaṃ//
1.025a saṃkaliya-payaṃ bi^uṇaṃ s-egaṃ vimakaliya-paya-vihīṇa dalaṃ/
1.025b vimakaliya-guṇaṃ jaṃ hui taṃ uvarāo ya vimakaliyaṃ//[19]
1.026a saya-saṃkaliya-dhaṇāo uvarāo tīi jāma vimakaliyaṃ/
1.026b tā kiṃ jāyaï taṃ bhaṇi jaï vimakaliyaṃ viyāṇāhi//

3. atha guṇākāram=āha/

1.027a ṭhavi gunna-rāsi hiṭṭhe kavāḍa-saṃdhī va uvari guṇa-rāsī/[20]
1.027b anuloma-viloma-gaī guṇijja su-kameṇa guṇa-rāsī//
1.028a vīsā saü battīsihi nava saï caüsaṭṭha sattavīsehiṃ/

---

[7] tah=aṭṭha] taha 'ṭṭha N.
[8] paṇayālīsāiṁ] paṇayālīsāi N.
[9] N puts this line between 1.016 and 1.017.
[10] Only one half of 1.016 is extant. It consists of 12 + 18 mātrās and is the first half of a gāthā stanza. N interchanged 16a and 16b.
[11] sama-diṇa-dalam=agga-guṇaṁ] sama diṇa dala magga guṇaṃ N.
[12] paṇh-akkharihiṃ] paṇha'kkharihiṃ N; paṇhu] panhu N.
[13] Aḍillā metre.
[14] ikk-ikkaṃ] ikkikaṃ N.
[15] ikk-ikkaṃ] ikkikaṃ N.
[16] das-ahiyassa] dasa'hiyassa N.
[17] sohevi] sāhevi N.
[18] sohivi] sohi vi N; kayaṃ] juyaṃ N.
[19] guṇaṃ] juyaṃ N.
[20] saṃdhī] saṃdhi N.

1.028b aḍ-ahiya saü saṭṭhi-guṇaṃ kiṃ kiṃ patteya hoṃti phalaṃ//
1.029a aha guṇa-rāsī khaṃḍivi ig-ega^aṃkeṇa guṇavi kari piṃ.ḍaṃ/
1.029b para-kami caḍaṃti paṃtī cheya karavi su-kami guṇiya puṇo//

dutīka-guṇākāra-satya-alīka-parijñāna/

1.030a guṇa-rāsi-gunna-rāsī pihu pihu piṃḍaṃ navassa sesa-guṇaṃ/
1.030b taha guṇiya-rāsi-piṃḍaṃ nava-sesa-samaṃ havaï suddhaṃ//
1.031a kheva-samaṃ khaṃ joe rāsi avigayaṃ kha-joḍaṇe hīṇe/[21]
1.031b sunna-guṇaṇ-āi sunnaṃ sunna-guṇe sunna sunneṇa//[22]
1.032a sunnassa ya guṇayāraṃ sunnassa ya bhāgaharaṁ tahā vaggaṃ/[23]
1.032b sunnassa vagga-mūlaṃ ghaṇ-āi bhaṇi jaï viyāṇāsi//

4. atha bhāgāharam=āha/

1.033a jassāo pāḍijjaï saṃharaṇīo ju haraï sujji haro/[24]
1.033b uvari lihi hāraṇīyaṃ hiṭṭhi haraṃ saṃbhave bhāyaṃ//[25]

5. atha vargaḥ/

1.034a paḍham-aṃku vaggu ṭhaviyaṃ avara kame vi^uṇa^āi^aṃkehiṃ/[26]
1.034b guṇi puvva-sahiya puṇa taha vagga-juyaṃ ṭhāṇ-ahiya vaggaṃ//[27]
1.035a jo aṃku tiṇa ya aṃke guṇijja so vaggu ahava iccha duhā/[28]
1.035b iṭṭh-ūṇa-juya guṇeviṇu tah=iṭṭha-vagg-ahiya iya vaggaṃ//[29]
1.036a eg-āi-nav-aṃtāṇaṃ solasa-caüvīsa^aṭṭhavīsāṇa/
1.036b patteya vagga-rāsī jaṃ jāyaï taṃ bhaṇaha sigghaṃ//

6. atha varga-mūlam=āha/

1.037a jaṃ havaï vagga-rāsī tass=aṃtāo gaṇijja jāva dhuraṃ/
1.037b visama-sama visama-ṭṭhāṇe vaggaṃ sohevi mūl-aṃkaṃ//[30]
1.038a vi^uṇu kari cāli bhāyaṃ phala paṃtī tassa vaggi sohi puṇo/[31]
1.038b puvva-vihi jāva carimaṃ jaṃ vi^uṇa tam=addhiyaṃ mūlaṃ//[32]

7. atha ghanam=āha/

1.039a dhurim-aṃka-ghaṇaṃ ṭhāviya tass=eva dhur-aṃka-vaggu tihu guṇiyaṃ/
1.039b vīy-aṃke guṇiūṇaṃ ṭhāṇa-ahiya su-kami joḍijjā//[33]
1.040a puṇu vīya^aṃka-vaggaṃ dhurim-aṃkihi guṇivi ti^uṇa kari juttaṃ/
1.040b puṇu tassa ya aṃkassa ya ghaṇaṃ karivi sahiya ghaṇam=eyaṃ//[34]

---

[21]avigayaṃ kha-joḍaṇe] avigayakkhaṃ joḍaṇe N.
[22]sunna-guṇaṇ-āi] sunna guṇāṇaï N; guṇe] gaṇe N.
[23]bhāgaharaṁ] bhāgaharaṃ N.
[24]haro] hare N.
[25]haraṃ saṃbhave] haraṃsaṃ bhave N.
[26]avara kame] avarakame N; āi^aṃkehiṃ] āi aṃkehiṃ N.
[27]ahiya vaggaṃ] ahiyavaggaṃ N.
[28]tiṇa ya] tiṇaya N.
[29]iṭṭh-ūṇa] daṭṭhūṇa N.
[30]-sama visama-] -sama-visama- N; sohevi] sāhevi N.
[31]phala paṃtī] phalapaṃtī N.
[32]jaṃ] ∅ N.
[33]guṇiūṇaṃ] gaṇiūṇaṃ N.
[34]aṃkassa] aṃsassa N.

1.041a icchiya^aṃku ti-hā ṭhavi uvaruppari guṇiya jaṃ havaï sa ghaṇo/
1.041b aggimu puvv-aṃki hayaṃ ti^uṇaṃ puvva-ghaṇa-juya s-egaṃ//[35]
1.042a eg-āī jāva navaṃ taha solasa-du-saya-paṃcavīsāṇa/
1.042b tinni-saya-nav-ahiyāṇaṃ patteyaṃ kiṃ havei ghaṇaṃ//[36]

8. atha ghana-mūlam=āha/

1.043a ghaṇa-paya do^aghaṇa-pae ghaṇa-paya-bhāyaṃ ghaṇeṇa pāḍijjaṃ/[37]
1.043b taṃ laddh-aṃkaṃ mūlaṃ cālivi taïy-aṃka-tali dijjā//[38]
1.044a tav-vaggu ti^uṇu tass=eva pacchae dharivi bhāu pāḍijjā/[39]
1.044b laddhaṃ paṃti ṭhavijjaï har-aṃka-vigamo ya kāyavvo//
1.045a paṃtissa aṃta-vaggaṃ ti^uṇaṃ puvv-aṃki guṇivi sohijjā/[40]
1.045b aṃti payassa ghaṇaṃ puṇa sohiya taṃ laddha puṇa evaṃ//

atha bhinna-parikarma-aṣṭakam=āha/[41]

1.046a bhinn-aṃku ima ṭhavijjaï rū uvari aṃsassa majjhi cheya tale/
1.046b hīṇ-aṃse biṃdu-cayaṃ accheyaṃ jattha tatth=egaṃ//
1.047a cheya-haya-rūva-rāsī aṃsā juya gaya savaṃnaṇaṃ havaï/
1.047b annonna-cheya-guṇiyā havaṃti kami sadisa-cchey-aṃsā//
1.048a sadisa-ccheya karevinu tā kīraï joḍa hīna aṃsāṇaṃ/
1.048b na havaï cheyāṇa juī kayā vi iya bhaṇiya satthehiṃ//
1.049a chey-aṃke vi^uṇa kae uvarima-rāsī havei addhīya/
1.049b savve vi pāyarehiṃ bhinna-ṭhiī esa nāyavvā//

9. atha bhinna-saṃkalitam=āha/[42]

1.050a sadisa-cchey-aṃsa-juī cheeṇa vihatta bhinna-saṃkaliyaṃ/
1.050b ti-cha-paṇa-nav-aṃsa-piṃḍaṃ taha paüṇa-ti diuḍha sa-ti-hāyaṃ//

10. atha bhinna-vyavakalitam=āha/[43]

1.051a ādāyassa vayassa ya savaṃnaṇaṃ karavi sadisa-cheya puṇo/
1.051b pihu pihu aṃsāṇa juī tay-aṃtare bhinna-vimakaliyaṃ//
1.052a addha ti-hāya khaḍ-aṃsā nav-aṃsu aṭṭhāu sohi kiṃ sesaṃ/
1.052b saḍḍha-tiya paṃca sa-ti-hā nav-aṃsa khaḍ-aṃsāu sohijjā//

11. atha bhinna-guṇākāram=āha/

1.053a aṃseṇa aṃsa guṇiyaṃ cheeṇa vi cheya guṇivi hariyavvaṃ/
1.053b jaṃ havaï laddham=aṃkaṃ taṃ jāṇaha bhinna-guṇayāraṃ//
1.054a pā^ūṇa-paṃca-dammā guṇijja sa-ti-hāya^aṭṭha-dammehiṃ/
1.054b addhaṃ khaḍ-aṃsi guṇiyaṃ pihu pihu kiṃ havaï tassa phalaṃ//

---

[35]-juya s-egaṃ] -juyasesaṃ N.
[36]nav-ahiyāṇaṃ] nava'hiyāṇaṃ N.
[37]do^aghaṇa-pae] doa ghaṇapae N.
[38]taïy-aṃka] taīyaṃka N.
[39]tav-vaggu] tavaggu N.
[40]aṃta] aṃka N.
[41]N puts the serial number 9 at the head of this line. bhinna] abhinna N.
[42]N puts the serial number 10 at the head of this line.
[43]N does not have this line.

12. atha bhinna-bhāgāharam=āha/

1.055a kariūṇa cheya aṃsā harassa vivarīya na haraṇīyassa/[44]
1.055b puvva-vihi guṇivi bhāyaṃ esa vihī bhinna-bhāyassa//[45]
1.056a aḍḍhāiehi bhāyaṃ harijjae paüṇa-satta-dammehiṃ/
1.056b cahu sa-ti-hāi vihattaṃ sa-vā-cha kiṃ tāṇa laddha-phalaṃ//

13. atha bhinna-vargam=āha/

1.057a aṃsāṇa vagga-rāsī hiṭṭhima-cheyāṇa vagga bhāeṇa/
1.057b pāḍevi jaṃ ji laddhaṃ taṃ jāṇahu bhinna-vagga-phalaṃ//[46]
1.058a aḍḍhāiyassa vaggaṃ sa-ti-hā-paṃcassa paüṇa-sattassa/
1.058b bhaṇi addha ti-hāya puṇo jaï vagga-vihī viyāṇāsi//

14. atha bhinna-varga-mūlam=āha/

1.059a aṃsassa vagga-mūle cheyaṇa-mūleṇa bhāu pāḍijjā/
1.059b visama-sama-visama-karaṇe hui mūlaṃ bhinna-vaggassa//

15. atha bhinna-ghanam=āha/

1.060a aṃsassa ghaṇaṃ kujjā cheyassa ghaṇāṇa bhāu hariūṇaṃ/
1.060b jaṃ kiṃ pi tattha laddhaṃ bhinna-ghaṇaṃ taṃ viyāṇāhi//[47]
1.061a saḍḍhaya-sattassa ghaṇaṃ sa-vāya-pannara sa-pā-ti-hāyassa/[48]
1.061b jaṃ jāyaï ghaṇa-rāsī patteyaṃ taṃ bhaṇijjāsu//

16. atha bhinna-ghaṇa-mūlam=āha/

1.062a aṃsa-ghaṇa-mūla-rāse cheyaṇa-ghaṇa-mūla bhāu pāḍijjā/[49]
1.062b ghaṇa-paya do^aghaṇa-pae iya karaṇe havaï ghaṇa-mūlaṃ//[50]

17. atha trairāsikam=āha/

1.063a āi^aṃt=eka-jāī ṭhavijjae anna-jāi majjheṇa/[51]
1.063b aṃteṇa majjhi guṇiyaṃ āima-bhāgaṃ tirāsiyagaṃ//
1.064a jā ikkārasa-daṃmihi dosiya kara-satta kappaḍo hoi/
1.064b tā caüvīsihi dammihi kaï hattha havaṃti te kahasu//
1.065a bhaṇisu hava nāṇa-vaṭṭaṃ nava muṃda lahaṃti damma paṇavīsaṃ/
1.065b iya aggha-pamāṇcṇaṃ solasa muṃdāṇa kaï mullaṃ//
1.066a caṃdaṇa-palaṃ sa-vāyaṃ sa-ti-hā-nava damma mullu pāvei/
1.066b tā cha-pala khaḍ-aṃs-ūṇā kittiya dammāiṁ pāvaṃti//[52]
1.067a dammi sa-vā-sattehiṃ pippali dui-sera chaṭṭham-aṃs-ahiyā/[53]
1.067b labbhaï tā nava-dammihi ti-hāya^ūṇehiṁ kiṃ havaï//[54]

---

[44] haraṇīyassa] hāraṇīyassa N.
[45] guṇivi bhāyaṃ] guṇi vibhāyaṃ N.
[46] jāṇahu N] jāṇa N/Ms.
[47] jaṃ] ja N.
[48] sa-vāya-pannara sa-pā-ti-hāyassa] savāya panarasa pā tihāyassa N.
[49] cheyaṇa-ghaṇa-mūla bhāu] cheyaṇaghaṇa mūlabhāu N.
[50] do^aghaṇa-pae] doa ghaṇapae N.
[51] Regard aṃ of aṃt=eka as two short syllables (a-ṃ).
[52] dammāiṁ] dammāiṃ N.
[53] chaṭṭham-aṃs-ahiyā] chaṭṭhamaṃsa'hiyā N.
[54] ūṇehiṁ] ūṇehiṃ N.

1.068a pā^uṇa-vīsa-saehiṃ dammihi sa-ti-hāya-paṃca-patthā ya/[55]
1.068b tā taṃdul-āi annaṃ kaï labbhaï ikki dammeṇa//
1.069a bāraha-vannī kaṇao sa-ti-hā-saya-dammi tolao ikko/
1.069b jaï hui ta ikki-māsaya-das-aṃsa-hīṇassa kaï mullo//
1.070a jaï joyaṇa-chaṭṭh-aṃsaṃ paṃgulao calaï satta-divasehiṃ/
1.070b tā saṭṭhi-joyaṇāiṃ kittiya kāleṇa gacchei//
1.071a aṃgula-satt-aṃso jaï diṇassa chaṭṭh-aṃsi kīḍao calaï/
1.071b gacchihaï aṭṭha joyaṇa niyattaī keṇa kāleṇa//

atha paṃca-rāsikam=āha/ sapta-nava-ekādasa-rāsiko ya/

1.072a hiṭṭhima-phal-aṃka vivariya pihu pihu kami do vi pakkha guṇiūṇaṃ/
1.072b thov-aṃka-rāsi-bhāyaṃ paṇa-satta-nav-āi-rāsīṇaṃ//

18. atha paṃca-rāśikam=āha/

1.073a māseṇa paṃcaga-sae varise saṭṭhissa kiṃ phalaṃ havaï/
1.073b aha no najjaï kālaṃ phala mūlaṃ taha pamāṇa-dhaṇaṃ//
1.074a māse ti-hāya^ūṇe saḍḍha-sae divaḍhu dammu vavahāro/
1.074b tā satarahi pā^ūṇihiṁ sa-vāya-nava-māsa kiṃ havaï//[56]
1.075a saḍḍh-aṭṭha-maṇahaṁ bhāḍaï joyaṇa-sa-ti-hāi damma-paüṇa-due/[57]
1.075b tā nava-sa-vā-maṇāṇaṃ kiṃ hui dasa-joyaṇe paüṇe//
1.076a jaï vārasa-kamma-yarā cahu divasihi tīsa-damma pāvaṃti/
1.076b paṇayālīsa-diṇehiṃ tā kiṃ pāvaṃti aṭṭha-jaṇā//
1.077a jaï kiri bhitti-suvanno guṃj-ūṇa-ti-māsa paüṇa-vīsa dhaṇe/
1.077b tā saḍḍha-dasī vannī guṃj-ahiya-du-māsa kaï mullaṃ//

19. atha sapta-rāśikam=āha/

1.078a chad-dīha ti-kara-vitthara dui kaṃvala navaï damma pāvaṃti/[58]
1.078b nava-dīha paṃca vitthari tā kaṃvala satta kaï mullaṃ//

20. atha nava-rāśikam=āha/

1.079a cīra vāraha paṃca vannehi/
1.079b te dīhaṇa satta-kara tinni-hattha-vitthāru acchaï/
1.079c tahaṃ savvahaṃ mullu kiu cha-saya damma dosiyahi nicchaï/
1.079d jaï cahuṁ vannihi aṭṭha kara dīhi paṃca vitthāri/[59]
1.079e tā nava-cīraha mullu kaï kahi dosiya viccāri//#[60]

21. atha ekādaśa-rāśikam=āha/[61]

1.080a du-cha-ti-du^iga patth-āī jā kara-puḍa muṃga saṭṭhi-dammehiṃ/[62]
1.080b tā nava-ti-du^iga-ti-kame patth-āī muṃga kaï mullaṃ//[63]

---

[55]vīsa] vīsā N.
[56]ūṇihiṁ] ūṇihiṃ N.
[57]maṇahaṁ] maṇahaṃ N.
[58]chad-dīha] cha dīha N.
[59]cahuṁ] cahuṃ N.
[60]Raḍḍā metre. kaï] kaïṃ N.
[61]rāśikam=āha] rāśayo āha N.
[62]kara-puḍa] kara puḍa N.
[63]du^iga] duga N.

22. atha vyasta-trairāśikam=āha/[64]

1.081a majjhaṃ ca āi-guṇiyaṃ aṃteṇa vihatta vittha-tiyarāsī/
1.081b aṃt-āi ega-jāī ṭhavi majjhe anna-jāīya//[65]
1.082a daha-seiyaṃmi patthe maviyā satt-ahiya-vīsa-patthāiṃ/
1.082b solasi seī patthe kaï pattha havaṃti te kahasu//[66]
1.083a chāsaṭṭhi-ṭaṃka-tulle tuliyā maṇa-vīsa vakkhar-aṃtaïyā/
1.083b jaï vāhattari-tulle tuliyaṃti havaṃti kitiya maṇā//[67]
1.084a saḍḍh-ikkārasa-vannī tolā-cālīsa-saḍḍha kaṇao ya/
1.084b tā dasa-sa-vāya-vannī pavaṭṭaṇe havaï kevaḍao//[68]
1.085a nava^āyāma ti-vitthara dui-saï vīs-ahiya kaṃbalā savve/
1.085b paṃc-āyāma du-vitthara kaï kaṃvala hoṃti te kahasu//

23. atha kraya-vikarayam=āha/

1.086a majjh-aṃta guṇiya mūlaṃ aṃt-āī guṇiya savvaüppattī/[69]
1.086b vikaya-kay-aṃtari bhāyaṃ nāijjaï mūla lāha-dhaṇaṃ//[70]
1.087a sataraha maṇa ṭaṃkeṇaṃ lijjahi pannarasa vikkiṇijjaṃti/
1.087b jaï dasa ṭaṃkā lāhe tā kahu ṭaṃkāṇa te mūle//
1.088a tihu dammi paṃca vatthū lijjahi navi dammi satta vikkijjā/
1.088b daṃma duvālasa lāhe kittiya dammāṇa sā mūle//
1.089a uvari damma tali vatthu ṭhavijjahi vaṃkaï vinni vi rāsi guṇijjaï/
1.089b āima-rāsi lāhi tāḍijjaï vihū rāsi^aṃtari pāḍijjaï//#[71]

24. atha bhāṃḍa-pratibhāṃḍakam=āha/

1.090a bhaṃḍa-paḍibhaṃḍa-karaṇe vivariya mullaṃ phalaṃ ca vivarīyaṃ/
1.090b kami guṇavi do vi rāsī harijja lahu-rāsiṇā bhāyaṃ//
1.091a saï dammi du-maṇa pippali tihu saya-dammehi paṃca-maṇa suṃṭhī/
1.091b tā pippali-satta-maṇe pāvijjaï soṃṭhi kitiya maṇā//

25. atha jīva-vikraya-karaṇam=āha/

1.092a jīvassa vikkaeṇa ya varisa vivarīya phal-aṃka vivarīyaṃ/
1.092b sesaṃ ca puvva-vihiṇā jāṇijjahu jīva-vara-mullaṃ//
1.093a dasa-varisā tiya karahā ṭaṃkā saü aṭṭha^ahiya pāvaṃti/
1.093b tā nava-varisā karahā kaï mullaṃ havaï paṃcāṇa//

// iti parama-jaina-śrī-candra-aṃgaja-ṭhakkura-pherū-viracitāyāṃ gaṇita-sāra-kaumudī-pāṭyāṃ paṃcaviṃśati-parikarmma-sūtra samāptāni//[72]

// iti prathamo 'dhyāyaḥ//

---

[64] trairāśikam=āha] trairāśikom=āha N.
[65] anna] aṃta N.
[66] kaï] kaha N.
[67] tuliyaṃti] tuliyaṃ ti N.
[68] kevaḍao (S. kiyat)] kevaïo N.
[69] guṇiya mūlaṃ] gaṇiya mūlaṃ N.
[70] vikaya-kay-aṃtari] vikaya kayaṃtari N.
[71] Aḍillā metre.
[72] aṃgaja] aṅgaja N; pherū] pheru N.

Chapter 2

Dvitīyo 'dhyāyaḥ

1. atha bhāga-jātau kalā-savarṇṇanam=āha/

2.001a sama-cheya karavi pacchā aṃsa-juī havaï bhāga-jāī ya/
2.001b addhassa addhu tassa ya paṇ-aṃsa-chaṭṭh-aṃsu kiṃ havaï//[1]

2. atha prabhāga-jātim=āha/

2.002a cheeṇa cheya guṇiyaṃ aṃse aṃsā pabhāga-jāī ya/
2.002b addhassa addhu tassa ya paṇ-aṃsa-chaṭṭh-aṃsu kiṃ havaï//[2]

3. atha bhāga-bhāga-jātim=āha/

2.003a cheeṇa rūva-guṇie cheya-game havaï bhāga-bhāga-vihī/[3]
2.003b aṃsāṇa juī bhāyaṃ dhaṇeṇa pihag=aṃsa guṇavi phalaṃ//[4]
2.004a egi ti-bhāya-du-bhāyaṃ egisu nava-bhāya-satta-bhāyaṃ ca/[5]
2.004b egi cha-bhāya-ti-bhāyaṃ kiṃ saya-dammāṇa pihagu phalaṃ//[6]
2.005a vāvi cha-kara caü nālaya bharaṃti kami diṇ-igi dala ti-caür-aṃso/[7]
2.005b jaï sama-kāli vimuccahi tā pūrahi keṇa kāleṇa//[8]

4. atha bhāga-anubaṃdham=āha/

2.006a aha-hari uvarimu haru guṇi sa^aṃsi hiṭṭhima-hareṇa guṇi rūvaṃ/[9]
2.006b jā havaï carima-cheyaṃ esā bhāga-aṇubaṃdha-vihī//
2.007a saḍḍha-tiya tassa pāyaṃ sahiyaṃ jaṃ tassa chaṭṭham-aṃsa-juyaṃ/
2.007b tass=addha-jutta kiṃ hui tah=addha sa-ti-hāya tassa pāya-juyaṃ//#[10]

---

[1]= 2.002b. addhu N] adhu N/Ms.
[2]= 2.001b. addhu N] adhu N/Ms.
[3]rūva-] rūvaṃ PM; cheya] chea PM.
[4]Cited in PM A33. aṃsāṇa juī] aṃsaṃ juie PM; bhāyaṃ] bhāgaṃ PM; pihag] pihug PM; guṇavi] guṇa PM; phalaṃ] phala PM.
[5]egi] ega PM; egisu] egi su N.
[6]Cited in PM A34. egi] ega PM; dammāṇa] damma PM; pihagu] pihaga PM.
[7]aṃso] aṃsaṃ PM.
[8]Cited in PM A40. vimuccahi] ti muccahi N, vimuccaï PM; pūrahi] pūraï PM; keṇa kāleṇa] kittieṇa kālena PM.
[9]haru guṇi] haru gaṇi N; sa^aṃsi] sa aṃsi N.
[10]Uggāhā metre.

5. atha bhāg-apavāham=āha/

2.008a hiṭṭhima-hari uvarima-haru guṇijja hiṭṭhima-hare gay-aṃseṇa/
2.008b uvarima-rūva guṇijjahi evaṃ bhāga-apavāhaṃ ca//
2.009a tiya addh-ūṇaṃ paüṇaṃ tassa khaḍ-aṃs-ūṇa taha ya addhaṃ ca/
2.009b t-aṃsa-caür-aṃsa-rahiyaṃ kiṃ kiṃ patteya hoṃti phalaṃ//

6. atha bhāga-mātṛ-jātau āha/

2.010a bhāg-āi-paṃca-jāī-samāsaṇaṃ taṃ ca bhāga-mattīya/[11]
2.010b pihu pihu jah-utta-karaṇaṃ karevi sama-cheya^aṃsa-juī//
2.011a addhaṃ payassa pāyaṃ ti-bhāya-bhāyaṃ tah=addha addh-ahiyaṃ/[12]
2.011b taïy-aṃsu addha-hīṇaṃ egaṭṭhaṃ kiṃ havei dhaṇaṃ//

7. atha vallī-savarṇane āha/

2.012a vallī-savannaṇa-vihī hiṭṭhima-cheeṇa guṇavi chey-aṃsā/
2.012b uvarima^aṃse riṇu dhaṇu pakīrae hiṭṭhim-aṃsāṇa//
2.013a dui tolā tiya māsā tah=eva caü guṃja paṃca visuvā ya/
2.013b te sattam-aṃsa-hīṇā savannaṇe kiṃ havaï vallī//

8. atha sthaṃbh-aṃsaka-jātau āha/

2.014a sama-cheya^aṃsa-piṃḍaṃ rūvāo sohi jaṃ havaï sesaṃ/
2.014b teṇa paccakkha bhāyaṃ laddh-aṃke thaṃbha-parimāṇaṃ//[13]
2.015a addha-khaḍ-aṃsa-duvālasa^aṃsā jala-paṃka-vāluya-ttha kame/
2.015b paccakkha tinni-kaṃviya bhaṇi paṇḍiya thaṃbha-parimāṇaṃ//
2.016a bhāū paṃcamu gayaü puv-addhi/
2.016b dakkhiṇa aṭṭhamaü solas-aṃsu pacchima paṇaṭṭhaü/
2.016c cāuddhu gaü uttaraha sīha-bhaïṇa ima chaṭṭhu naṭṭhaü/
2.016d talaï rahiu paṃḍiya nisuṇi gorū saü igayālu/[14]
2.016e te ikkaṭṭhā jaï karahi kaï loḍaï thaṇa-vālu//#[15]
2.017a addhu sa-ti-hāu viṃjjhe khaḍ-aṃsu satt-aṃsa^ahiu jala-tīre/[16]
2.017b aṭṭh-aṃsu nav-aṃsa-ahiu thali gaya caü sesa kiṃ jūhe//[17]

// iti parama-jaina-śrī-candra-aṅgaja-ṭhakkura-pherū-viracite gaṇita-sāre kaumudī-pāṭyāṃ aṣṭau bhāga-jātayaḥ//

// iti dvitīyo 'dhyāyaḥ//

[11] bhāg-āi-paṃca] bhāgāī paṃca N.
[12] pāyaṃ] bhāya N.
[13] Regard pac of pacchakkha as two short syllables (pa-c).
[14] igayālu] paṇayālu N. For this emendation see under GSK 2.16 in Part IV.
[15] Raḍḍā metre.
[16] addhu] adhu N.
[17] nav-aṃsa-ahiu] navaṃsahiu N.

Chapter 3

Tṛtīyo 'dhyāyaḥ

1. atha vyavahāra-gaṇanāyāṃ miśraka-vyavahāre āha/

3.001a niya-kāli pamāṇa-dhaṇaṃ phaleṇa para-kālu guṇavi taj-joyaṃ/[1]
3.001b missi guṇiūṇa donna vi joya-vihattammi phala-mūlaṃ//[2]
3.002a māseṇa paṃcaga-sae cahu māsihi damma paṃca-saï vīsā/
3.002b tassa phalaṃ kiṃ mūlaṃ jaï muṇasi ta bhaṇasu sigghenạ//

2. atha bhāvyake āha/

3.003a niya-kāli pamāṇa-dhaṇaṃ guṇijja phala-kāli kami phal-āīṇi/
3.003b aṃsāṇa juī-bhāyaṃ missi guṇavi laddha mūl-āī//
3.004a māse sayassa paṇa phalu egaṃ vippassa addhu vittī ya/[3]
3.004b lehaga pāyaṃ varise nava-saya paṃc-ahiya missa-dhaṇaṃ//

3. atha eka-patrī-karaṇe āha/

3.005a gaya-kāla-phala-samāse māsa-phal-akkeṇa bhāi kālo ya/
3.005b māsa-phalu piṃḍu saya guṇi dhaṇa-piṃḍe hari sayassa phalaṃ//
3.006a dugi tigi caü-paṃcaga-saï māse dhaṇu dinnu ega du ti cha sayaṃ/
3.006b cahu chahi das-aṭṭha-māsihi egaṃ pattaṃ kahaṃ havaï//

4. atha prakṣyepake aha/[4]

3.007a sama-cchey-aṃsa-juī hara missaṃ patteya aṃsi guṇiūṇa/
3.007b pakkheva-karaṇam=eyaṃ missāu phalaṃ muṇijjei//
3.008a duṃni-tiya-paṃca-caü-maṇa bīyaṃ pakkhaviya taṃ ca nippannaṃ/
3.008b bi-saya daha-uttara hala-hari dinnāṇa vi kim=aha bhinna-phalaṃ//
3.009a ṭaṃk-aṭṭha chahi vuṇijjahi mulle dasa ṭaṃka paṃci dammehiṃ/
3.009b ṭaṃka chayāsī paṭṭaṃ kiṃ vuṇiyaṃ kiṃ vuṇāvaviyaṃ//
3.010a kaṃcolu satta udae satta=uvari egu hiṭṭhi vikkhaṃbhā/
3.010b dasi dammi bhariu caṃdaṇi aṃguli ikk-ikki kaï mullo//

[1] guṇavi] vi N.
[2] joya-vihattammi] joyāvihatammi N.
[3] vittī ya] vittīya N.
[4] N suggested prakṣepake for prakṣyepake.

5. atha sama-viṣama-krayayoḥ āha/

3.011a mulle vatthu vihattho pihag=aṃsa-guṇe ya aṃsa-juva-bhāvaṃ/[5]
3.011b davveṇa aṃsa guṇio sama-visama-kayaṃ ti-rāsi-vihi puvvaṃ//#[6]
3.012a damm-ikki seru haraḍaï tinni baheḍā cha-sera āmalayā/
3.012b bho vijja dehi phakkiya sama-mattā ikka-dammassa//
3.013a tihuṁ addhu seru pippali sa-ti-hā-navi dammi miriya seru igo/
3.013b cahuṁ paüṇu seru suṃṭhī igassa ti^uḍū samaṃ dehi//
3.014a daṃmi nava-sera taṃdula ikkārasa muṃga seru ikku ghio/
3.014b ti-du^iga^aṃsa vaṇiya kami sa-vāya-dammassa me dehi//

6. atha suvarṇṇa-vyavahāre āha/

3.015a jasu vannā jaṃ tullaṃ taṃ teṇa guṇevi kīrae piṃḍaṃ/[7]
3.015b tulli vihatte vannī vannī-bhāe havaï tullaṃ//
3.016a nava dasa aṭṭh=ikkārasa vannī tolā ya tiya cha paṇa juyalaṃ/[8]
3.016b egattha gāliyaṃ taṃ kerisa vannī havaï kaṇayaṃ//

7. atha suvarṇṇe bhinna-udāharaṇam=āha/

3.017a aṭṭha sa-vā nava paüṇa cha vannī tulle ti paṃca dui māsā/[9]
3.017b tiya-cha-paṇa^aṃsa-sahiyā āvaṭṭe kiṃ havaï kaṇayaṃ//

8. atha pakva-suvarṇṇasya āha/

3.018a vanna-suvanna-guṇ-ikkaṃ vipakka-kaṇae vihatta vannāya/[10]
3.018b icchā-vannī-bhāe pakka-suvannassa tull-aṃkaṃ//
3.019a cha-paṇ-aṭṭha-satta-tolaya nava-satta-das-aṭṭha-vanna pakkāya/[11]
3.019b saṃjāya vīsa-tolā kerisa-vannī havaï kaṇayaṃ//

dutīkaḥ/

3.020a satt-aṭṭha-nava-cha-vannā caü-paṃca-ti-satta-tolayā kamaso/
3.020b ikkārasī ya vannī tulle kiṃ havaï pakkavio//[12]

9. atha naṣṭa-suvarṇṇa-varṇṇam=āha/

3.021a uppanna-vannaeṇaṃ suvanna-piṃḍaṃ guṇevi sohijjā/[13]
3.021b vanna-suvanna-vah-ikkaṃ gaya-vanna-suvannae bhāyaṃ//
3.022a tiya-paṃca-satta-māsā nav-aṭṭha-dasa-vanna aṭṭha-mās=anne/
3.022b uppannā dasa-vannā kā vannī aṭṭha-māsāṇaṃ//[14]
3.023a uppanna-vanna-tāḍiya-kaṇaya-juī vanna-kaṇaya-vaha-piṃḍaṃ/[15]
3.023b sohivi bhāyaṃ gaya-kaṇaya-vanni uppanna-vann-ūṇe//

---

[5] vihattho] vi hattho N; guṇe ya] guṇeya N; aṃsa-juva] aṃsa juva N.
[6] Uggāhā metre. aṃsa guṇio] aṃsaguṇio N.
[7] jasu vannā] ja suvannā N.
[8] tolā ya] tolāya N.
[9] tulle ti] tulleti N.
[10] vihatta vannāya] vihattavannāya N.
[11] vanna pakkāya] vannapakkāya N.
[12] ikkārasī ya] ikkārasīya N.
[13] uppanna] upanna N.
[14] uppannā] upannā N.
[15] uppanna] upanna N.

3.024a ahiyassa hīṇa ccheyaṃ hīṇassa ya ahiya iccha-vannīo/
3.024b chey-aṃka tulla-bhāgā iya icchā-karaṇa-vanna-vihī//
3.025a paṇa-satta-nava^igārasa-vannīo pihagu pihagu kiṃ lijjā/
3.025b jeṇa hui dasī vannī tulle tol=ikku taṃ bhaṇasu//

// iti miśraka-vyavahāram//

1. atha seḍhī-vyavahāro yaḥ[16]

3.026a gacch=eg-ūṇ-uttara-haya saḥ-āi aṃta-dhaṇu puṇa vi āi-juyaṃ/
3.026b du-vihatta majjhima-dhaṇaṃ gaccha-guṇaṃ havaï savva-dhaṇaṃ//
3.027a vīs=āi paṃca uttara satta diṇe turiya-haraḍaī-māṇaṃ/
3.027b taṃ bhaṇi taha naṭṭh-āī^uttara-gacchaṃ puṇo bhaṇasu//

2. atha naṣṭa-ādy-ānayane karaṇam=āha/

3.028a naṭṭh-āi-jāṇaṇ-atthe savva-dhaṇaṃ gaccha-bhatta laddhāo/[17]
3.028b eg-ūṇa-gacchi uttaru guṇevi dali sohi ses=āī//[18]

3. atha naṣṭa-uttara-ānayane karaṇa-sūtram=āha/

3.029a uttara-naṭṭh-āṇayaṇe gaccheṇa vihatta savva-dhaṇa-rāsī/
3.029b āi-vihīṇaṃ kāuṃ nir-ega-gaccha-dala-laddha cayaṃ//[19]

4. atha naṣṭa-gaccha-ānayane āha/

3.030a aḍaüttara-haya-gaṇiyaṃ du-guṇ-āī-vuḍḍhi-hīṇa-vagga-juyaṃ/[20]
3.030b mūlaṃ dhaṇa-vi^uṇ-ūṇaṃ sa-cayaṃ caya-vi^uṇa hari gacchaṃ//[21]

5. atha saṃkaliti-aikya-ānayane āha/

3.031a iga-caya-saṃkaliy-aṃkaṃ vi-jueṇa paeṇa guṇivi tihu bhāyaṃ/
3.031b laddhaṃ saṃkaliya-juī na saṃsayaṃ ittha nāyavvaṃ//
3.032a saṃkaliya vagga taha ghaṇa pihu pihu paṃcāṇa kiṃ havaï ikkaṃ/[22]
3.032b gaṇiūṇa bhaṇasu sigghaṃ jaï gaṇiya-vihiṃ viyāṇāsi//[23]

6. atha varga-eka-ghana-aikya-ānayanam=āha/

3.033a iccha-paya bi^uṇa s-egaṃ ti-hariya saṃkaliya-guṇiya vagga-juı/
3.033b saṃkaliya-vaggu jaṃ hui taṃ ghaṇa-piṃḍaṃ viyāṇehi//

7. atha saṃkalita-vargga-ghana-aikya-ānayane āha/

3.034a s-ega-bi^uṇa-paya paya-guṇa-s-ega-pay-addheṇa guṇiya hui jaṃ taṃ/
3.034b saṃkaliya vagga taha ghaṇa tinhāṇa juī muṇeyavvaṃ//

// iti seḍhī-vyavahāre sūtra-gāthā sammattā//

---

[16]yaḥ N/Ms] yathā N.
[17]gaccha-bhatta laddhāo] gaccha bhattaladdhāo N.
[18]eg-ūṇa-gacchi uttaru] egūṇa gacchiu tarū N.
[19]nir-ega] nirare N.
[20]gaṇiyaṃ] guṇiyaṃ N.
[21]dhaṇa-vi^uṇ-ūṇaṃ] dhaṇa vi uṇūṇaṃ N.
[22]ghaṇa] ghaṇaṃ N.
[23]jaï] jaya N. Cf. kaï in 4.049e.

atha kṣetra-vyavahāram=āha/

3.035a caüraṃsa dīha-caürasa vikkhaṃbh-āyāmu guṇiya taṃ khettaṃ/
3.035b caüraṃse cha-kara bhuyā ti-paṃca-kara dīha-caüraṃse//
3.036a bhuva-piṃḍ-addhaṃ caühā kameṇa bhuva-hīṇa sesa guṇa su-kame/
3.036b tassa pae taṃ khittaṃ ti-bhue a caü-bbhue jāṇa//
3.037a muha-bhuva kara-paṇavīsaṃ bhūmi-bhuvaṃ saṭṭhi vāma vāvannaṃ/
3.037b dāhiṇa uṇayālīsaṃ kiṃ jāyaï tassa khitta-phalaṃ//
3.038a bhūmi-bhuva hattha-caüdasa terasa egaṃ ca bīya pannarasaṃ/
3.038b eyaṃ visama-ti-koṇaṃ khitta-phalaṃ assa kiṃ havaï//
3.039a sayalāṇa caürasāṇaṃ bhū-muha-joy-addha laṃba-guṇa khittaṃ/
3.039b t-aṃsāṇa bhū-bhuv-addhaṃ laṃba-guṇaṃ havaï khitta-phalaṃ//
3.040a bhuva-juva terasa panarasa bhū-bhuva igavīsa paṃca hattha muhe/
3.040b majjhe laṃbu duvālasa erisa-khittassa kiṃ māṇaṃ//
3.041a ti-kkoṇa-phalaṃ vi^uṇaṃ bhū-bhattaṃ majjha-laṃbao havaï/[24]
3.041b bhuva-laṃba-vagga^aṃtari sesassa pae havaï ahavā//
3.042a bhuva-laṃba-vagga-piṃḍaṃ tassa pae havaï nicchayaṃ kannaṃ/
3.042b savvattha khitta-gaṇaṇe esa vihī havaï nāyavvā//[25]
3.043a vikkhaṃbha-vagga daha-guṇa tam-mūle vaṭṭa-khitta-parihi dhuvaṃ/
3.043b vikkhaṃbha-pāya-guṇiyā parihī tā havaï khitta-phalaṃ//
3.044a dasa-vikkaṃbhe khitte sama-vaṭṭe kiṃ pi jāyae parihī/
3.044b gaṇiūṇa bhaṇahi paṃḍiya tasu khitta-phalassa kiṃ havaï//[26]
3.045a vaṭṭassa ya vikkhaṃbhaṃ ti^uṇaṃ taha chaṭṭham-aṃsa-juya parihī/
3.045b vikkhaṃbh-addhe guṇiyaṃ parihi-dalaṃ tassa khitta-phalaṃ//[27]
3.046a jīvā-sara-piṃḍ-addhaṃ sara-guṇiyaṃ vagga daha-guṇaṃ kāuṃ/
3.046b nava-bhāe jaṃ laddhaṃ tassa pae havaï dhaṇuha-phalaṃ//
3.047a dhaṇu-piṃḍe igavīsaṃ jīvā panarasa cha havaï jassa saraṃ/[28]
3.047b bhaṇi paṃḍiya gaṇiya-phalaṃ kiṃ jāyaï tassa dhaṇu-khittaṃ//
3.048a sara-vaggaṃ cha-guṇa-kiyaṃ jīvā-vagg-ahiya mūla dhaṇu-piṃḍaṃ/
3.048b dhaṇu-vaggāo jīvā-vagg-ūṇa cha-bhāya mūla saraṃ//
3.049a dhaṇu-sara-juy-addha-hīṇaṃ dhaṇuhāo vagga caüṇa paya jīvā/
3.049b patteya gaṇiya-māṇaṃ eyāṇa phalaṃ havaï nūṇaṃ//
3.050a bāl-iṃde ti-bhuva-dugaṃ muruje do dhaṇuha caürasaṃ majjhe/
3.050b do dhaṇuha jav-ākāre kulise caü-bhuva du kappijjā//
3.051a ti-bhuvaṃ gaya-daṃta-uvama caü-bbhuvaṃ sagaḍa-cakka-vaṭṭa-samaṃ/
3.051b caṃdassa sarisa dhaṇuhaṃ vaṭṭaṃ paripunna-caṃda-samaṃ//
3.052a bāl-iṃda-uvama-khittaṃ vitthāre paṃca vīsa kara dīhe/
3.052b dala laṃbaṃ tinni dharā gaya-daṃte kiṃ havei phalam//
3.053a nimm-āgāre khitte ubhaya-muhe ti kara paṃca kara laṃbe/
3.053b dharā-muhe paṇa hatthaṃ ti majjhi daha laṃba kulis-uvamo//[29]

// iti kṣetra-vyavahāra-sūtraṃ samāptaṃ//

1. atha khāta-vyavahāram=āha/

[24] vi^uṇaṃ] vi^ulaṃ N.
[25] vihī] vihi N.
[26] gaṇiūṇa] guṇiūṇa N.
[27] guṇiyaṃ parihi-dalaṃ] guṇiyā parihi dalaṃ N.
[28] cha havaï] chaka chaka N.
[29] Regard rā of dhārā as two short syllables (ra-a).

3.054a tala-muha-majjhe visamaṃ uṃḍḍattaṃ ahava dīha visamaṃ vā/
3.054b taṃ egaṭṭhaṃ kāuṃ visama-ṭṭhāṇehiṁ hariya samaṃ//[30]
3.055a sama-vitthara-dīha-guṇaṃ uṃḍḍatte guṇiya havaï khatta-phalaṃ/[31]
3.055b khatte sama-bhuva-vehe ghaṇa-uvamaṃ jāyae gaṇiyaṃ//[32]
3.056a du ti caü kara uṃḍḍatte pukkharaṇī paṃca hattha vitthāre/
3.056b solasa hatth=āyāme kiṃ jāyaï tassa khatta-phalaṃ//
3.057a dīha-kara saḍḍha-solasa vitthāre dasa sa-vāya aṭṭh=udae/[33]
3.057b aha vittharu dīh=udae sama nava kara kim=iha pihagu phalaṃ//

2. atha kūpasya phala-ānayanam=āha/

3.058a kuva-vitthāraṃ vaggaṃ ti^uṇa khaḍ-aṃs-ahiya vehi guṇiyavvaṃ/
3.058b cahuṁ bhāe jaṃ laddhaṃ taṃ kara-saṃkhā havaï savvaṃ//[34]
3.059a kūvassa ya vikkhaṃbhaṃ cha-hattha kara-vīsa jassa uṃḍḍattaṃ/
3.059b kūvassa tassa paṃḍiya khatta-phalaṃ kiṃ havei dhuvaṃ//
3.060a ti-kkoṇay-āi-khittā puvv-uttā-khitta-phala-samā jāṇa/[35]
3.060b te vi guṇiyaṃti vehe havaṃti ghaṇa-hattha khatta-phale//[36]

3. atha pāṣāṇa-phala-ānayana-karaṇa-sūtram/

3.061a dīh-aṃgulāṇi vitthara-piṃḍ-aṃgula-tāḍiyāṇi vibhaehiṃ/
3.061b jiṇa^aṭṭha-terasehiṃ havaṃti pāhāṇa-ghaṇa-hatthā//
3.062a saḍḍha-tiya hattha vitthari kar-addhu piṃḍe silā sahe jassa/
3.062b sa-ti-hāya-paṃca dīhe kam=ittha hui tassa gaṇiya-phalaṃ//
3.063a jaṃ havaï viviha-rūvaṃ vaṭṭa-ti-koṇ-āi-sayala-pāhāṇaṃ/
3.063b khitta-phalu vva gaṇeviṇu piṃḍa-guṇaṃ havaï tassa phalaṃ/[37]
3.064a dasa-hatthe vikkhaṃbhe gharaṭṭa-paṭṭa vva vaṭṭa-pāhāṇe/
3.064b divaḍha-kara-māṇa-piṃḍe kiṃ hoi imassa gaṇiya-phalaṃ//
3.065a golass=udaya-ghaṇ-addhaṃ sa-nav-aṃse ahiya taṃ havaï selaṃ/
3.065b parihi-caütthaṃ bhāyaṃ haya parihi nav-aṃs-ahiya khittaṃ//[38]
3.066a cha-kara-dīh-udaya-vitthara sama-vaṭṭaṃ golayassa pāhāṇaṃ/[39]
3.066b kiṃ gaṇiyaṃ kiṃ khittaṃ jaṃ hui taṃ bhaṇahi patteyaṃ//

4. atha pāṣāṇasya taulyam=āha/

3.067a ghaṇa-kaṃbiya^ikkeṇaṃ ḍhilliya-saṃbhūya-pāhaṇaṃ savvaṃ/
3.067b paṃcasa-maṇaṃ jayaï tulio caüvisa-tullo ya//[40]
3.068a vaṃsī aḍayālīsaṃ mammāṇī saṭṭhi kasiṇu bāsaṭṭhī/
3.068b jajjāvaya kannāṇaya uṇavanna kuḍukkaḍo saṭṭhī//[41]

---

[30]ṭṭhāṇehiṁ] ṭṭhāṇehiṃ N.
[31]guṇiya] gaṇiya N; khatta-phalaṃ] khittaphalaṃ N.
[32]khatte] khāttaṃ N. Presumably, this is a corruption caused by the '*pṛṣṭhamātrā* e' of Jaina Nāgarī.
[33]aṭṭh=udae] addhudae N.
[34]cahuṁ] cahuṃ N.
[35]ti-kkoṇay-āi-khittā] tikkoṇayāiṃ khittā N.
[36]guṇiyaṃti vehe] guṇiyaṃ tivehe N.
[37]gaṇeviṇu] guṇeviṇu N.
[38]= 5.025b.
[39]Regard dī as two short syllables (di^i).
[40]= 5.026. ya added by N.
[41]= 5.027.

// iti khāta-vyavahāra-sūtra-gāthā 15 sammattā//

atha citi-vyavahāram=āha/

3.069a goṃmaṭṭa pāya-sevaṃ caüraṣa vaṭṭaṃ munārayaṃ tākaṃ/
3.069b sovāṇa pulaṃ kūvaṃ vāvī iya nava-vihā bhittī//
3.070a paḍhama mavi suddha-bhittī vitthara dīh-udaya-guṇiya jaṃ havaï/[42]
3.070b tassāu vāra-vārī ālaya-kaṭṭāu sohijjā//
3.071a sesāo dasam-aṃsaṃ divaḍḍhayaṃ maṭṭiyassa ghaṭṭei/
3.071b sesā pāhaṇa-saṃkhā havaṃti ghaṇa-hattha-māṇeṇa//
3.072a paṃca kara bhitti^udaye dasa dīha du vitthare ya tam-majjhe/
3.072b bāru ti udaï du vitthari kā saṃkhā havaï pāhāṇe//[43]

atha iṭṭānāṃ gaṇanā/

3.073a dīhe vitthari piṃḍe addhu ti-hā aṭṭham-aṃsu iṭṭa kame/
3.073b caür=udaï divaḍhu vitthari daha dīhe bhitti ke iṭṭā//[44]

1. atha goṃmaṭam=āha/

3.074a goṃmaṭṭa-mūla-parihī^addhaṃ pā-parihi-guṇiya sa-nav-aṃsaṃ/
3.074b bhiti-gabbhāo cayaṇaṃ bāhira-majjhāu taṃ khittaṃ/[45]
3.075a bhiti-gabbhāo parihī uṇavīsa cha vitthar=assa kiṃ cayaṇaṃ/[46]
3.075b bāhira-parihī paṃḍiya caüvīsaṃ kiṃ havaï khettaṃ//
3.076a parihī vikkhaṃbh-addhe guṇiya nav-aṃs-ahiya khallu guṃmaṭṭe/
3.076b aṇubhava-sahiyaṃ bhaṇiyaṃ na saṃsayaṃ ittha nāyavvaṃ//

2. atha pāya-sevam=āha/

3.077a caüraṃsa-pāya-sevaṃ bāhira-bhittī ya majjhimaṃ thaṃbhaṃ/
3.077b vittharu dīh-udaya-guṇaṃ jaṃ hui taṃ kaṃviyā jāṇa//
3.078a bhitti taha thaṃbha aṃtari kam-ucca-maggaṃ phiraṃta taṃ dīhaṃ/
3.078b tala^uvara-juy-addh=udayaṃ vitthara guṇiyaṃ havaï pūraṃ//

3. atha vaṭṭaṃ/

3.079a taha vaṭṭa-pāya-seve thaṃbhaṃ bhittī ya gaṇahu kūvu vva/
3.079b pūr-aṃtara chat-ti-dalaṃ taṃ caüraṃsu vva jāṇeha//

4. atha munārayā/

3.080a vaṭṭa-pā-seva-sarisā munārayā hoṃti sayala-majjhāo/[47]
3.080b puṇu ittiyaṃ visesaṃ ti-koṇa-dala vaṭṭa-dala bhittī//

5. atha tāka/

3.081a vāriss=uvarima-tākaṃ dīh=udae guṇiya bhitti-piṃḍa-guṇaṃ/

[42] paḍhama mavi] paḍhamamavi N; vitthara dīh-udaya-guṇiya] vittharadīhudaya guṇiya N.
[43] bāru] bārū N.
[44] caür=udaï] caü rudaï N.
[45] bhiti N] bhitti N/Ms.
[46] bhiti N] bhitti N/Ms.
[47] Regard pā as two short syllables (pa^a).

3.081b satt-aṃsa-divaḍḍh-ūṇaṃ sihā-juyaṃ jāyae khallaṃ//
3.082a satta-kara tāka-dīhaṃ sihā-sahiya hattha-cāri jass=udayaṃ/
3.082b hatth-ega bhitti-piṃḍaṃ kiṃ jāyaï tassa khalla-phalaṃ//

6. atha sopānam/

3.083a sovāṇa-hiṭṭa^uvarima-joy-addhaṃ udaya-vitthare guṇiyaṃ/
3.083b nava hiṭṭhi uvari egaṃ du pihula chaha udaï kim=iha phalaṃ//

7. atha pula-baṃdham=āha/

3.084a vittharu dīhaṃ udae guṇiyaṃ tāka-vihīṇaṃ bhuva-juva-sahiyaṃ/
3.084b niggama^ahiyaṃ taha khall-ūṇaṃ jala-pula-baṃdhaṃ taṃ hui nūṇaṃ//#[48]

8. atha kūpa/

3.085a kuva-bhitti-majjhi parihī vitthara^udaeṇa guṇiya havaï phalaṃ/
3.085b dasa udaï du kara vitthari aṭṭhārasa parihi kiṃ cayaṇaṃ//

atha vāpī ṣaṭ bhedi/

3.086a caüraṃsa dīha vaṭṭā khaḍ-aṃsa aṭṭh-aṃsa saṃkha-vatt-āī/
3.086b bahu chaṃdi hoṃti vāvī te diṭṭha-pamāṇi gaṇiyaṃti//[49]

// iti citi-vyavahāra-sūtra-gā 18 sammattā//

atha krakaca-vyavahāra-karaṇa-sūtram=āha/[50]

3.087a dāru jah-acchiya-māṇe tassāu jah-iccha-phalaha kīraṃti/[51]
3.087b duṇha dalu dīhu vittharu guṇijja phalahehi bhāgu tti//[52]
3.088a aṭṭha-kara-dīhu dārū kar-addhu vitthāri dali ti-hāu kare/[53]
3.088b dīh=egu pāu vitthari nav-aṃsu dali kim=iha phalaheṇa//

atha kara-vatte dāru-cchedita-gaṇanā/

3.089a kara-vatta-līha je hui te dīhiṇa guṇiya hoṃti hatthāiṃ/[54]
3.089b vitthara-vasāu koḍī cirāvaṇī aggha-māṇeṇa//
3.090a iga divaḍha visuva saï gaji du ti vitthari gaji asīhi koḍī ya/[55]
3.090b cahu visuvahi saṭṭhi gaje paṃc-āi-nav-aṃti cālīse//[56]
3.091a dass-ai java terasa visuvā vitthāri tāva tīsehiṃ/
3.091b uvaraṃ te jā solasa tā vīsi gajehi koḍī ya//[57]
3.092a uvari jā vīsa visuvā tā koḍī dasi gajehi jāṇeha/[58]
3.092b uvari kara-vattu na calaï iya bhaṇiyaṃ sutta-hārīhiṃ//
3.093a dāru gaja satta dīhe visovagā aṭṭha satta vitthāre/

---

[48] Aḍillā metre.
[49] vāvī N] vākī N/Ms; gaṇiyaṃti] guṇiyaṃti N.
[50] karaṇa-sūtram] sūtrakaraṇam N.
[51] phalaha] phaliha N.
[52] phalahehiṁ] phalahehiṃ N.
[53] dārū] dāro N.
[54] hui] huiṃ N.
[55] asīhi] asīhiṃ N.
[56] cahu] cahuṃ N.
[57] uvaraṃ te] uvaraṃte N.
[58] Regard jā as two short syllables (ja^a).

3.093b dasa-līha phalaha-gārasa cīriya kaï koḍiyā homti//
3.094a aṭṭha-java kaṃviy-aṃguli jav-egu kara-vatta-līha phalahi ige/
3.094b vaṭṭassa khaṃḍa-karaṇe piṃḍaṃ taṃ dīhu jāṇeha//
3.095a mahuva-vaḍa-sāla-sīsama-niṃva-sirīs-āi sama-cirāvaṇiyaṃ/
3.095b khayar-aṃjaṇa-kīra sa-vā seṃvalu sura-dāru guṇi paüṇaṃ//

// iti krakaca-vyavahāro samatto// gāhā 9//

atha rāśi-vyavahāram=āha/

3.096a sama-bhuvi kay=anna-rāsī tap-parihi-khaḍ-aṃsa-vaggu udaya-guṇe/
3.096b jaṃ hui te ghaṇa-hatthā ghaṇa-hatthe ikki patto ya//
3.097a tila-kuddava-dhannāṇaṃ nav-aṃsu udao ya rāsi-parihīo/
3.097b dasam-aṃsu mugga-gohuma vora-kulatthā igārasamo//
3.098a siharu vva vaṭṭa-rāsī caür=udayaṃ tassa parihi chatīsaṃ/
3.098b bhitti-saṃlagga addhā kūṇ-aṃtari pāya parihī ya//[59]
3.099a bāhira-kūṇe paüṇaṃ parihī udao sa eha jāṇeha/
3.099b kiṃ jāyaï kara-saṃkhā pihu pihu rāsīṇa taṃ bhaṇasu//
3.100a dala-pāya-paüṇa-parihī guṇivi kame du-caü-sa-tti-hāeṇa/
3.100b puvvu vva phalaṃ pacchā niya-niya-guṇayārae bhāyaṃ//

// iti rāśi-vyavahāra-sūtraṃ sammattaṃ// gāthā 5//

atha cchāyā-vyavahāra-karaṇa-sūtram=āha/[60]

3.101a thaṃbh-āi-bhitti-chāyā daṃḍi miṇavi guṇahu daṃḍa-māṇeṇa/[61]
3.101b tass=eva daṃḍa-chāyā harijja bhāyaṃ phaleṇ=udayaṃ//[62]
3.102a caüvīs-aṃgula-daṃḍe cchāyā thaṃbhassa tinni-daṃḍa sa-vā/[63]
3.102b daṃḍa sa-vā^aṭṭhārasa^aṃgula kiṃ thaṃbhu uccattaṃ//[64]

atha sādhana-ānayana-karaṇam/

3.103a sama-bhūmi du-kara-vitthari du-reha-vaṭṭassa majjhi ravi-saṃkaṃ/
3.103b paḍham-aṃta-chāya-gabbhe jam-uttarā addhi uday-atthaṃ//
3.104a caü caü iga mayar-āī paṇa tiya iga kakkaḍ-āi dhuva-rāsī/
3.104b satt-aṃgula paha-muṇi-juva-phala rū-gaya-jutta divasa-gaya-sesaṃ//#[65]

// iti cchāyā-vyavahāra-sūtraṃ sammattaṃ// gāthā 4// ekatra gāthā 104//

// iti parama-jaina-śrī-candra-aṃgaja-ṭhakkura-pherū-viracitāyāṃ gaṇita-kaumudī-pāṭyāṃ aṣṭau vyavahārāṇi samāptaḥ//[66]

// iti tṛtīyo 'dhyāyaḥ//

---

[59] Regard saṃ as two short syllables (sa-ṃ).
[60] karaṇa-sūtram] sūtra-karaṇam N.
[61] chāyā] cchāyā N; daṃḍi] daṃḍe PM; guṇahu] guṇaha PM; māṇeṇa] māṇena PM.
[62] Cited in PM A1. tass=eva] tassa va PM; daṃḍa-chāyā] daṃḍacchāyā N, PM; harijja] harija PM.
[63] tinni] tinti PM.
[64] Cited in PM A2. daṃḍa sa-vā^aṭṭhārasa] daṃḍassa ya addhārasa PM; thaṃbhu] thaṃbha PM.
[65] Uggāhā metre.
[66] aṃgaja] aṅgaja N.

Chapter 4

Caturtho 'dhyāyaḥ

atha desa-adhikāram=āha/

4.001a ḍhilliya-rāyaṭṭhāṇe kajjaṃ bhūya-karaṇa-majjhaṃmi/
4.001b jaṃ desa-leha-payaḍī taṃ pherū bhaṇaï caṃda-suo//
4.002a jasu jasu vaṃṭivi dijjaï tasu tasu jīvalaï jaṃ bhave davvo/
4.002b so guṇivi laddha dammihi savvāṇa ya jīvalaï bhāyaṃ//
4.003a sẹva rāyaha paṃca jaṇa gae ya/[1]
4.003b taha savvaha jīvalaï tīsa sahassa egattha-rāsiṇa/
4.003c tiya terasa paṃca dui satta sahasa iya bhinna-rūviṇa/
4.003d vaya-kāraṇi jā nava-sahasa te savvi vi pāvaṃti/[2]
4.003e niya-niya-jivalā kaḍḍha tahaṁ kiṃ kiṃ kasu āvaṃti//#[3]
4.004a uppakkhaï jaṃ davvaṃ hui taṃ paṃḍiya karijja saya-guṇiyaṃ/[4]
4.004b caṭṭī harevi bhāyaṃ jaṃ labbhaï taṃ saī hoi//[5]
4.005a gāmi nayari dese jaï navi lakhi paṃcāsa-sahasi caṭṭī ya/
4.005b sattari-sahasa upakkhaï tā tassa kisā saī hoi//
4.006a jissa saī bheijjaï jittiya dhaṇa kaḍḍha tahi sa vaṭṭijjā/[6]
4.006b juyal-aṃt-aṃka phusivi taha paṇa-bhāge hoṃti visuvā ya//[7]
4.007a sahase ti^aṃtim-aṃkā phusavi kame lihasu du-caür-aṭṭha-guṇā/[8]
4.007b te visuv-āī jāṇaha evam dasa-sahasi lakkhe vā//
4.008a jaï caṭṭı-mula-dhaṇaṃ du-lakkha nava-sahasa paṃca-saya tısa/[9]
4.008b caüka saī bheijjaï tāma dhaṇaṃ kittiyaṃ havaï//

1. atha deśa-aṃke/[10]

4.009a dhaṇa-rāsi^aṃtim-aṃko phusijja taṃ bi^uṇa visuva dasam-aṃse/[11]

---

[1] seva rāyaha] seva rāyaha [seva rāyaha] N (the 2nd 'seva rāyaha' added by N).
[2] savvi vi] savvivi N.
[3] Raḍḍā metre. kaḍḍha tahaṁ] kaḍḍhatahaṃ N.
[4] uppakkhaï] upakkhaï N.
[5] caṭṭī harevi] caṭṭīhare vi N.
[6] jissa] jisā N.
[7] juyal-aṃt-aṃka] juyalaṃ taṃka N.
[8] sahase ti^aṃtim-aṃkā] sahaseti aṃtimaṃkā N.
[9] saya] iga N, i N/Ms.
[10] N puts the serial numbers, 1 to 5, in parentheses.
[11] dasam-aṃse] dasamaṃso N.

4.009b do^aṃtim-aṃka phussie paṇ-aṃsi taha visuva sayam-aṃso//[12]
4.010a rāsissa aṃtim-aṃke visuvā visuvaṃsag-āi sesa kamā/
4.010b āima^aṃkāṇ=addhe dammā jāṇeha vīs-aṃse//

2. atha mukātayam=āha/

4.011a mukkātaï jaṃ varise taṃ gaya-diṇa guṇavi varisa-diṇi bhāyaṃ/[13]
4.011b paṃci sahassi mukātaï navi diṇi caü-māsi kiṃ havaï//
4.012a jittā damma maseliya dijjahi mās-ikki te ti-bhāg-ūṇā/
4.012b sesa havaṃti visovā divase divase muṇeyavvā//

3. atha dhāvaka-gatau/

4.013a lahu-gaï diṇa-saṃkha-guṇaṃ lahu-dīha-gaïssa aṃtare bhāyaṃ/[14]
4.013b laddha-diṇehi milaṃtī appa-gaī bahu-gaī do vi//[15]
4.014a caü-joyaṇīya pacchā navama-diṇe satta-joyaṇī calio/[16]
4.014b tassa vahoḍaṇa-heū milei so kaï ya divasehiṃ//[17]
4.015a paṃc-āi-du-vaḍḍhaṃtā joyaṇa divaseṇa callae karaho/[18]
4.015b joyaṇa-caüdasa-karahī kittiya divasehi so milaï//[19]
4.016a āi-majjh-aṃta-rāsī aṃtāo āi-hīṇa majjheṇa/[20]
4.016b bhāe laddhaṃ bi^uṇaṃ ega-juyaṃ karaha-diṇa-māṇaṃ//[21]

4. atha saṃvatsara-ānayanam=āha/

4.017a vikkam-āi je varisa māsa citt-āi karivi diṇa/
4.017b cha-muṇi-naṃda-laddh=ahiya-māsa te vacchara-juya puṇa/
4.017c nava-nihāṇa-rasa-varisa māsa-dui dui-diṇa ūṇaya/
4.017d tājiya-vaccharu havaï māsa muharama māīṇaya/
4.017e tājikku puṇ=evaṃ karivi para ahiya-māsa sohevi puṇi/
4.017f nava-muṇi-cha-varisa dui-diṇa ahiya paṃḍiya vikkama-samaü bhaṇi//#[22]

// iti deśa-adhikāra-karaṇa-sūtraṃ sammattaṃ//

atha vastra-adhikāram=āha/[23]

4.018a juja-paṭṭolaya^atalasa-sār-āī paṭṭa-vattha em-āī/
4.018b kara-vāsaka-tāṇ-āī iya suhamā thūla sāḍ-āī//

---

[12] sayam-aṃse] sayamaṃso N.
[13] varisa-diṇi bhāyaṃ] varisa diṇibhāyaṃ N.
[14] bhāyaṃ] bhāgaṃ PM.
[15] Cited in PM A3. laddha-diṇehi] labdhadiṇehiṃ PM; milaṃtī] milaṃti PM; bahu] lahu N.
[16] joyaṇīya] joaṇīu PM; joyaṇī] joaṇī PM; calio] caliu PM.
[17] Cited in PM A4. kaï ya] kahaï PM.
[18] vaḍḍhaṃtā-joyaṇa] vaḍḍhaṃto joaṇa PM.
[19] Cited in PM A5. joyaṇa] joaṇa PM; kittiya] kittia PM; divasehi] divasehiṃ PM; so] sā PM.
[20] Regard maj of majjhaṃta as two short syllables (ma-j). majjh-aṃta] mabbhaṃta PM; aṃtāo] aṃtāu PM; majjheṇa] mabbheṇa PM.
[21] = 4.054. Cited in PM A6. bhāe] bhāyaṃ PM; laddhaṃ] labdhaṃ PM; bi] vi PM; juyaṃ] juaṃ PM; māṇaṃ] mānaṃ PM.
[22] Chappaya metre.
[23] N puts the serial number (5) at the head of this line.

4.019a saya-hatthi sayala-kappaḍi sīvāṇi divaḍhu kar-egu kattaraṇe/[24]
4.019b iga du tiya kora dhuvaṇe ghaṭṭaï paṭṭ-aṃsuy-āi kame//
4.020a sayala-khīmehiṁ kappaḍa-sama-saṃkha navāra kiṃci hīṇ-ahiyā/[25]
4.020b dahalīja viṇā savve thaṃbhāu sa-vāiyā udae//[26]
4.021a udayassa vāra visuvā kamara-tale aṭṭha uvari savvehi/
4.021b iga-thaṃbhi du-thaṃbhe vā itto siya-kappaḍaṃ bhaṇimo//
4.022a savvāṇa paḍa-tala-uvara-juy-addha udae guṇijja jā kamaraṃ/
4.022b piṭṭhī-vitthara dīhaṃ haya aṭṭh-aṃs-ahiya juya vatthaṃ//[27]
4.023a majjhima-ḍaṃḍassāo caüṇaṃ khīmassa kaḍa-yala-paveso/
4.023b tassa divaḍḍhā parihī bārasam-aṃs-ūṇa caüraṃse//
4.204a caü-kara majjhima-ṭhaṃbhaṃ solasa kamaraṃ ca parihi bāvīsaṃ/[28]
4.024b tassa khīmassa paṇḍiya kiṃ jāyaï vattha-parimāṇaṃ//[29]
4.025a aṭṭh-aṃsa taha ya vaṭṭe ti^uṇaṃ kamaraṃ tay-addha-juya daüraṃ/
4.025b iya dhara-haya sīmāṇaya thaṃbhāu tanāva caü-guṇiyaṃ//
4.026a taṃgoṭī iga-thaṃbhā hiṭṭh-uvara-juy-addha udaya-guṇa vatthaṃ/
4.026b thaṃbhā parihi paṇaṃ guṇa duga-thaṃbhā majjha-paḍa^ahiyā//
4.027a kharigaha maṃḍava^uvaraṃ ubhaya-dise ji kara tassa addh=udayaṃ/
4.027b tatto paṇa-guṇa parihī parihi-dalaṃ udaya-guṇa vatthaṃ//
4.028a bhitti-valaya-paḍa donni vi duvāra-paḍa be vi udaya dīha-guṇā/
4.028b iya vatthaṃ addh=addhaṃ maṃḍava saha vāra taha bhittiṃ//
4.029a vārigaha khaḍ-aṭṭh-aṃsā cchatt-āgārā ya maṇḍav-āgārā/
4.029b eyāṇaṃ ca tarakkā-hiṭṭh-uvara-juy-addha udaya-guṇā//
4.030a iga-thaṃbha cha-guṇa parihī du-thaṃbha parihī ya majjha-paḍa^ahiyaṃ/
4.030b aṭṭh-aṃsa-jutta udae thaṃbhāu tanāva paṃca-guṇā//
4.031a mīrāṇa vārigaha hui cilaṃga caürasa du^ega-thaṃbhā ya/[30]
4.031b samaüdaya caüṇa kamaraṃ vi^uṇa parihi aṭṭham-aṃs-ahiyaṃ//[31]
4.032a vallahali patta-vallī-jhuṃbukkā-kaliya majjha-jhallariyā/
4.032b eyāṇa ya kappaḍao taha taïya puḍassa puṇa ahio//
4.033a chāyā-paḍa caṃdovaya sarāi cājamaṇiy-āi bhitti-paḍā/
4.033b vittharu dīhe guṇiyā sujjhaṃti viṇoya-citta viṇā//
4.034a dahalīja rūi tākā chajjaya kuvvāya carakha paḍirūvā/[32]
4.034b chatt-ālaṃva nisāṇā te ṭippa-pamāṇi nāyavā//
4.035a uddesa siyāvaṇiyaṃ saï gaji nāvāra damma-solasagaṃ/
4.035b cittaṃ gaj-ikki pacchā-dahalīja sarāi ceti dugaṃ//[33]
4.036a kimisaṃ gaj ikki citte suhame caüvīsa thūli vīsā ya/
4.036b cattāri ṭaṃka ḍorī iga suttaṃ aruṇa nīlaṃ vā//
4.037a nāvāra saraja cammaṃ nīla-aruṇa-kasiṇa-vattha taṃ payaḍaṃ/
4.037b suttaṃ navāra saï gaji niva paüṇaṃ iyara ser-addhaṃ//

// iti vastra-adhikāre gāhā 20 sammattā//[34]

[24]divaḍhu kar-egu] kara divaḍhu egu N.
[25]khīmehiṁ] khīmehiṃ N. Regard khī as two short syllables (khi-i).
[26]dahalīja viṇā] dahalī javiṇā N.
[27]aṭṭh-aṃs-ahiya] aṭṭhaṃsa hiya N.
[28]parihi] parihī N.
[29]Regard khī as two short syllables (khi^i).
[30]vārigaha] vāriga N.
[31]aṭṭham-aṃs-ahiyaṃ] aṭṭhamaṃsa hiyaṃ N.
[32]dahalīja rūi] dahalī jarūi N.
[33]dahalīja sarāi] dahalī jasarāi N.
[34]20] 21 N.

atha jaṃtra-adhikāra-karaṇa-sūtraṃ=āha/

4.038a diṇayar=aggi rasa tera caüras=iṃdiya juga īsara/[35]
4.038b iya kuṭṭhihi cha^igāi igigi samahiya lihi maṇahara/[36]
4.038c kara nihi solasa taha ya uvahi vasu tihi disi sasihara/[37]
4.038d icchā dali rū harivi kamiṇa ṭhavi jaṃtu muṇahi para/
4.038e jā sunnu vāri tā=aṇukkamihi jaṃtari tav-vivarīu dhuya/
4.038f jā savvi gehi visam=ahava sama-sama-visamāi sam-aṃka juya//#[38]

ṣaṭ-gṛhe jaṃtra/[39]

4.039a dāhiṇa-kann=eg-āī satt-ahiya khaḍ-āi paṃc-ahiya vāme/[40]
4.039b daṃta caütīsa sura sara muṇi uṇavīs=aṭhara paṇavīsaṃ//[41]
4.040a paṇatīsa ti caü du ravī teraha jiṇa tīsa rikkha sagavīsaṃ/[42]
4.040b maṇu sataraha dasa nava nakha tevisa puvv-āi jaṃta cha-gihaṃ//[43]
4.041a lihi dhurāu pā^oli addha aha meli puṇ=uvarima/
4.041b pāya^oli iya kamihiṃ jāva pā-jaṃtu havaï ima/
4.041c majjhim-addhu uvakamihi carima-pā-jaṃtu puṇu vi kami/
4.041d caü-gih-āi-caü-vuḍḍhi-jaṃta iya hui iga-caya-kami//#[44]
4.042a tihi nihi rasa juya vasu kara tera gāra/[45]
4.042b sasi muṇi ravi maṇu disi kala guṇa sara/[46]
4.042c adh-upaü cahu cahuṭhe caüsaṭhi-gihi/[47]
4.042d rū-ti-du-caü-kami 'ṇukam=eg-āi lihi//#[48]

atha viṣama-jaṃtra-ānayane/

4.043a eg-āi jah-icch-oliṃ giha-saṃkh=iga-juya sa-puvva paḍham-oliṃ/[49]
4.043b tatto majjhima majjhima-gihāu giha-jutt=asu-kamehiṃ//[50]
4.044a dhuri paṃti carima^aṃkāu jattha ahiy-aṃku havaï tattha gihe/[51]
4.044b savva-giha-saṃkha sohivi lihijja iya visama-giha-jaṃtaṃ//

---

[35] caüras N/Ms] caüdas N. N superscribes the eight numbers, 12, 3, 6, 13, 14, 5, 4, and 11, above the eight words of this line.

[36] cha^igāi] kha(0) igāi N. See the footnote to verse 43a below.

[37] N superscribes the eight numbers, 2, 9, 16, 7, 8, 15, 10, and 1, above the eight words of this line.

[38] Chappaya metre. visam=ahava] visama hava N.

[39] ṣaṭ-gṛhe jaṃtra/] // ṣaṭ gṛhe jaṃtra// N.

[40] satt-ahiya] sattahi ya N; paṃc-ahiya] paṃcahi ya N.

[41] uṇavīs=aṭhara] uṇavīsa ṭhāra N. N superscribes the eight numbers, 32, 34, 33, 5, 7, 19, 18, and 25, above the eight words of this line.

[42] N superscribes the ten numbers, 35, 3, 4, 2, 12, 13, 24, 30, 28, and 27, above the ten words of this line.

[43] N superscribes the six numbers, 14, 17, 10, 9, 20, and 23, above the first six words of this line.

[44] Rolā metre.

[45] N superscribes the eight numbers, 15, 9, 6, 4, 8, 2, 13, and 11, above the eight words of this line.

[46] N superscribes the eight numbers, 1, 7, 12, 14, 10, 16, 3, and 5, above the eight words of this line.

[47] adh-upaü] adhu paü N.

[48] Metre? N superscribes the four numbers, 1, 3, 2, and 4, above the first four words of this line.

[49] eg-āi] kha igāi N. See the footnote to verse 38b above.

[50] giha-jutt=asukamehiṃ] giha jutta sukamehiṃ N.

[51] tattha] tittha N.

4.045a juga gaha loyaṇa hara-nayaṇa iṃdiya muṇi aṭṭhehi/[52]
4.045b sasi rasa jaṃtu ig-āi lihi ikkāsī-kuṭṭhehi//#[53]

// iti jantra-adhikāro sammatto// gāhā 8//

atha prakīrṇaka-adhikāram=āha/[54]

1. kusuma-ānayanam=āha/[55]

4.046a du-guṇā du-guṇa ji uvvarahi vāra vāra tihu jutta/[56]
4.046b jaï ko kusumu na uvvaraï tā dhuri tinni nirutta//#[57]
4.047a ikku sura-gihu cahu duvārehiṃ/
4.047b patteya tahi jakkhu igu vāra-tulla tasu majjhi sura-vaï/
4.047c dhammiu kusumāṇa vi vahala-sayala biṃba addh-addhasu ṭhavaï/[58]
4.047d jaṃt-āvaṃta igegu de savihi vāri jakkhassa/[59]
4.047e sesa vīsa jahi uvvarahi savve kaï hui tassa//#[60]

2. atha āmra-ānāyanam=āha/

4.048a je pattā te aṃsi guṇijjahi āi-hīṇa kari vuḍḍhi harijjahi/[61]
4.048b laddhā bi^uṇa rūva-saṃjuttā paṃḍiya te jaṇa gayā niruttā/[62]
4.048c seḍhiya-saṃkaliye phala-saṃkhā laddha-vihatte hui jaṇa-saṃkhā//#[63]
4.049a aṃsu aṭṭhamu kaṭaka-majjhāu/[64]
4.049b gaü aṃba-toḍaṇa-vaṇihi bhakkhaṇ-attha āesi rāṇaya/[65]
4.049c caür-ādi vaḍḍhaṃta chahaeṇa parihi savvehi āṇiya/[66]
4.049d jaṃ kaṭakku thiu laddha tihi vīsa vīsa savvehi/[67]
4.049e kaï jaṇa gaya kittaü kaṭaku kaï aṃb=āṇiya tehi//#[68]

3. atha jamātrika-varisola-ānayanam=āha/

---

[52]aṭṭhehi] aṭṭhehiṃ N. N superscribes the seven numbers, 4, 9, 2, 3, 5, 7, and 8, above the seven words of this line.

[53]Dohā metre. N superscribes the two numbers, 1 and 6, above the first two words of this line.

[54]In the Nahatas' text the nineteen verses of this section have been given their serial numbers, 1 to 19, separately from the foregoing sections, but we have renumbered them for the sake of convenience.

[55]N puts the serial numbers, 1 to 11, in parentheses.

[56]uvvarahi] uvvaraïṃ PM; tihu] tihiṃ PM.

[57]Dohā metre. Cited in PM A7. jaï] aha jaï N; kusumu] kusuma PM; uvvaraï] uccaraï PM; tinni] tinti PM; nirutta] niruttā PM.

[58]addh-addhasu] addhaddha suṭhavaï N.

[59]jaṃt-āvaṃta] jaṃ tāvaṃta N.

[60]Raḍḍā metre. Paraphrased in prose Sanskrit in PM A8. kaï] kaïṃ N.

[61]guṇijjahi] guṇijjaïṃ PM; harijjahi] harijjaï PM.

[62]paṃḍiya] paṃḍia PM.

[63]Aḍillā metre with six feet? Cited in PM A9. seḍhiya-saṃkaliye] seḍhīsaṃkalie PM; jaṇa] phala N, jaṇa PM.

[64]aṃsu] aṃsa PM; aṭṭhamu] aḍḍhama PM; majjhāu] majjāu PM.

[65]gaü] gaya PM; vaṇihi] vaṇihiṃ PM; āesi rāṇaya] āpasa rāyaṇa PM.

[66]caür-ādi] caürādiṃ PM; vaḍḍhaṃta] vuḍḍhi PM; parihi savvehi āṇiya] parihiṃ savvehiṃ āṇia PM.

[67]kaṭakku thiu] kaṭaka tthaü PM; tihi] tahiṃ PM; savvehi] savvehiṃ PM.

[68]Raḍḍā metre. Cited in PM A10. kaï] kaya N (cf. jaï in 3.032b), kaï PM; kaṭaku] kaṭaka PM; aṃb=āṇiya] aṃbāṇiaṃ PM; tehi] tehiṃ PM.

4.050a guṇaka thappivi kamiṇa eg-āi/[69]
4.050b uvaruppari guṇi vi guṇi vāra vāra ikk-ikku dījaï/[70]
4.050c varisolā je havaï savvi tei paḍhamaha bhaṇijjahi/[71]
4.050d te vi aṃka rūvāha viṇu puvva parihi guṇiyaṃti/[72]
4.050e hui ji ti bhakkhahi savvi jaṇa paṃḍiya iu pabhaṇaṃti//#[73]
4.051a gaya jamāiya paṃca sāsuraï/
4.051b varisolā=aṇukkamihi diyaï sāsu taṭṭhiya bhareviṇu/[74]
4.051c tahaṃ bhuṃjiya rahahi ji te bi^uṇa ti-caü-paṇa-guṇa kareviṇu/
4.051d aṃtima sahi bhakkhahi avari bhaṇahi eṇa bahu khaddha/
4.051e savihi egu sā bhakkhiyā kaï thākaï kaï khaddha//#[75]

4. atha vastra-phala-ānayanam=āha/

4.052a je jaṇa gahaṃti hatthaṃ te caüṇa guṇijja laddha-vattha-kare/[76]
4.052b taṃ vatthu dīhu vittharu kara-jaṇa-guṇa caüṇa savvi jaṇā//[77]
4.053a vara-vatthu igu caüd-disi igegu karu ṭhahiu tihu tihu jaṇehiṃ/[78]
4.053b nava nava kari jaṇi pattā kaï jaṇa vara-vatthu kaï hatthā//[79]

5. atha karabha-gatyām=āha/

4.054a āi-majjh-aṃta-rāsī aṃtāo āi-hīṇa majjheṇa/[80]
4.054b bhāe laddhaṃ bi^uṇaṃ ega-juyaṃ karaha-diṇa-māṇaṃ//[81]
4.055a caü-joyaṇ-āi tiya-tiya-vaḍḍhaṃto nicca callae karaho/
4.055b solasa-joyaṇa-karahī kittiya divasehi sā milaï//

6. atha viparīta-uddeśakam=āha/

4.056a ses=ūṇa-jutta vaggaṃ gaya^ahiyaṃ tassa mūla bhāya-guṇaṃ/[82]
4.056b guṇayāreṇa vihattaṃ so amuṇiya-rāsi nāyavvo//
4.057a paṃca-guṇa nava-vihattaṃ tav-vaggaṃ nav-ahiyassa mūlaṃ ca/[83]
4.057b do-hīṇa tinni sesaṃ viviriya^uddesago rāsī//

7. atha para-cintā-jñānam=āha/[84]

---

[69] thappivi] thapya vi PM; eg-āi] egād PM.

[70] uvaruppari] avaruppara PM; guṇi vi] guṇivi N; guṇi vāra] guṇiṃ vāra PM; ikk-ikku] ikkikka PM; dījaï] dijjaï PM.

[71] varisolā] varasolāṃ PM; havaï] havaïṃ PM; paḍhamaha] paṭhamaha PM; bhaṇijjahi] bhaṇijjaïṃ PM.

[72] te vi] tevi N; parihi] paraïṃ PM; guṇiyaṃti] guṇiaṃti PM.

[73] Raḍḍā metre. Cited in PM A11. hui] huiṃ PM; ji ti] ti ti N.; bhakkhahi] bhakhaïṃ PM; savvi] savva PM; paṃḍiya] paṃḍia PM; iu] ia PM.

[74] varisolā=aṇukkamihi] varisolā'ṇukkamihi N.

[75] Raḍḍā metre. Paraphrased in Sanskrit prose in PM A12.

[76] caüṇa] caüguṇaṃ PM.

[77] Cited in PM A13. vatthu dīhu vittharu] vattha dīha vitthara PM; guṇa] ∅ PM; caüṇa] caügu PM.

[78] ṭhahiu tihu tihu jaṇehiṃ] ṭhāhiu tihuṃ tihu jaṇehiṃ N.

[79] A solution in prose Sanskrit is given in PM A14.

[80] Regard maj of majjhaṃta as two short syllables (ma-j).

[81] = 4.016.

[82] ses=ūṇa-jutta] sesūṇa jutta N; gaya^ahiyaṃ] gaya ahiyaṃ N; mūla bhāya-guṇaṃ] mūlabhāya guṇaṃ N.

[83] tav-vaggaṃ] tavaggaṃ N.

[84] para] patra N.

4.058a sattari guṇa ti^unehiṃ paṃcahi igavīsa panara sattehiṃ/[85]
4.058b piṃḍeṇa saü paṇ-uttaru devi harivi muṇaha para-cittaṃ//[86]
4.059a ciṃtiya-suya kara-sahiyaṃ bi^uṇ=igi juya paṃca-guṇa suyā-sahiyaṃ/[87]
4.059b daha-guṇa kha-paṇa-kar-ūṇaṃ sesa kame muṇaha sunna viṇā//[88]

8. atha marddita-aṃka-jñānam=āha/

4.060a sayal-aṃka-piṃḍu sohivi rāsiss=aṃtāu sesa-piṃḍāo/
4.060b jaṃ hīṇu navasu pāḍaï pūraï maliy-aṃku sunnu navaṃ//

9. atha sadṛśa-aṃka-ānayanam=āha/

4.061a eg-āī ya nav-aṃtā aṭṭha viṇā icchiy-aṃku navi guṇio/
4.061b puvv-aṃka-rāsi-tuliyā havaṃti eg-āi-saris-aṃkā//[89]

10. atha go-saṃkhyā-ānayanam=āha/

4.062a uvarāo jā hiṭṭhi hui tā=aṇukkamihi ṭhavijja/[90]
4.062b uvaruppari savve vi guṇi gāvi ema jāṇijja//#[91]
4.063a cahuṃ duvārihiṃ gāvi nīsariya/
4.063b gaya pāṇī paṃca sari satta rukkha-tali te baïṭṭhiya/
4.063c āvaṃti vārihi navihi païsi chac=ca vāḍihi niviṭṭhiya/
4.063d rakkhahi aṭṭha gu-vāla tahaṁ sāricchiya sahi te vi/[92]
4.063e paṃḍiya kittiya gāvi huiṁ tammi nayari savve vi//#[93]

11. atha go-dugudha-vaṃṭanam=āha/

4.064a go jaṇa sama-bhāgeṇaṃ jaṇa-saṃkha ig-āi ṭhavi kam-ukkamaso/[94]
4.064b jā aṃtima go^aṃkaṃ sama-paṇhe duddhu aṃka-samaṃ//

iti śrī-candra-aṃgaja-ṭhakkura-pherū-viracite gaṇita-sāre deśa-adhikāra-ādyāḥ catvāri adhikārāni sammattā//[95] gāhā 64//

---

[85] ti^unehiṃ] ti^ūṇehiṃ PM; paṃcahi] paṃcahiṃ PM.

[86] Cited in PM A18. saü paṇ-uttaru] saya paṇuttara PM; harivi] haravi PM; cittaṃ] cintaṃ PM.

[87] ciṃtiya-suya kara-sahiyaṃ] ciṃtiya suyakarasahiyaṃ N, ciṃtiasua karasahiaṃ PM; igi juya] iga jua PM; suyā-sahiyaṃ] suāsahiaṃ PM.

[88] Cited in PM A17. daha] dasa PM; daha-guṇa kha-paṇa-kar-ūṇaṃ] daha guṇa kha paṇaka rūvaṃ N; sesa kame] sesaṃke PM; sunna] sunti PM.

[89] rāsi-tuliyā] rāsi guṇiyā N.

[90] uvarāo] uvarāuṃ PM; hui] huiṃ N; tā=aṇukkamihi] tāṇukkamihiṃ PM.

[91] Dohā metre. Cited in PM A15. gāvi] gāvī PM; ema] ima PM.

[92] tahaṁ] tahaṃ N.

[93] Raḍḍā metre. Paraphrased in prose Sanskrit in PM A16. huiṁ] huiṃ N.

[94] kam-ukkamaso] kamu kkamaso N.

[95] aṃgaja] aṅgaja N.

Chapter 5

Pañcamo 'dhyāyaḥ

atha uddeśa-paṃcagaṃ sūtram=āha/

5.001a paṇameviṇu siṭṭhi-karaṃ bhaṇāmi nippatti-paṃcag-uddesaṃ/
5.001b dhann-ikkhu-cuppaḍāṇaṃ desa-kar-agghāṇa māṇāṇaṃ/
5.002a savvattha anna nippaï bhūmi-visesena aṃtaraṃ bahuyaṃ/
5.002b ḍhilliya āsiya narahaḍa varuṇa-paesā imaṃ jāṇa//
5.003a khittassa dīha vitthara viggahayā guṇiya havaï bhū-saṃkhā/
5.003b vīsa-kami dīha-vitthari aha kaṃviya saṭṭhi vīgahao//
5.004a annassa phalaṃ jāyaï nippanne vīsa-visuva-vīgahao/
5.004b saṭṭhi maṇa dhanna-kuddava caüvīsa maüṭṭha jāṇeha//
5.005a caülā maṇa-bāvīsaṃ tila solasa mugga-māsa aṭṭhāraṃ/
5.005b vīsa kaṃguṇiya cīṇaya panaraha kūr-īsa vāīyā//[1]
5.006a solasa maṇa kappāsā cālīsa juvāri dasa saṇo taha ya/
5.006b ikkhu savāṇiya-sāhā itto āsāḍhiyaṃ jāṇa//
5.007a gohuva paṇayālīsaṃ kalāva massūra caṇaya vattīsaṃ/
5.007b java chappaṃna maṇāiṃ sarisama alasīi karaḍa dasaṃ//
5.008a vaṭula tori kulattha caüdasa maṇa hoṃti savva kaṇa tuliyā/
5.008b jīrā dhaṇiyā dasa-maṇa para sikkaya-majjhi gaṇiyaṃti//
5.009a savve vi vesavārā hālima metthī ya saggavattī ya/
5.009b kora-dhann-āi seṃkkaya saü-damma-karassa viggahae//[2]

// iti dhānya-utpatti-phalam//

5.010a nava khāri pacāsa-maṇī ikkhu-raso tassa paṃcam-aṃsu gulo/
5.010b sakkara chaṭṭh-aṃse hui solasam-aṃse ya khaṃḍā ya//
5.011a tassa divaḍḍhā ravvā hīṇa-ahiya puṇa havei nīravasā/
5.011b puṇu ittiyaṃ navi calaï jā bhaṇiyaṃ diṭṭha-patteṇaṃ//
5.012a khaṃḍāu ti-bhāg-ūṇā nivāta varisolagā bhave paüṇā/
5.012b aï-cukkha sesa sīro iga-vārā hoi khaṃḍa-samā//

// iti ikṣu-rasa-phalam//

5.013a tila-sarisama-karaḍa-maṇe tillaṃ nava-satta-paṃca-visuva kame/

---

[1]Regard kaṃ of kaṃguṇiya as two short syllables (ka-ṃ). kūr-īsa vāīyā] kūrī savāīyā N.
[2]Regard dhan of dhann-āi as two short syllables (dha-n).

5.013b duddhi aḍ-aṃsu nav-aṃso lūṇiu tatto ya paüṇa ghio//

// iti sneha-phalam//

5.014a dasi chālīe gāvī mahisī tav-vi^uṇa cahu vayalli halo/[3]
5.014b culhi pavāṇe kuḍhiyā nāviya valahāra mahara viṇā//
5.015a devaï kanna-calā taha nīlī kavilīya go adaṃtīya/
5.015b vippa sa-vāsaṇi ya puṇo karaṃ caraṃ n=atthi eyāṇaṃ//
5.016a ṭaṃkā vattīsa halo ti-viha kuḍhī ega-dīvaḍha-du-ṭaṃkā/[4]
5.016b mahis=ikku gāvi addho vuḍḍhiya-vasahassa ṭaṃko ya//
5.017a iya bhaṇiyaṃ uddesaṃ hīṇa-ahiya hoṃti caṭṭiy-aṇusāre/
5.017b addha ti-hā pā annaṃ tiṇa cara pā hīṇa bhāsa-karaṃ//[5]

// iti deśa-kara-phalam//

5.018a je pāī damm-akkihi bhavaṃti te ti^uṇa nicchae seī/
5.018b anne vi vi^uṇa pāī ṭaṃkaï ikke vi jāṇijjā//
5.019a ji ki vi sera bhaṇiyahi damm-ikkihi te vi sa-vāyā-maṇa ṭaṃk-ikkihi/[6]
5.019b maṇaha bhāu paṃcamu pāḍijjahu sesa sera damm-ikki muṇijjahu//#[7]
5.020a jittihi kittihi dammihi paṃḍiya maṇu egu vakkharo hoi/
5.020b tass=addhehiṁ visoihi sero ikko viyāṇāhi//[8]
5.021a gaṇima-vatthūṇa jittihi dammehiṃ hoi koḍiyā ikkā/[9]
5.021b tāvaïya visovehiṃ labbhaï egā gaṇima-vatthū//

// iti arghasya phalam//

atha mānāni/

5.022a vaṭṭassa ya vikkhaṃbhaṃ ti^uṇaṃ taha chaṭṭham-aṃsa-juya parihī/
5.022b sā pāya-vitthare guṇi jaṃ jāyaï taṃ ji khitta-phalaṃ//

darśanaṃ (6) paridhi 19 kṣetra-phalaṃ 28 || iti vṛttam//[10]

5.023a vaṭṭāo caüraṃsaṃ bārasa visuvā havei sa-visesaṃ/
5.023b caüraṃsāo vaṭṭaṃ taha vaṭṭaḍa paṃcam-aṃs-ūṇaṃ//
5.024a ti-kkoṇāo vaṭṭaṃ saḍḍha-duvālasa-visova hui khittaṃ/[11]
5.024b vaṭṭāo ya ti-koṇaṃ visovagā satta addh-ahiyā//

viśeṣa eṣāṃ darśanam/[12]

[3] chālīe] chālīehi N.
[4] ega-dīvaḍha-du-ṭaṃkā] ega dīvaḍha du ṭakīya N.
[5] bhāsa-karaṃ] bhā sakaraṃ N.
[6] damm-ikkihi] dammikihi N.
[7] Aḍillā metre.
[8] addhehiṁ] addhehiṃ N.
[9] Regard vat of vatthūṇa as two short syllables (va-t).
[10] 28 ||] 28 N.
[11] ti-kkoṇāo] tikkoṇayāo N.
[12] N puts this prose line, with āha at its end (··· darśanamāha), immediately before verse 25 (after // iti kṣetra-mānam//).

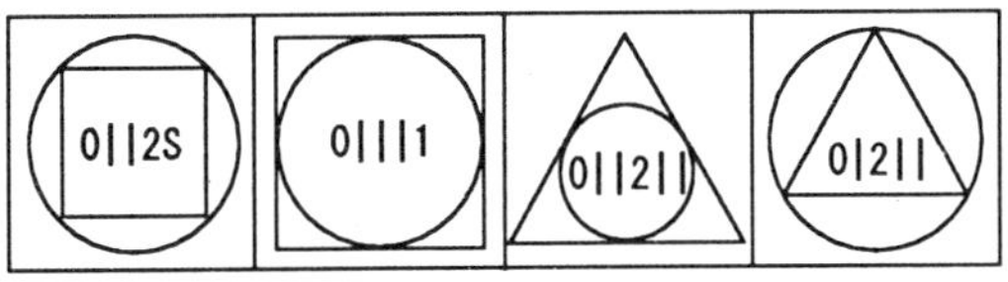

[13]

// iti kṣetra-mānam//

5.025a golassa ya udaya-ghaṇaṃ paüṇaṃ paüṇaṃ va havaï pāhāṇaṃ/
5.025b parihi-caütthaṃ bhāyaṃ haya parihi nav-aṃsa-juya khittaṃ//[14]

nyāsa (6) labdhaṃ golaka-phalaṃ 121 || kṣetra-phalaṃ 100SS6/[15]

ghani 216 paüṇaṃ 162 puṇu paüṇaṃ 121 || phalaṃ//[16]

parihi-caüttha-bhāyaṃ 4 ||| guṇita 19 jāta 90 | /[17]

asya nava-aṃsa 10 savisesam/[18]

evaṃ 100SS6 kṣetra-phalaṃ//[19]

5.026a ghaṇa-kaṃviya^ikkeṇaṃ ḍhilliya-saṃbhūya-pāhaṇaṃ savvaṃ/
5.026b pannāsa maṇaṃ jāyaï tulio caüvīsa-saya-tulle//[20]
5.027a vaṃsī aḍayālīsaṃ saṭṭhi mamāṇīya kasiṇu bāsaṭṭhī/
5.027b jajjāvara kannāṇaya uṇavanna kuḍukkuḍo saṭṭhī//[21]
5.028a maṭṭī maṇa paṇavīsaṃ tus-aṃnna maṇa^aṭṭha bārasa vaṇ-aṃnnaṃ/[22]
5.028b daha maṇa tilla-ghayaṃ taha solasa maṇa lavaṇa uddesaṃ//
5.029a rāju igu ti-jaṇa-sahio vārasa-gaja-bhitti-pāhaṇe ciṇaï/
5.029b caüdasa-sayāiṁ iṭṭā udesa jala-gaggarī tīsā//[23]
5.030a sagavīsa-maṇā hakkaṃ nava cunnaṃ bi^uṇu khoru ikki gaje/
5.030b pāhāṇa bhitti cijjaï nava-maṇaï ime va jāṇeha//
5.031a leve kevaṇa-cunnaṃ paüṇa-maṇaṃ pāya-sera-saṇa-sahiyaṃ/
5.031b taïy-aṃsa-khora-juttaṃ talavaṭṭe addhu jala-ṭhāṇe//
5.032a chāṇaya maṇa-cālīsaṃ taha kakkara saṭṭhi pakka hui cunnaṃ/
5.032b rakkha-pavāhiya saṭṭhī arakkha cālīsa kaliyā ya//
5.033a uddesa-paṃcagam=imaṃ caṃdā-suya-pheruṇā ao bhaṇiyaṃ/
5.033b jaha desa-kar-uppattī caṭṭiya samae muṇijjci//

[13]The figures in N:

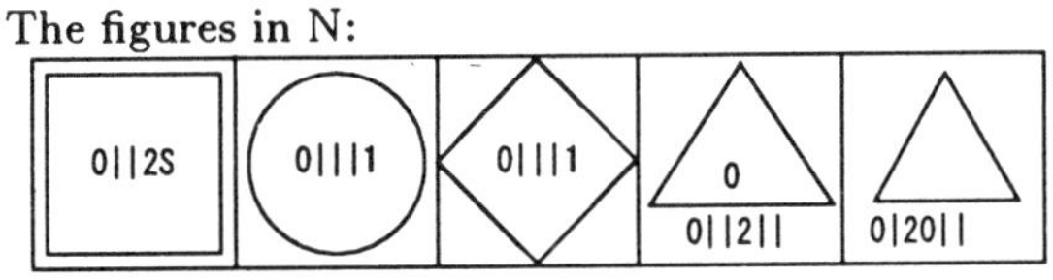

[14]= 3.065b.
[15]121 ||] 120 N. 'SS' stands for two avagraha-like symbols, which seem to be used for separating 100 and 6.
[16]121 ||] 120 N.
[17]caüttha-bhāyaṃ] ∅ N; 90 |] 90 N.
[18]savisesam/] ∅ N.
[19]100SS6] 100 N.
[20]= 3.067.
[21]= 3.068.
[22]maṇa^aṭṭha bārasa] maṇa aṭṭhabārasa N.
[23]sayāiṁ] sayāiṃ N.

// iti uddesa-paṃcagaṃ sammattaṃ//

// iti parama-jaina-śrī-candra-aṃgaja-ṭhakkura-pherū-viracita-gaṇita-sāra-kaumudī-pāṭyāṃ sūtraṃ samāptaḥ//[24]

// sarve vastu-baṃdha tathā gāthā miśrita 311//

likhitaṃ caitra sudi 5 saṃvat 1404//

[24] aṃgaja] aṅgaja N.

Chapter 6

Sūtrānukramaṇikā

6.000 sūtra saṃ° gāthā tathā vastu-baṃdha gaṇita 311//[1]

6.001 gā° 15 mūla prabandha sthāpanā/
6.002 gā° 78 parikarmāṇi pāṭī 25/[2]
6.003 gā° 5 saṃkalita utpatti vidhi 1/
6.004 gāha 6 vimala kalita gaṇanā 2/
6.005 gāha 6 guṇākāra bheda 2 gaṇanā 3/[3]
6.006 gāha 1 bhāgāhāra gaṇanotpatti 4/
6.007 gāha 3 varggasaṃ° utpatti gaṇanā 5/
6.008 gāha 2 varggamūla saṃ° utpatti gaṇanā 6/
6.009 gāha 4 ghana utpatti gaṇanā 7/
6.010 gāha 3 ghanamūlotpatti gaṇanā 8/
6.011 gāha 4 bhinna parikrama gaṇanā 9/[4]
6.012 gāha 3 bhinna saṃkalita gaṇanā 10/
6.013 gāha 2 bhinna guṇākāra gaṇanā 11/
6.014 gāha 2 bhinna bhāgāhara gaṇanā 12/
6.015 gā° 2 bhinna varggasya gaṇanā 13/
6.016 gā° 1 bhinna varggamūla gaṇanā 14/[5]
6.017 gā° 2 bhinna ghanasya gaṇanā 15/
6.018 gā° 1 bhinna ghana mūla gaṇanā 16/
6.019 gā° 9 trairāśika gaṇanā 17/
6.020 gā° 6 paṃca rāśika gaṇanā 18/
6.021 gā° 1 sapta rāśika gaṇanā 19/
6.022 gā° 1 nava rāśika gaṇanā 20/
6.023 gā° 1 ekādaśa rāśika gaṇanā 21/
6.024 gā° 5 vyasta trairāśika gaṇanā 22/
6.025 gā° 4 kraya vikraya bheda gaṇanā 23/
6.026 gā° 2 bhāṃḍa pratibhāṃḍa gaṇanā 24/

---

[1]We reproduce here the list of contents put after the colophon of the scribe, preserving the word forms in N.
[2]N omits '/' up to 6.027.
[3]gāha] " N up to 6.014.
[4]bhinna] abhinna N.
[5]gā°] " N up to 6.027.

6.027 gā° 2 jīva vikraya gaṇanā 25/
6.028 iti gāthā 78 parikarmāṇi 25 sūtrasya bījakaṃ yathā śubhamastu//
6.029 aparabhāga jāti 8 aṣṭa nāmāni/[6] sūtra gāthā° 17/
6.030 1 kalāsavarṇanu gāthā 1/
6.031 2 prabhāgajāti gāthā 1/
6.032 3 bhāga bhāgajāti gā° 3/
6.033 4 bhāgānubaṃdhājā° 2/
6.034 5 bhāga pravāha gaṇanā 2/
6.035 6 bhāga mātṛ jāti gā° 2/
6.036 7 vallī savarṇanu gāthā 2/
6.037 8 staṃbhoddesa jāti gāthā 4/
6.038 apara vyavahāra 8 gaṇanā/ sūtra gā° 104/[7]
6.039 1 prathama miśraka vyavahāra/ gā° 25/
6.040 gā° 2 miśraka gaṇanā pratha. 1/
6.041 gā° 2 bhāvyaka gaṇanā dutī 2/[8]
6.042 gā° 2 egapatrī karaṇa sūtra 3/
6.043 gā° 4 prakṣepaka 4/[9]
6.044 gā° 4 sama visama 5/
6.045 gā° 2 suvaṇṇī vyava° 6/
6.046 gā° 4 suvarṇa bhinno 7/
6.047 gā° 5 naṣṭa suvarṇṇa varṇṇa aṣṭama 8/
6.048 2 dutīka seḍhī vyavahāra/ gāthā 9/
6.049 gā° 2 seḍhī vyavahāra gaṇita 1/
6.050 gā° 1 naṣṭādyānayana 2/[10]
6.051 gā° 1 naṣṭottarānayana 3/
6.052 gā° 1 naṣṭagacchānayana 4/
6.053 gā° 2 saṃkalitaikyānayana 5/
6.054 gā° 1 vargaikyaghanaikyānayana 6/[11]
6.055 gā° 1 saṃkalita vargga ghanaika° 7/
6.056 3 kṣetra vyavahāra/ sūtra gāthā 19/
6.057 1 samacaüraṣa/
6.058 2 dīrgha caürasa/
6.059 3 ekādi sālaṃba/
6.060 4 trikoṇa kṣetra/
6.061 5 paṃcakoṇa kṣetra/
6.062 6 trikoṇa vikaṭa/
6.063 7 vṛttamaṃḍala/
6.064 8 dhaṇuhākāra/
6.065 9 gajadaṃtākāra/
6.066 10 vajrākāra/
6.067 11 mṛdaṃgākāra/
6.068 12 nānāvidhi/
6.069 4 khāta vyavahāra/ gāthā 15/

---

[6] N omits '/' up to 6.037.
[7] N omits '/' up to 6.097.
[8] gā°] " N up to 6.042.
[9] N omits gā° up to 6.047.
[10] gā°] " N up to 6.055.
[11] vargaikyaghanaikyānayana] varggaikaghanānayana N.

6.070 gā° 4 khātanānāvidhi gaṇana 1/
6.071 gā° 3 kūpasya phalānayana 2/[12]
6.072 gā° 6 pāṣāṇa phalānayana 3/
6.073 gā° 2 pāṣāṇa tolya gaṇana 4/
6.074 5 citi vyavahāra/ gāthā 19/
6.075 gā° 1 mūlaprabaṃdha gā° 1/
6.076 gā° 4 bhitti īṭapāṣāṇa 2/[13]
6.077 gā° 3 gomaṭa ciṇaṇa ga° 3/
6.078 gā° 2 pāyasevabhitti 4/
6.079 gā° 2 munārā gaṇita saṃkhyā 5/
6.080 gā° 2 tāka gaṇanā suddhi 6/
6.081 gā° 1 sopāna gaṇanā 7/
6.082 gā° 1 pulabaṃdhanaga° 8/
6.083 gā° 1 kūpa saṃgaṇanā 9/
6.084 gā° 1 vāpī saṃgaṇanā 10/
6.085 6 krakaca vyavahāra/ gāthā 9/
6.086 ° kāṣṭha dīrgha vi°/
6.087 karavattī cheda°/
6.088 nānākāṣṭha/
6.089 etā gāthā 9/
6.090 7 rāsi vyavahāra/ gāthā 5/
6.091 anna rāsi/
6.092 dīrghodaya/
6.093 vistara/
6.094 gaṇitasāri/
6.095 8 chāyā vyavahāra/ gāthā 4/
6.096 ° chāyā sādhanā/
6.097 ° digsādhanā/
6.098 iti aṣṭa vyavahāra sūtra gāthā 104//

[12]gā°] " N up to 6.073.
[13]gā°] " N up to 6.084.

Part III

English Translation

Chapter 1

Twenty-five Fundamental Operations

[Note: In this translation, the titles of the chapters and sections accompanied by serial numbers have been supplied by us. A double daṇḍa, '//', separates the translations of two consecutive verses. Original Apabhraṃśa terms are put in a pair of parentheses, (...), in the translation, and other information in footnotes. A brief explanation of a word in the translation also is put in a pair of parentheses, (...), immediately after the word in question, while additions to the translation are put in a pair of angular brackets, ⟨...⟩. Word numerals (the so-called *bhūtasaṃkhyās*) are put in a pair of single quotes, '...'. N indicates the Nahatas' edition or the editors themselves.]

1.1 Weights and Measures

1.1–2. Having saluted ⟨Jina⟩ the lord of the three worlds, the knower of the lord of Lakṣmī, the lord of the mountain⟨s⟩ and all other gods, and having written (copied) the algorithm of computations (*gaṇaṇa-pāḍī*) as it was expounded by the ancient teachers// — some was taken from there (i.e. writings of the ancient teachers), some gained from direct experience, and some after having heard ⟨it from others⟩ — Pherū, son of Caṃda, expounds this for the sake of all people.

⟨*Monetary units*⟩

1.3. *Paḍikāiṇi* and *kāiṇi*, *paḍivissaṃsa* and *vissaṃsa*; it should be known that these ⟨units⟩ up to *visova* are twenty times ⟨the preceding unit of currency⟩.

1.4a. Twenty *visovas* ⟨make⟩ one *damma* and fifty *dammas* one ⟨silver⟩ *ṭaṃkaya*.

⟨*Area measures*⟩

1.4b. Twenty *kamas* or sixty *kaṃvīs* in length and breadth make one *vīgahaya*.[1]

[1]This line, which deals with area measures, should occur after verse 7. The line, with *kami* for *kama* and *kaṃviya* for *kaṃvī*, occurs again in GSK 5.3b.

⟨*Linear measures*⟩

1.5–6. Twenty-four *pavv-aṃgulas* (joint-digits) should be known ⟨as equal to⟩ thirty-two *kar-aṃgulas* (hand-digits). Know that eight *javas* (barleycorns) spread breadthwise are equal to one *pavv-aṃgula*.[2] // Twenty-four *aṃgulas* ⟨make⟩ one *hattha* ('hand' or cubit). O learned man, four *hatthas* become one *ḍaṃḍa* (a stick), two thousand *ḍaṃḍas* one *kosa* (cry of a cow), and four *kosas* one *joyaṇa* (yoking).

1.7. This has been said about the linear (*sara*) *hattha*. Breadth (*vikkhaṃba*) multiplied by length (*āyāma*) is the square (*paḍa*) *hattha*. Know that area (*vitthāra*)[3] multiplied by height (*udaya*) is the cubic (*ghaṇa*) *attha* (hand).

⟨*Volume measures*⟩

1.8. Four *karapuḍas* (the hands joined and hollowed) ⟨make⟩ one *pāī*, four *pāīs*, it is said, ⟨make⟩ one *māṇaya*, four *māṇayas* one *seī*, and sixteen *seīs* become one *pattha*[4].

⟨*Weight measures*⟩

1.9. Six *gumja* seeds ⟨make⟩ one *māsaya*, four of these one *ṭaṃka*, ten *ṭaṃkas* one *pala*, six *palas* one *sera*, and forty *seras* one *maṇa*.

1.10. Sixteen *javas* ⟨make⟩ one *māsaya*, four of the same one *ṭaṃka*, and thrice this is one *tola*. Sixteen *javas* make one *vannī* and that which has twelve *vannīs* is pure gold (*mahā-kaṇaya*, lit. great gold).[5]

⟨*Time units*⟩

1.11. Sixty *palas* ⟨make⟩ one *ghaḍiyā*, sixty *ghaḍiyās* one nychthemeron (*diṇu-rayaṇī*), thirty nychthemerons make one month (*māsa*), and twelve months one year (*varisa*).

⟨*Names of decimal places*⟩

1.12–14. Know that *ega* (one), *dasa* (ten), *saya* (hundred), *sahasa* (thousand), *dasahasa* (ten thousand), *lakkha*, *dasa-lakkha* (ten *lakkha*), *koḍi*, *dasa-koḍī* (ten *koḍi*) , *avva*, *dasa-avva* (ten *avva*),// *khavva*, *dasa-khavva* (ten *khavva*), *saṃkha*, *dasa-saṃkha* (ten *saṃkha*), *paüma*, *dasa-paüma* (ten *paüma*), *nīla*, *dasa-nīla* (ten *nīla*), *nīla-saya* (hundred *nīla*), *nīla-sahasa* (thousand *nīla*),// *dasa-sahasa-nīla* (ten thousand *nīla*), *nīla-lakkha* (*lakkha nīla*), *nīla-dasa-lakkha* (ten *lakkha nīla*), *koḍi-nīla*, and so on, are the names of ⟨the notational places for⟩ the digits (*aṃka*) ⟨used for expressing⟩ numbers (*saṃkha*).

[2] Pherū mentions two kinds of *aṅgulas*; he calls the standard *aṅgula* (i.e. one-twenty-fourth part of *hasta*) *parvāṅgula*. In addition, he has a new *aṅgula* which he terms as *karāṅgula*. See under GSK 1.5–6 in Part IV. On *aṅgula*, see Michaels 1978, 156–57.

[3] S. *vistāra*, which usually means width.

[4] S. *prastha*. See GSK 3.96, where *patta* is defined as one cubic cubit of grain. Pherū's *pattha* is much larger than the traditional *prastha*. See Srinivasan 1979, 71–73, etc. Bhāskara I seems to have used the same *prastha* as Pherū did. See GSK under 1.8 in Part IV.

[5] The first line gives the gold weights, as distinct from the other weights in verse 8. The second line gives the grades of the fineness of gold. This is mentioned also in Pherū's *Dravya-parīkṣā*. On the fineness of gold, see Sarma 1983.

Thus 25 digits for computation (*gaṇitāṅka*).

⟨*Forty-five gateways to mathematics*⟩

1.15. ⟨There are⟩ twenty-five fundamental operations (*parikamma*), eight classes (*jāī*) ⟨of reduction of fractions⟩, eight kinds of procedures (*vivahāra*), and four topics (*ahigāra*). Thus ⟨there are⟩ forty-five gateways (*dāra*) ⟨to mathematics⟩.[6]

1.2 Eight Fundamental Operations of Integers

1. Now he tells[7] the sum (*saṃkalita*).[8]

1.16a. Add unity to the requisite (*icchā*) and halve it. Multiply it by the requisite. This is the summation of the natural series (*saṃkaliya*).

1.16b. (Missing.)[9]

1.17. One half of an even number of days (*samadiṇa*) is multiplied by the next term (*agga*); an odd number (*visama*) ⟨of days⟩ is multiplied by half of the next term. The result ⟨in each case⟩ is the summation of the natural series. There is no doubt here. This should be known.

1.18. The requisite is multiplied by ⟨the number of⟩ the letters in question (*paṇh-akkhara*), increased by the 'question', again multiplied by the requisite, and ⟨finally⟩ divided by twice the 'question'. The summation of the natural series is told by the digits obtained.

1.19. Tell the sums of one up to ten terms each multiplied by ten, where terms regularly increase by one ⟨from one⟩, and also the number of terms (*mūla*, lit. root) from these ⟨sums⟩.

1.20. The sum of the natural series being multiplied by eight and then increased by one, its square root (*paya*) decreased by one and halved ⟨is the number of terms⟩. Or, the number of terms (*mūla*) in the summation of a natural series is the square root (*mūla*) of twice ⟨the sum of the series⟩, which is the same as the remainder (*sesa*) ⟨left in the sum after the square root is extracted⟩.

2. Now he tells the difference (*vyavakalita*).

1.21. Just as the number of terms (*paya*) in the summation of the natural series increases one by one from unity, so ⟨the number of terms⟩ decreases one by one from the root-quantity (*mūla-rāsi*) in the case of the difference of the natural series (*vimakaliya*).

1.22. The number of terms in the subtrahend series (*vimakaliya*) increased by one is added to the number of terms of the summation (*saṃkaliya*). Then

[6] At the end of the text, all topics are listed serially but not the last four *adhikāras*.

[7] Henceforth, we render *āha* in the present tense, taking its contextual meaning into account.

[8] In this section, instead of ordinary addition, Pherū deals with the summation of series, like PG and GSS.

[9] Verse 16b should read, as in PG 14a, 'when the first term and the common difference ⟨of the series⟩ are each unity.'

multiplied by the difference of the periods of the two ⟨series⟩ and halved, it becomes the remainder of subtraction (*vimakaliya-sesa*) ⟨of the two given series⟩.

1.23. O learned man, after having subtracted from the sum of one thousand terms ⟨starting from one and increasing by one⟩ the sums of terms starting from ten and increased each time by ten (i.e., 10, 20, 30, ...), severally, tell what will be the numbers representing the respective remainders after subtractions.

1.24. Having subtracted the remainder of subtraction from the sum of the summation, twice the remainder, its square root, which must be equal to the remainder ⟨left after extracting the square root⟩, becomes the original number of terms of the subtrahend series.

1.25. Twice the number of terms of the summation, increased by one, decreased by the number of terms of the subtrahend series, halved and multiplied by ⟨the number of terms of⟩ the subtrahend series, is ⟨the remainder of⟩ the subtraction from the highest (*uvara*) ⟨term of the summation⟩.

1.26. Tell, if you know the subtraction of the natural series, what will be produced when up to three terms from the top are subtracted from the sum of the ⟨first⟩ hundred terms of the natural series.

3. Now he tells the multiplication (*guṇākāra*).

1.27. Having placed the multiplicand (*gunna-rāsi*) below and the multiplier (*guṇa-rāsī*) above as in the door-junction (*kavāḍa-saṃdhi*) (i.e., junction of two doors), the multiplier should be multiplied successively in the direct or inverse order (*anuloma-viloma-gaī*).

1.28. ⟨When⟩ two thousand ⟨is multiplied⟩ by thirty-two, nine hundred and sixty-four by twenty-seven, ⟨and⟩ one hundred and eight multiplied by sixty, what is the product (*phala* lit. fruit, result) in each case?

1.29. Now, having broken the multiplier ⟨into the digits standing in different notational places or into two or more parts whose sum is equal to it⟩, and having multiplied ⟨the multiplicand⟩ by each digit ⟨separately⟩, make their sum (*piṃḍa*). ...[10]

How to judge whether ⟨the result obtained by⟩ the second (*dutīka*) ⟨rule of⟩ multiplication[11] is true or false (*satya-alīka*).

1.30. Multiplier and multiplicand. Take the sum ⟨of the digits of each⟩ separately ⟨or⟩ the remainders ⟨after dividing⟩ by nine. Multiply ⟨the two sums or the two remainders, and apply the same operation to this product also⟩. ⟨If the result is⟩ equal to the sum of ⟨the digits of⟩ the product (*guṇiya-rāsi*) or to the remainder ⟨after dividing it⟩ by nine, ⟨the product is⟩ correct (*suddha*).

1.31. In addition ⟨of a number to zero⟩, zero (*kha*) becomes equal to the additive (*kheva*). When 'sky' (0) is added to or subtracted from ⟨a number⟩, the number

---

[10] The latter half of verse 29 is not understood.

[11] The 'second ⟨rule of⟩ multiplication' seems to refer to verse 29.

(*rāsi*) remains unchanged (*avigaya*). In multiplication, etc. with zero (*sunna*), ⟨the result is⟩ zero. In multiplication of zero by zero, ⟨the result is⟩ zero.

1.32. Say, if you know, ⟨what will be the results of⟩ multiplication (*guṇayāra*) of zero, and division (*bhāgahara*) of zero, the square (*vagga*), the square root (*vagga-mūla*), the cube (*ghaṇa*), etc. of zero.

4. Now he tells the division (*bhāgāhara*).

1.33. That from which ⟨a number⟩ is taken off is the dividend (*saṃharaṇīya*) ⟨and⟩ that which takes off ⟨itself⟩ after multiplying is the divisor (*hara*). Write down the dividend above and the divisor below, and the division (*bhāya*) is possible.

5. Now the square (*varga*).

1.34. Place the square of the first digit. Multiply the rest in order by twice the first digit.[12] And along with the previous results again add the square ⟨of the next digit⟩, always increasing (moving) by one place. It is the square.

1.35. A number multiplied by the same number is the square. Or, place the requisite number at two places, subtract an arbitrary number ⟨from one⟩, and add ⟨the same to the other⟩. Multiply ⟨the sum by the difference⟩ and further add the square of the arbitrary number. This is the square ⟨of the given number⟩.

1.36. From one to nine, sixteen, twenty-four, and twenty-eight. Tell ⟨me⟩ quickly the square quantity of each.

6. Now he tells the square root (*varga-mūla*).

1.37–38. Of that which is a square number, from the last ⟨place⟩ up to the first, count odd and even ⟨places⟩. Subtract a square number from the ⟨first (highest)⟩ odd place. The square root ⟨of that square⟩,// make it doubled, move ⟨the result to the next place⟩, and divide ⟨that place by it⟩. ⟨Put twice⟩ the quotient *phala* in the row ⟨of the square root⟩ and subtract its square again. ⟨Repeat⟩ the previous operation up to the last ⟨place⟩. That which has been doubled is halved. This is the square root ⟨of the given number⟩.

7. Now he tells the cube (*ghana*).

1.39–40. Having set down the cube of the first digit, the square of the same first digit, multiplied by three and further multiplied by the second digit, should be added in the proper sequence, by moving one place.// Again, the square of the second digit, multiplied by the first digit and further multiplied by three, should be added. Again, that ⟨second⟩ digit should be cubed and added. ⟨And so on.⟩ This is the cube ⟨of the given number⟩.

1.41. Set down the desired digit (number) three times ⟨one below the other⟩. Multiply upwards. The result is the cube. The last ⟨number⟩ multiplied by

[12]Pherū's terminology for the order of the digits expressing a number is different from others': he calls the unit place 'the last' (*aṃta* in verse 37 and *carima* in verse 38; cf. also *juyal-aṃt-aṃka* in 4.6, *ti^aṃtim-aṃkā* in 4.7, etc.) and the highest place 'the first' ('*paḍhama*' and '*āi*' in the present verse, *dhura* in verses 37 and 39, and *dhurima* in verses 39 and 40).

the preceding digit (number) and by three and increased by the cube of the preceding ⟨number⟩ and by one ⟨gives the cube.⟩

1.42. What is the cube of one to nine, of sixteen, of two hundred and twenty-five, and of three hundred and nine?

8. Now he tells the cube root (*ghana-mūla*).

1.43–45. ⟨Digits in the given number are divided, from the units' place on, into periods of⟩ one cube place and two non-cube places. Diminish (divide) the portion up to the ⟨first (highest)⟩ cube place by the cube ⟨of an appropriate number⟩. The ⟨cube⟩ rootof what has been obtained, having been moved, should be placed below the third ⟨place from that cube place⟩.[13]// Set down ⟨temporarily⟩ thrice its square behind that (i.e., below the second place from that cube place) and divide ⟨that place by it⟩. The quotient *phala* is placed in the line and the digits of the divisor ⟨which has been temporarily set down⟩ should be removed.// Thrice the square of the last ⟨digit⟩ of the line ⟨of the cube root⟩, multiplied by the preceding digits ⟨of the line⟩, is subtracted ⟨from the third place from the cube place⟩ and further the cube of the last place ⟨of the line of the cube root⟩ is subtracted ⟨from the next cube place⟩. Do the same to what has been obtained (i.e., to the remaining digits).

1.3 Eight Fundamental Operations of Fractions

Now he tells the eight fundamental operations of fractions (*bhinna*).

⟨*Expression of fractions*⟩

1.46. The digits of a fraction (*bhinna*) are placed in this way: the integer (*rū*) above, the numerator (*aṃsa*) in the middle, and the denominator (*cheya*) below. A dot (*biṃdu*) is piled (*caya*) on a negative numerator (*hīṇ-aṃsa*). When there is no denominator, one is ⟨put down⟩.

⟨*Reductions of fractions*⟩

1.47. Multiply the integral number (*rūva-rāsī*) with the denominator and add or subtract the numerator. This becomes the reduction to the same colour (*savaṃnaṇa*). ⟨The numerators and the denominators of fractions⟩, when mutually multiplied by the denominators, become numerators having a common (*sadisa*) denominator ⟨and the common denominator⟩, respectively.[14]

1.48–49. Reduce ⟨the fractions⟩ to ⟨those having⟩ the common denominator. Then addition or subtraction of the numerators is made. Addition of the denominators must not be done. This has been told in ⟨mathematical⟩ treatises (*sattha*).// If the denominator be doubled, the upper quantity would become half ⟨of the correct value⟩. That this is the state of fractions should be known to all *pāyaras* (wise people?).

[13]For 'below the third place' see PG 29b: '*saṃyojya tṛtīyapadasyādhas*,' which has been fully explained by Shukla, tr., p. 13.

[14]For Pherū's terminology for the reductions of fractions see §2.2.10 of Part I.

9. Now he tells the sum of fractions.

1.50a. The sum (*juī*) of numerators having a common denominator, when divided by the denominator, is the sum (*saṃkaliya*) of the fractions.

1.50b. The sum of one-third, one-sixth, one-fifth, and one-ninth, as well as ⟨the sum of⟩ three less by a quarter, one and a half, and one-third ⟨should be told⟩.

⟨**10.** *Difference*⟩

1.51. Make the reduction of positive and negative ⟨mixed numbers⟩ (*ādāyassa vayassa ya*)[15] into fractions (*savaṃnaṇa*) ⟨severally⟩ and then the reduction ⟨of the fractions obtained⟩ to a common denominator (*sadisacheya*). Add the numerators ⟨coming from the positive and the negative mixed numbers⟩ separately, take their difference (*aṃtara*), ⟨and divide the difference by the denominator⟩. This is the subtraction (*vimakaliya*) of fractions.

1.52. Subtract a half, one-third, one-sixth, and one-ninth ⟨all at once (or separately)⟩ from eight. What is the remainder? Or, three and a half, five and one-third, and one-ninth should be subtracted from one-sixth.[16]

11. Now he tells the multiplication of fractions.

1.53. Multiply the numerator ⟨of one fraction⟩ with the numerator ⟨of the other⟩ and also the denominator with the denominator ⟨and then the first product⟩ should be divided ⟨with the second⟩. The number which is obtained, know that as ⟨the result of⟩ the multiplication of fractions.

1.54. Five less by a quarter *dammas* is multiplied by eight and one-third *dammas* and one-half is multiplied by one-sixth. What will be the result in each case?[17]

12. Now he tells the division of fractions.

1.55. Having interchanged the numerator and the denominator of the divisor and not of the dividend, multiply and divide as in the previous process (*vihi*) ⟨of the multiplication of fractions⟩. This is the process (*vihī*) of the division of fractions.

1.56. Seven less one-quarter should be divided (*bhāyaṃ harijjae*) by one and a half. Six and one-quarter is divided by four and one-third. What are their resulting quotients *phala*?[18]

13. Now he tells the square of fractions.

1.57. Divide the square of the numerator by the square of the denominator ⟨which is written⟩ below. Whatever is obtained, know that to be the result of

---

[15] Lit., 'of the income and the expenditure'.

[16] Or, 'Or, three and a half should be subtracted from five and one-third, and one-ninth from one-sixth.'

[17] This verse is carelessly worded. Perhaps it should mean 'Multiply five less by a quarter *dammas* by eight and one-third, and half a *damma* by one-sixth. What will be the result in each case?'

[18] Here it is not clear which is the divisor and which is the dividend in the first problem, for both are in the instrumental case.

squaring the fraction.

1.58. Tell the squares of one and a half, of five and one-third, of seven less by a quarter, and also ⟨of⟩ half and one-third if you know the method (*vihī*) of squaring.

14. Now he tells the square root of fractions.

1.59. The square root of the numerator should be divided by the square root of the denominator, ⟨after having extracted the square roots at first⟩ by marking the odd-even-odd ⟨places⟩. The root of the square of a fraction is obtained.

15. Now he tells the cube of fractions.

1.60. The cube of the numerator should be made and divided by the cube of the denominator. Whatever is obtained there, know that to be the cube of the fraction.

1.61. The cube of seven and a half, of fifteen and a quarter, of a quarter and one-third. Tell separately the cubic quantities produced.

16. Now he tells the cube root of fractions.

1.62. The cube root of the numerator should be divided by the cube root of the denominator, ⟨after having extracted the cube roots at first⟩ by marking one cube place and two non-cube places. The cube root is obtained.

1.4 Proportion

17. Now he tells the rule of three (*trairāśika*).

1.63. The first and the last ⟨terms⟩ are of the same denomination (*jāī*). The term of a different denomination is placed in the middle. The middle term is multiplied by the last term and divided by the first term. This is the rule of three (*tirāsiyaga*).[19]

1.64. O cloth merchant (*dosiya*), if for eleven *dammas* seven *karas* of cloth (*kappaḍa*) is sold, tell, for twenty-four *dammas* how many *hatthas* (= *karas*) are sold.

1.65. I will tell (*bhaṇisu*) now the exchange of coins (*nāṇavaṭṭa*). Nine coins (*muṃda*) fetch twenty-five *dammas*. By this price standard (*aggha-pamāṇa*), what is the price (*mulla*) of sixteen coins?[20]

1.66. If one and a quarter *palas* of sandalwood (*caṃdaṇa*) fetches the price of nine and one-third *dammas*, how much will six *palas* less by one-sixth of the same (sandalwood) fetch?

1.67. For seven and a quarter *dammas*, two *seras* increased by one-sixth of long

[19]= DP 60.

[20]Pherū gives a rule designed specifically for this type of problems in DP 49–50, the first sentence of which is exactly the same as that of this verse. See under GSK 1.65 in Part IV.

pepper (*pippali*) are obtained. How much of the same can be obtained for nine *dammas* less by one-third?

1.68. For one hundred and twenty *dammas* less by a quarter, five and one-third *patthas* of food such as rice (*taṃdula*), etc. ⟨are obtained⟩. How much rice can be obtained for one *damma*?

1.69. If one *tola* of gold (*kaṇaya*) of twelve *vannīs* (or of pure gold) ⟨can be obtained⟩ for one hundred and one-third *dammas*, then what will be the price of one *māsaya* less by one-tenth?

1.70. If a lame man (*paṃgulaya*) walks one-sixth *joyaṇa* in seven days, in what time will he walk sixty *joyaṇas*?

1.71. If a worm (*kīḍaya*) crawls one-seventh *aṃgula* in one-sixth day, in what time will it go eight *joyaṇas* and return?

Now he tells also the rule of five; the rules of seven, of nine, and of eleven.

1.72. Having transposed the fruit-digits at the bottom ⟨of the two columns of numerical data⟩ and having multiplied ⟨the terms⟩ on both sides (*pakkha*) separately in succession, divide ⟨the product of the larger set of digits⟩ by the product of the smaller set of digits (*thov-aṃka-rāsi*[21]) in the case of the rules of five, seven, nine, etc.

18. Now he tells the rule of five.

1.73. If ⟨the rate is⟩ five per cent (*paṃcaga-saya*) in a month, what is the interest (*phala*) on sixty in one year? Tell us the time (*kāla*), knowing the interest and principal (*mūla*); also the principal (*pamāṇa-dhaṇa*), ⟨knowing the time and interest⟩.

1.74. In a month less by one-third the interest (*vavahāra*) on one hundred and a half *dammas* is one and a half *dammas*. Then, what will be on seventeen less by a quarter ⟨*dammas*⟩ in nine and a quarter months?

1.75. The wages (*bhāḍaya*) for carrying eight and a half *maṇas* to a distance of one and one third *joyaṇas* is two less by a quarter *dammas*. Then, what will be ⟨the wages⟩ for ⟨carrying⟩ nine and a quarter *maṇas* to a distance of ten less by a quarter *joyaṇas*?

1.76. If twelve workers (*kamma-yara*, artisans) earn in four days thirty *dammas*, what will eight men earn in forty-five days?

1.77. If, assuming (*kiri*) pure gold (*bhitti-suvanna*) ⟨of twelve *vannīs*⟩ weighing four *māsas* less by one *guṃja* fetches twenty less by a quarter ⟨units of⟩ money, what price will gold of ten and a half *vannīs*, weighing two *māsas* increased by one *guṃja*, ⟨fetch⟩?

19. Now he tells the rule of seven.

1.78. Two blankets (*kaṃvala*), each measuring six ⟨*karas*⟩ in length (*dīha*) and

[21]Lit. the quantity of smaller number of digits.

three *karas* in breadth, fetch ninety *dammas*. What then is the price of seven blankets, each nine ⟨*karas*⟩ long and five ⟨*karas*⟩ wide?

20. Now he tells the rule of nine.

1.79. Twelve pieces of cloth (*cīra*) having five colours; each of them covers (*acchaï*) ⟨an area of⟩ seven *karas* long and three *hatthas* (= *karas*) wide. The price of all these is fixed (*nicchaï kiu*) by the cloth merchants at six hundred *dammas*. If so, tell, after thinking ⟨well⟩, O cloth merchants, what is the price of nine pieces of cloth, having four colours, each eight *karas* long and five broad?

21. Now he tells the rule of eleven.

1.80. If *mūṃg* pulses measuring two *patthas*, six *seīs*, three *māṇas*, two *pāīs*, and one *karapuḍa* (lit. two, six, three, two, and one from *pattha* up to *karapuḍa*) cost sixty *dammas*, what is the price of nine *patthas*, three *seīs*, two *māṇas*, one *pāī*, and three *karapuḍas* (lit. nine, three, two, one, and three from *pattha*)?[22]

22. Now he tells the inverse rule of three (*vyasta-trairāśika*).

1.81. The middle term multiplied by the first and divided by the last is the inverse rule of three (*vittha-tiya-rāsī*). The last and the first terms are of the same denomination, when the term of a different denomination is placed in the middle.

1.82. When ⟨a certain quantity of grain⟩ is measured in *patthas* of ten *seīs*, it is twenty-seven *patthas*. How many *patthas* will it be, if measured in *patthas* containing sixteen *seīs*?

1.83. ⟨Grain⟩ contained in a container (*vakkhara*), when weighed by ⟨*sera*⟩ consisting of sixty-six *ṭaṃkas*, comes to twenty *maṇas*. How many *maṇas* will it be, if weighed by ⟨*sera*⟩ equalling seventy-two ⟨*ṭaṃkas*⟩? [23]

1.84. There is a gold alloy of eleven and a half *vannīs* weighing forty and a half *tolas*. How much gold of ten and a quarter ⟨*vannīs*⟩ can be had in exchange (*pavaṭṭaṇa*) for that?

1.85. ⟨When a certain quantity of wool can be made into⟩ two hundred blankets (*kaṃbala*), each nine ⟨*hatthas*⟩ long and three ⟨*hatthas*⟩ broad, how many blankets, each five long and two broad, can be made ⟨out of the same quantity of wool⟩, do tell.

23. Now he tells the purchase and sale (*kraya-vikraya*).

1.86. The product of the middle and the last terms is the capital (*mūla*) and the product of the last and the first terms is the total product (*savva uppattī*), when they are ⟨separately⟩ divided by the difference between the selling and the

[22]Actually, there are only three terms here, because *seī*, *māṇa*, *pāī*, and *karapuḍa* are submultiples of *pattha*. Convert all either into *karapuḍa* or into *pattha* and then apply the rule of three.

[23]This and the previous examples contain assumptions which do not agree with the weight systems of the GSK. See under GSK 1.83 in Part IV.

buying rates (*vikaya-kaya*). When one knows the capital ⟨instead of the profit⟩, the value of the profit (*lāha*) ⟨can be calculated⟩.

1.87. ⟨A certain thing⟩ is obtained (*lijjahi*) ⟨at the rate of⟩ seventeen *maṇas* per one *ṭaṃka* and sold ⟨at the rate of⟩ fifteen ⟨*maṇas*⟩ per one *ṭaṃka*. If the profit is ten *ṭaṃkas*, then tell how many *ṭaṃkas* are the principal.

1.88. Five objects (*vatthū*) are obtained for three *dammas*; they should be sold at the rate of seven objects for nine *dammas*. If the profit is forty-two *dammas*, how many *dammas* is the principal?

1.89. The *dammas* are placed (written) above and the number of objects below. The two sets of terms are multiplied cross-wise (*vaṃkaï*). The first quantity (product) is multiplied by the profit and divided by the difference between the two quantities (products).

24. Now he tells the barter (*bhāṃḍa-pratibhāṃḍaka*).

1.90. In the procedure for barter (*bhaṃḍa-paḍibhaṃḍa-karaṇa*), having transposed the price (*mulla*) and the fruit (*phala*) and multiplied successively the two ⟨sets of⟩ quantities, divide ⟨the product of the larger set of terms⟩ by the product of the smaller ⟨set of terms⟩.

1.91. If for one hundred *dammas* two *maṇas* of long pepper (*pippali*) are obtained and for three hundred *dammas* five *maṇas* of dry ginger (*suṃṭhī*), how much of dry ginger will then be obtained for seven *maṇas* of long pepper?

25. Now he tells an operation for the sale of living beings (*jīva-vikraya*).

1.92. In the sale of living beings (*jīva*), years also should be transposed and the fruit-digits should be transposed. The rest is by means of the previous rule (i.e., the rule of five). Thus you know the price of the best living beings.

1.93. If three camels (*karaha*) of ten years ⟨in age⟩ fetch one hundred and eight *ṭaṃkas*, then what is the price of five camels of nine years ⟨in age⟩?

Thus the rules for twenty-five fundamental operations (*parikarman*) in ⟨the book for⟩ the procedural mathematics (*pāṭī*), Gaṇita sāra kaumudī (Moonlight of the Essence of Mathematics), composed by the most ⟨devout⟩ Jain, son of the respected Candra, Ṭhakkura Pherū, are completed.

Thus ⟨ends⟩ the first chapter.

Chapter 2

Eight Classes of Reduction of Fractions

1. Now he tells the homogenization of fractions (*kalā-savarṇana*) for the part class (*bhāga-jāti*).

2.1a. After reducing ⟨the fractions⟩ to ⟨those having⟩ a common denominator, addition of the numerators is done afterwards. This is the ⟨process for reduction of⟩ part class (*bhāga-jāī*).

2.1b.[1]

2. Now he tells the the multi-part class (*prabhāga-jāti*).

2.2a. The denominator is multiplied by the denominator and the numerator by the numerator. This is the ⟨process for reduction of⟩ multi-part class (*pabhāga-jāī*).

2.2b. What is the half of a half and its one-fifth and one-sixth ⟨severally⟩?[2]

3. Now he tells the part-part class (*bhāga-bhāga-jāti*).

2.3. Multiply unity (*rūva*) by thc dcnominator (*cheya*) ⟨of the fraction⟩ and remove the denominator. This is the process of part-part ⟨class⟩ (*bhāga-bhāga-vihī*). After multiplying each part (*aṃsa*, i.e. investment) by the total sum (*dhaṇa*), ⟨each product is⟩ divided by the sum of the parts. The fruit (*phala*, i.e. share) ⟨is obtained⟩.[3]

2.4. What will be the result (*phala*, i.e. shares) of one hundred *dammas* severally ⟨when divided as follows⟩: two parts among three parts for one, seven parts among nine parts for another, and three parts among six parts for another?

---

[1] Verse 1b is wrongly placed. It is an example for the multi-part class (*pabhāga-jāī*) and the right place for it is at 2b, where also it occurs. Here Pherū does not speak of subtraction for the part-class. A better definition is in GSK 1.48.

[2] Two alternative interpretations are possible: (1) 'What is the half of a half and one-sixth of one-fifth of it (i.e., of a half) ⟨severally⟩?' (2) 'What is the half of a half and its (the product's) one-fifth and further one-sixth?'

[3] For the second half of this verse, cf. GSK 3.7 below.

2.5. A cistern (*vāvi*) of six ⟨cubic⟩ *karas* ⟨in capacity⟩ is filled by four pipes (*nālaya*) in one day, half a day, one-third, and one-fourth part of a day respectively ⟨when they are opened separately⟩. If they are opened at the same time, what time will the cistern take to fill?

4. Now he tells the part-addition ⟨class⟩ (*bhāga-anubaṃdha*).

2.6. Having multiplied by the lower denominator (*hara*) the upper denominator, and having multiplied by the lower denominator increased by the ⟨lower⟩ numerator (*aṃsa*) the whole number (*rūvaṃ*, actually the upper numerator), that which has a division at the end is the process of part-addition ⟨class⟩ (*bhāga-aṇubaṃdha-vihī*).

2.7. ⟨Reduce to a single fraction⟩ three and a half, plus its quarter, along with its one-sixth, added to its half. What is the result? Likewise, a half is increased by its one-third and ⟨further⟩ by its quarter. ⟨What is the result?⟩

5. Now he tells the part-subtraction ⟨class⟩ (*bhāga-apavāha*).

2.8. Having multiplied by the lower denominator the upper denominator, multiply by the lower denominator diminished by the ⟨lower⟩ numerator the upper whole number (*rūva*, actually the upper numerator). Thus the part-subtraction ⟨class⟩ (*bhāga-apavāha*).

6. Now he tells the part-mother class (*bhāga-mātṛ-jāti*).

2.9. Three diminished by ⟨its⟩ half, this less by ⟨its⟩ quarter, this less by its one-sixth. Likewise, a half deprived of ⟨its⟩ one-third, this of ⟨its⟩ one-fourth. What are the results separately?

2.10. The summation (*samāsaṇa*) ⟨of two or more⟩ of the five classes starting with the part ⟨class⟩ is ⟨the class called⟩ the mother of fractions (*bhāga-mattīya*[4]). After performing ⟨the operation for reducing each class by the respective rule⟩ separately as told before, the reduction to a common denominator and the addition of the numerators ⟨should be carried out⟩.

2.11. A half, a quarter of a quarter, one divided by one-third, a half increased by its half, one-third diminished by its half: all these ⟨are added⟩ together. What will be the amount of money?

7. Now he tells about the chain-reduction ⟨class⟩ (*vallī-savarṇana*).

2.12. The operation (*vihī*) for reducing a chain ⟨of measures⟩ to a common denominator (*vallī-savannaṇa*) ⟨is as follows⟩. Having multiplied ⟨the pair of⟩ denominator and numerator by the lower denominator, the lower numerator should be added to, or subtracted from, the upper numerator.

2.13. Two *tolas*, three *māsas*, and also four *guṃjas* and five *visuvas*: all these are diminished by one-seventh. What will be the amount obtained by reducing this chain ⟨to *tolas*⟩?

[4]Here, *mattīya* must be equivalent to S. *mātṛkā*, although, according to the PSM, *mattīya* = S. *mṛttikā* and *māua* = S. *mātṛkā*.

8. Now he tells about the pillar-part class (*sthaṃbh-aṃsaka-jāti*).

2.14. Reduce the fractions to ⟨those having⟩ a common denominator, add the numerators, and subtract the result from one. Whatever is the remainder, with that divide the visible quantity (*paccakkha*). The quotient *phala* is the measure of the pillar (*thaṃbha*).

2.15. One half, one-sixth and one-twelfth ⟨of a pillar⟩ are respectively situated under water, mud, and sand, and three *kaṃviyas* are visible.[5] Tell, O learned man, the measure of the pillar.

2.16. ⟨Of a herd of cows⟩, one-fifth went to the eastern side[6] and one-eighth to the south; one-sixteenth disappeared in the west; one-fourth (*cāuddha*) went to the north; one-sixth was lost by the fear of lions. O scholar, listen carefully. One hundred and forty-one cows (*gorū*) are left in the valley (*tala*). If they are collected in one place, how many cows (*thaṇa-vāla*) will go (*loḍaï*) ⟨there⟩?

2.17. ⟨Of a herd of elephants⟩, one-half increased by one-third ⟨of itself⟩ are near Mt. Vindhya, one-sixth increased by one-seventh ⟨of itself⟩ are near the water, and one-eighth along with one-ninth ⟨of itself⟩ are on dry grounds. Four are remaining. How many are there in the herd?

Thus the eight classes (*jāti*) of fractions (*bhāga*) in the procedural mathematics (*pāṭī*), Gaṇita-sāra-kaumudī (Moonlight of the Essence of Mathematics), composed by the most ⟨devout⟩ Jain, son of the respected Candra, Ṭhakkura Pherū, ⟨are completed⟩.

Thus ⟨ends⟩ the second chapter.

---

[5]For *kaṃviya* see under GSK 1.5–6 in Part IV.

[6]*puvaddhi*, S. *pūrva-ardhe*, lit. 'in/to the eastern half'.

Chapter 3

Eight Types of Procedures

3.1 Procedure for Mixture

1. Now he tells about the procedure for mixture (*miśraka-vyavahāra*) in the procedural mathematics (*vyavahāra-gaṇanā*).

3.1. Having multiplied the capital (*pamāṇa-dhaṇa*) by its time, and the other time by the interest (*phala*), take their sum. Having multiplied both ⟨products⟩ by the mixed amount (*missa*, i.e., capital plus interest), they are separately divided by the sum. The interest and the capital (*mūla*) ⟨are obtained⟩.

3.2. ⟨The rate of interest⟩ being five percent per month, ⟨a certain capital becomes⟩ in four months five hundred and twenty *dammas*. What are its interest and capital? If you know, tell quickly.

2. Now he tells about the commission of surety (*bhāvyaka*).

3.3. Having multiplied the capital by its time and the interest, etc. (interest, commission, fees, etc.) respectively by the time of interest (*phalakāla*), divide ⟨each of these products⟩ by the sum of the products (*aṃsāṇa*, lit. of the parts), and then having multiplied ⟨each⟩ by the mixed amount, the capital, etc. are obtained.

3.4. In a month, for one hundred, the rate of interest is five, ⟨commission⟩ of the brāhmaṇa one, the fee (*vittī*) ⟨of the calculator⟩ a half, ⟨the fee of⟩ the scribe (*lehaga*) a quarter, and the mixed amount in one year is nine hundred and five. ⟨Find the capital, the interest, and the shares of the brāhmaṇa, the calculator and the scribe⟩.

3. Now he tells about the conversion of several bonds into one (*eka-patrī-karaṇa*).

3.5. The sum of the interests ⟨accruing on the given bonds⟩ for elapsed time ⟨in months⟩, when divided by the sum of the interests ⟨on the same bonds⟩ for one month, ⟨yields⟩ time ⟨in months for an equivalent single bond⟩. The sum of the interests for a month, multiplied by one hundred and divided by the sum of the capitals ⟨gives⟩ the ⟨rate of⟩ interest per hundred ⟨per month for a single bond⟩.

3.6. The amounts of money, one hundred, two hundred, three hundred, and six hundred, were lent at the rate of two, three, four, and five percent per month, ⟨respectively⟩. Four, six, ten, and eight months ⟨having passed since the execution of the respective bonds⟩, how can a single bond be made ⟨out of these⟩?

4. Now he tells about the proportionate distribution (*prakṣyepaka*).

3.7. Divide the mixed amount (harvest) by the sum of the parts (investments) having equal denominators, after having multiplied by each part (investment). This is the method of investments, ⟨by means of which⟩ one should know the fruits (shares) from the mixed amount (harvest).

3.8. Two, three, five, and four *maṇas* of seeds were thrown together (or invested). This produced two hundred and ten ⟨*maṇas*⟩. What is the share (*bhinna-phala*) of the harvest[1] of each one?

3.9. (Not understood.)

3.10. ⟨There is a piece of⟩[2] sandalwood ⟨whose⟩ height is seven ⟨*aṃgulas*⟩ and ⟨whose⟩ width is seven ⟨*aṃgulas*⟩ at the top and one ⟨*aṃgula*⟩ at the bottom. It costs (*bhariya*, lit., is maintained with) ten *dammas*. What is the price of each *aṃgula*?

5. Now he tells about the purchase in proportion (*sama-viṣama-kraya*[3]).

3.11. The rate prices divided by the quantities separately should be multiplied by the parts (*aṃsa*) ⟨or the ratios in which these are purchased⟩. Divide (*bhāva*) these ⟨severally⟩ by the sum (*juva*) of the ⟨resulting⟩ parts, and the ⟨resulting⟩ parts should be multiplied by the ⟨total⟩ price. ⟨This gives the price of each item for⟩ the equal or unequal purchase. ⟨From each price, the amount of each item can be calculated⟩ by the rule of three, as ⟨told⟩ before.

3.12. For one *damma*, ⟨one can get⟩ one *sera* of yellow myrobalan (*haraḍaï*) ⟨or⟩ three ⟨*seras*⟩ of vibhītikā (*baheḍā*, *Terminalia bellerica*) ⟨or⟩ six *seras* of myrobalan (*āmalaya*). O physician (*vijja*), give for one *damma* equal amounts ⟨of all the three⟩ after grinding (*phakkiya*[4]) them.

3.13. Half a *sera* of long pepper ⟨can be obtained⟩ for three ⟨*dammas*⟩, one *sera* of pepper for nine and one-third *dammas*, and one less by a quarter *sera* of dry ginger for four *dammas*. Give for one *damma* all the three equally.

3.14. For one *damma*, nine *seras* of rice ⟨or⟩ eleven *seras* of *mūṃg* pulse ⟨or⟩ one *sera* of clarified butter (*ghiya*) ⟨can be obtained⟩. O merchant, give me three, two, and one parts respectively of these for one and a quarter *dammas*.

6. Now he tells about the procedure for gold (*suvarṇṇa-vyavahāra*).

---

[1] *halahari dinnāṇa*, lit. 'of those given by the plough-holder (farmer).'
[2] The first word, *kaṃcolu*, is not understood.
[3] Lit. 'the purchase of equal or different ⟨quantities⟩.'
[4] Or, a deśī meaning 'quickly'? Cf. *kṣipraṃ* in L 99 and *āśu* (twice) in PG (73)–(74) = Tr (71)–(72), which treat problems of the same type.

3.15. Whatever be the ⟨purity in⟩ *vannas* and whatever be the weight (*tulla*) ⟨of each gold piece⟩, having multiplied one by the other ⟨severally⟩, their sum is taken. ⟨This⟩, when divided by the ⟨total⟩ weight, ⟨gives⟩ the ⟨purity in⟩ *vannīs*, and ⟨when divided⟩ by the ⟨purity in⟩ *vannīs*, becomes the weight ⟨of the gold piece produced from them⟩.

3.16. ⟨Four pieces of gold of⟩ nine, ten, eight, and eleven *vannīs* ⟨and weighing⟩ three, six, five, and two *tolas* ⟨respectively⟩ were melted together (*egattha gāliya*). Of what *vannīs* will the gold (*kaṇaya*) become?

7. Now he tells an example with fractions for gold.

3.17. ⟨Three pieces of gold of⟩ eight, nine and a quarter, and five and three-quarters *vannīs* and weighing three, five, and two *māsas* increased by one-third, one-sixth, and one-fifth ⟨respectively⟩ when melted together (*āvaṭṭa*), what type of gold will it become?

8. Now he tells ⟨a rule⟩ for refined gold (*pakva-suvaṇṇa*).

3.18. The sum of the products of the ⟨purities in⟩ *vannas* and the ⟨weights of several pieces of⟩ gold (*suvanna*), when divided by ⟨the weight of⟩ the refined (*vipakka*) gold, gives the ⟨purity in⟩ *vannas* ⟨of the refined gold⟩. And ⟨the same sum⟩, when divided by the desired ⟨purity in⟩ *vannīs*, gives the digits for the weight of the refined (*pakka*) gold.

3.19. ⟨Four pieces of gold weighing⟩ six, five, eight, and seven *tolas* and having nine, seven, ten, and eight *vannas* ⟨respectively, when melted together⟩ for refinement (*pakka*), have become twenty *tolas* ⟨in weight⟩. Of what *vannī* is the ⟨refined⟩ gold?

Second ⟨example⟩:

3.20. ⟨Four pieces of gold having the purities⟩ seven, eight, nine, and six *vannas* and ⟨weighing⟩ four, five, three, and seven *tolas* respectively, ⟨when melted together for refinement, become⟩ eleven *vannīs*. What is the weight of the refined (*pakkaviya*) ⟨gold⟩?

9. Now he tells the lost (unknown) purities (*vaṇṇa*) of gold pieces.

3.21. Having multiplied the sum of ⟨the weights of all pieces of⟩ gold by the ⟨purity in⟩ *vannayas* of their alloy, ⟨this product⟩ should be diminished by the sum of the products of the ⟨purities in⟩ *vannas* and the weights ⟨of the pieces of gold whose weights and purities are both given⟩ and divided by the weight of the piece of gold of lost *vannas*. ⟨This gives the lost purity in *vannīs*.⟩

3.22. ⟨There are three pieces of gold weighing⟩ three, five, and seven *māsas* and having nine, eight, and ten *vannas* ⟨respectively⟩, and another piece weighing eight *māsas* ⟨but of unknown purity⟩. ⟨These, when melted together⟩, become ten *vannas*. What is the *vannīs* of the gold piece weighing eight *māsas*?

3.23. The sum of the ⟨given⟩ weights of pieces of gold, multiplied by the ⟨purity in⟩ *vannas* of their alloy produced, diminished by the sum of the products of the ⟨purities in⟩ *vannas* and the weights ⟨of those pieces whose weights and purities are both given⟩ and then divided by the ⟨purity in⟩ *vannas* of the piece of gold of lost weight which is diminished by the ⟨purity in⟩ *vannas* of the alloy, ⟨gives the lost weight⟩.

3.24. The denominator for ⟨a piece of gold whose purity is⟩ greater than the desired purity is the subtrahend ⟨by which the purity becomes the desired purity⟩ and that for ⟨a piece of gold whose purity is⟩ smaller than the desired purity is the additive ⟨by which the purity becomes the desired purity⟩. ⟨The unit fractions⟩ that have the digits of the denominators ⟨obtained thus⟩ are parts (shares) for the weights ⟨to be employed in proportionate distribution⟩. This is a rule concerning purities for producing the desired ⟨purity⟩.

3.25. Tell, how much of gold should be taken of ⟨purities⟩ five, seven, nine, and eleven *vannīs* respectively, so that we get an alloy one *tola* in weight and ten *vannīs* ⟨in purity⟩.

Thus the procedure for mixture.

3.2 Procedure for Series

1. Now what is ⟨called⟩ the procedure for series (*seḍhī-vyavahāra*) is ⟨as follows⟩.

3.26. The common difference (*uttara*) multiplied by the number of terms (*gaccha*) minus one and increased by the first term (*āi*) ⟨gives⟩ the value (*dhaṇa*) of the last term (*aṃta*). This again, increased by the first term and divided by two, gives the value of the middle term (*majjhima*). This, multiplied by the number of terms, becomes the value of all the terms (*savva-dhaṇa*) (i.e., the sum of the series).

3.27. Twenty is the first term and five is the common difference for seven days. Tell the amount of yellow myrobalan for a horse (*turiya*). Likewise, tell again the lost (unknown) first term, common difference, or number of terms ⟨respectively from the rest by means of the rules given below⟩.

2. Now he tells an operation (*karaṇa*) for calculating the lost (unknown) first term (*ādya*).

3.28. For knowing the lost first term, divide the sum of the series by the number of terms. From the quotient *phala*, subtract the common difference, multiplied by the number of terms minus one, and halved. The remainder is the first term.

3. Now he tells an operational rule (*karaṇa-sūtra*) for calculating the lost (unknown) common difference (*uttara*).

3.29. For calculating the lost common difference, having divided the sum of the series by the number of terms, having made it less by the first term, ⟨divide⟩ it by the half of the number of terms minus one. The quotient is the common difference (*caya*).

4. Now he tells ⟨an operational rule⟩ for calculating the lost (unknown) number of terms (*gaccha*).

3.30. Multiply the sum of the series[5] by eight and by the common difference. Add to it the square of twice the first term as diminished by the common difference (*vuḍḍhi*). Its square root is diminished by twice the first term (*dhaṇa*[6]) and increased by the common difference. By dividing this by twice the common difference, the number of terms of the series is obtained.

5. Now he tells ⟨an operational rule⟩ for calculating the sum of the sums.

3.31. The digits for the sum of the natural series (*saṃkaliy-aṃka*) whose common difference ⟨and the first term are both⟩ unity are multiplied by the number of terms (*paya*) plus two and divided by three. The quotient is the sum (*juī*) of the ⟨successive⟩ sums of natural series (*saṃkaliya*). There is no doubt. Thus it should be known.

3.32. If you know the method of calculation (*gaṇiya-vihi*), calculate and tell quickly, what will be the sums of the sums, of the squares, and of the cubes, severally, of the ⟨first⟩ five terms ⟨of the natural series⟩.

6. Now he tells how to calculate the sum of the squares (*varga-aikya*) and the sum of the cubes (*ghana-aikya*).

3.33. Twice the desired number of terms of the series plus one, having been divided by three and multiplied by the sum of the natural series (*saṃkaliya*), ⟨gives⟩ the sum of the square series (*vagga-juī*). Whatever is the square of the sum of the natural series, know that to be ⟨equal to⟩ the sum of the cubic series (*ghaṇa-piṃḍa*).

7. Now he tells ⟨an operational rule⟩ for calculating the sum of the sum, the square and the cube.

3.34. Twice the number of terms plus one, multiplied by the product of half the number of terms plus one and the number of terms, should be known as the sum of the sum of the natural series ⟨up to the desired term⟩ and the square and the cube ⟨of the desired term⟩.

Thus the verses (*gāthā*) for the rules in the procedure for series are completed.

3.3 Procedure for Plane Figures

Now he tells the procedure for plane figures (*kṣetra-vyavahāra*).

3.35a. ⟨In the case of⟩ a square (*caüraṃsa*) and a rectangle (*dīha-caürasa*), the length (*āyāma*) and the breadth (*vikkhaṃbha*), when multiplied ⟨by one another, gives⟩ the area (*khetta*[7]).

[5] *gaṇiyaṃ* (S. *gaṇitaṃ*), lit. calculated.

[6] S. *dhana*, which is unusual for the 'first term'. A more appropriate word is *muha* (S. *mukha*).

[7] S. *kṣetra*, lit. a field or a geometric figure. A more accurate expression of the area is *khitta-phala* (S. *kṣetra-phala*, lit. the field-fruit), which is used in verses 37, 38, 39, etc.

3.35b. In a square, the sides (*bhuyā*) are six *karas* ⟨each⟩. In a rectangle, ⟨the shorter sides are⟩ three *karas* and ⟨the longer sides are⟩ five *karas*. ⟨What are their areas?⟩

3.36. Half the sum of the sides (*bhuva*) ⟨is put down⟩ at four places. From it subtract severally the four sides and multiply the remainders successively. Know that its square root is the area of a trilateral (*ti-bhuya*) and also of a quadrilateral (*caü-bbhuya*).

3.37. ⟨In a quadrilateral⟩, the face side (*muha-bhuva*) is twenty-five *karas*, the base side (*bhūmi-bhuva*) sixty ⟨*karas*⟩, the left ⟨side⟩ fifty-two, and the right ⟨side⟩ thirty-nine ⟨*karas*⟩. What will be its area (*khitta-phala*)?

3.38. The base side is fourteen *hatthas*, one ⟨lateral side⟩ thirteen ⟨*hatthas*⟩, and the second ⟨lateral side⟩ fifteen ⟨*hatthas*⟩. This is a scalene triangle (*visama-tikoṇa*). What will be its area?

3.39. Of all quadrilaterals, half the sum of the base and the face, multiplied by the perpendicular (*laṃba*) ⟨gives⟩ the area. Of a trilateral (*taṃsa*), half the base multiplied by the perpendicular gives the area.

3.40. The two ⟨lateral⟩ sides are thirteen and fifteen ⟨*hatthas*⟩, the base twenty-one ⟨*hatthas*⟩, and the face five *hatthas*. In the middle, the perpendicular is twelve ⟨*hatthas*⟩. What is the measure of this figure?

3.41. Twice the area (*phala*) of a triangle (*ti-kkoṇa*), divided by the base, becomes the perpendicular in the middle. Take the difference between the squares of ⟨each lateral⟩ side (*bhuva*) and the perpendicular. The square roots (*paya*) of the remainders are the two segments of the base (*ahavā*).

3.42. Take the sum of the squares on the side (*bhuva*) and the upright (*laṃba*[8]) ⟨of a right-angled triangle⟩. Its square root certainly becomes the hypotenuse (*kanna*). It should be known that this rule (*vihi*) is applicable everywhere in geometric computation (*khitta-gaṇana*).

3.43. Multiply the square on the diameter (*vikkhaṃbha*) ⟨of a circle⟩ by ten. Its square root is certainly the circumference (*parihi*) of the circular figure (*vaṭṭa-khitta*). The circumference, multiplied by a quarter of the diameter, becomes the area.

3.44. In a uniformly (*sama*) circular figure having the diameter ten, whatever be produced as the circumference, calculate and tell. O learned man, what will be its area?

3.45. The diameter of a circle, multiplied by three and increased by one-sixth ⟨of the diameter⟩, is the circumference. Half the circumference, multiplied by half the diameter, is its area.

3.46. Half the sum of the chord (*jīvā*) and the arrow (*sara*) is multiplied by the arrow, squared, multiplied by ten, and divided by nine. The square of the quotient becomes the area (*phala*) of a bow (*dhaṇu*, i.e. of a segment of a circle).

[8]Usually called *koṭi*.

3.47. The length (*piṃḍa*) of a bow (i.e., of an arc) is twenty-one and the chord is fifteen ⟨in a bow-like figure⟩ whose arrow is six. Tell, O learned man, the calculated result. What will become the area of this bow⟨-like figure⟩?

3.48. The square root of the square on the arrow made six-fold and increased by the square on the chord is the length of the bow. When the square on the chord is subtracted from the square on the bow, the square root of one-sixth of the remainder is the arrow.

3.49. Half the sum of the bow and the arrow is subtracted from the bow. The square root of four times the square ⟨of the remainder⟩ is the chord. Each quantity obtained by the computation is indeed the fruit (length) of those ⟨parts of a bow-like figure in a circle⟩.

3.50. In the moon's crescent shape (*bāl-iṃda*) one should imagine a pair of trilaterals (*ti-bhuva*), in a *muraja* drum two bows with a rectangle in the middle, in a barleycorn shape (*java-ākāra*) two bows, and in a thunderbolt shape (*kulisa*) two quadrilaterals (*caü-bhuva*).

3.51. A shape like the elephant's tusk (*gaya-daṃta*) is ⟨assumed to be⟩ a trilateral, a shape like a rim of a cartwheel (*sagaḍa-cakka-vaṭṭa*) a rectangle, a shape like the moon a bow, and a full moon shape (*paripunna-caṃda*) a circle.

3.52. An area in the shape of the crescent moon has the ⟨central⟩ width of five ⟨*karas*⟩ and the ⟨central⟩ length of twenty *karas*. ⟨An area in the shape of⟩ the elephant's tusk has the altitude (*laṃba*) of one half ⟨of that of the previous case⟩ and the base (*dharā*) of three ⟨*karas*⟩. What is the area ⟨of each figure⟩?

3.53. In an area in the shape of the rim of a wheel (*nimma-āgāra*), both faces (*muha*) are three *karas* and the altitude is five *karas*. ⟨A figure⟩ like a thunderbolt has five *hatthas* for ⟨the width at⟩ the base and face and three at the middle and ten ⟨*hatthas*⟩ for the altitude. ⟨What is the area of each figure?⟩

Thus the rules in the procedure for plane figures are completed.

3.4 Procedure for Excavations

1. Now he tells the procedure for excavations (*khāta-vyavahāra*).

3.54. Take irregular (*visama*, unequal) ⟨measures of breadths⟩ at the bottom (*tala*), top (*muha*) and middle (*majjha*) ⟨of a pit⟩, or of heights (*uṃḍḍatta*), or unequal lengths (*dīha*). Having added them up, divide ⟨the sum⟩ by the number of unequal places ⟨where unequal measures were taken⟩. This gives the mean (*sama*, equal) ⟨measure⟩.

3.55. The product of the mean breadth and length, multiplied by the height, becomes the capacity of a pit (*khatta-phala*). In the case of a pit with equal sides and depth (*veha*), the calculation ⟨of the capacity⟩ becomes the same as the cube ⟨of a number⟩.

3.56. The height ⟨at three different places⟩ is two, three and four *karas* in a pond (*pukkharaṇī*), ⟨which has⟩ five *hatthas* (= *karas*) for width and sixteen *hatthas* for length. What is its capacity?

3.57. ⟨There is a rectangular pond whose⟩ length is sixteen and a half *karas*, width ten and a quarter ⟨*karas*⟩, and height (*udaya*) eight ⟨*karas*⟩.[9] And, ⟨in another pond⟩, the breadth, the length, and the height are equally nine *karas*. What is the capacity of each here?

2. Now he tells how to calculate the capacity (*phala*) of a cylindrical well (*kūpa*).

3.58. The square on the diameter (*vitthāra*) of a cylindrical well (*kuva*) should be multiplied by three, increased by one-sixth ⟨of the square⟩, and multiplied by the depth. By dividing it by four, whatever is obtained is the entire ⟨capacity⟩ in *karas* (*kara-saṃkhā*, lit. the number of *karas*) ⟨of the well⟩.

3.59. Of a well whose diameter (*vikkhaṃbha*) is six *hatthas* and height twenty *karas* (= *hatthas*), of that well, O learned one, what is the capacity, for sure?

3.60. Know that the areas of figures like triangle, etc. are the same as the areas told before. Those ⟨areas⟩ also, multiplied by the depth, become the capacities in cubic *hatthas* (*ghaṇa-hattha*).

3. Now an operational rule for calculating the volumes of stones (*pāṣāṇa-phala*).

3.61. The length in *aṃgulas*, multiplied by the breadth and the height in *aṃgulas* and divided by 'Jinas, eight, thirteen' (13824), gives the solid volume (*pāhāṇa*, lit. a stone) in cubic *hatthas*.[10]

3.62. O friend, ⟨there is⟩ a stone slab, whose breadth is three and a half *hatthas*, thickness half a *kara* (= *hattha*), and length five and one-third ⟨*kara*⟩. What will be its calculated result (volume) here?

3.63. Of all stones, whatever the shape be, whether of a circle or of a triangle, etc., after having calculated as in the case of the area, ⟨the result⟩ multiplied by the thickness (*piṃḍa*) will be its volume (*phala*).

3.64. In a circular stone slab like a grindstone (*gharaṭṭa-paṭṭa*), the diameter is ten *hatthas* and the thickness one and a half *karas* (= *hatthas*). What will be its calculated result (volume)?

3.65. Half the cube of the height (*udaya*, i.e. diameter) of a sphere, increased by one-ninth part of itself, becomes the solid volume (*sela*, lit. stone) ⟨of the sphere⟩. One-fourth part of the circumference, multiplied by the circumference and increased by one-ninth part ⟨of itself⟩, is the surface area ⟨of the sphere⟩.

3.66. Of a sphere, which is uniformly round, and whose length, height, and width are six *karas* ⟨each⟩, what is the solid volume (*pāhāṇa*) calculated and what is the surface area? Whatever be the results, tell them separately.

---

[9] The Nahatas' text reads 'and height one-half ⟨*kara*⟩.' See under 3.57 in Part IV.

[10] Or, by taking the three numerals separately, 'and divided by 'Jinas' (24), by eight and by thirteen, gives the volume in stone cubic *hatthas*.'

4. Now he tells the weight (*taulya*) of stones.

3.67. A cubic *kaṃbiya* of all varieties of stone originating in Ḍhilliya (Delhi) weighs fifty *maṇas* and twenty-four *tullas* (*tolas*).[11]

3.68. *Vaṃsī* stone weighs forty-eight ⟨*maṇas*⟩, marble (*mammāṇī*) sixty, black stone (black granite) sixty-two, *jajjāvaya* and of Kannāṇā forty-nine, and *kuḍu-kkaḍa* sixty.

Thus the 15 verses for the rules in the procedure for excavation are completed.

3.5 Procedure for Piling

Now he tells the procedure for bricks (*citi-vyavahāra*).

3.69. There are nine types of walls (*bhitti*): dome (*goṃmaṭṭa*), square and circular spiral stairways (*pāya-seva*), minaret (*munāraya*), arch (*tāka*), staircase (*sovāṇa*), bridge (*pula*), well (*kūva*), and stepwell (*vāvī*).

3.70–71. First measure the correct (total) wall space (*suddha-bhittī*) and multiply the breadth by the length and height. Whatever is obtained, from it repeatedly subtract ⟨the volume of⟩ the wood (*kaṭṭha*) used in the house (*ālaya*).// From the remainder, reduce (subtract) one and a half times its one-tenth part, ⟨being the volume⟩ of the clay (*maṭṭiya*) ⟨used as mortar⟩. The rest is the solid volume (*pāhaṇa-saṃkhā*) in terms of cubic *hatthas*.

3.72. Five *karas* are the height of the wall, ten the length, and two the width. In the middle, there is a door (*bārū*) three ⟨*karas*⟩ in height and two ⟨*karas*⟩ in width. What is the solid volume (*saṃkhā pāhāṇe*)?

Now computation of bricks —

3.73. The length, width, and height of a brick (*iṭṭa*) are a half, one-third, and one-eighth ⟨*hattha*⟩, respectively. ⟨If a wall measures⟩ four ⟨*hattha*⟩ in length, one and a half in breadth, and ten in length, how many bricks ⟨are there⟩?

1. Now he tells the dome (*goṃmaṭa*).

3.74. Half the circumference of a dome at its foot, multiplied by one-quarter of the circumference and increased by one-ninth ⟨of itself⟩, is the piling (*cayaṇa*) ⟨when the circumference is measured⟩ from the inside of the wall (*bhiti-gabbhāo*). ⟨When the circumference is measured⟩ at the middle of the outside(? *bāhira-majjhāu*), it is the area (*khitta*).

3.75. The inner circumference of a wall (*bhiti-gabbhāo parihī*) is nineteen and the breadth six. What is its piling? The outer circumference is, O learned man, twenty-four. What will be the area (*khetta*)?

3.76. The circumference multiplied by half the diameter and increased by one-ninth ⟨of itself⟩ is ⟨the volume of⟩ the empty space (*khalla*) in a dome. This has

[11] This verse is repeated in GSK 5.26, which reads *caüvīsasaya tulle* for *caüvīsa tullo*.

been told according to experience. There is no doubt. It should be known thus.

2. Now he tells the spiral stairway (*pāya-seva*).

3.77–78. In a square spiral stairway, there is a wall outside and a pillar in the middle. The product of the breadth, length and height, know it to be ⟨the volume in⟩ *kaṃviyas.*// In between the wall and the pillar, ⟨there is a stairway?⟩, which goes up at regular incline, turning around the pillar. Add the lengths at the bottom and top and halve. Multiplying this by the height and by the breadth, this becomes the volume (*pūra*) ⟨of the steps⟩.

3. Now the circular (*vaṭṭa*) ⟨stairway⟩ —

3.79. Likewise, in a circular spiral stairway, calculate ⟨the volume of⟩ the pillar and wall as in the case of a cylindrical well. The volume of the intermediate space (*pūraṃtara*), know it to be a rectangular ⟨solid⟩ of ⟨sides⟩, six, three, and a half (?).

4. Now the minaret (*munārayā*) —

3.80. The minarets are like the circular spiral stairways with regard to everything in the middle. But this is the difference: the walls are like half triangle and half circle.[12]

5. Now the arch (*tāka*) —

3.81. In the case of an arch above a door, multiply the length and height and further multiply the thickness of the wall. This, diminished by one and a half times its one-seventh, is the volume of the empty space (*khalla*) along with its apex (*sihā*).

3.82. Of an arch, the length is seven *karas*, the height along with the apex is four *hatthas* (= *karas*). The thickness of the wall is one *hattha*. What is the volume of the empty space?

6. Now the staircase (*sopāna*) —

3.83a. In the case of a staircase, add ⟨the lengths of⟩ the bottom and the top ⟨steps⟩. Its half multiplied by the height and by the width ⟨gives the volume⟩.

3.83b. The length at the bottom is nine, at the top one, the width two, and the height six ⟨*hatthas*⟩. What is the volume?

7. Now the bridge structure (*pula-baṃdha*) —

3.84. Multiply the width, the length, and the height ⟨with each other⟩. Subtract ⟨the volume of empty space of⟩ arches, add ⟨the volume of⟩ two sides (*bhuvajuva*) and the exits (*niggama*), and subtract the empty space (*khalla*). That is indeed ⟨the volume of⟩ the bridge structure above water (*jala-pula-baṃdha*).

[12] 'The wall is half triangular and half circular': this apparently means that the cross-section of the wall is as follows:  , as in the Qutub Minar.

8. Now the cylindrical well (*kūpa*) —

3.85a. The inner circumference (*majjhi parihī*) of the wall of a well,[13] multiplied by the breadth and by the height, becomes the volume (*phala*) ⟨of the wall⟩.

3.85b. ⟨Of the wall of a cylindrical well⟩, the height is ten ⟨*karas*⟩, the width two *karas*, and the circumference eighteen ⟨*karas*⟩. What is the piling (*cayaṇa*)?

Now the stepwell (*vāpī*) of six kinds —

3.86. The stepwells are many in shape (*chaṃda*): square (*caür-aṃsa*), rectangular (*dīha*, lit. long), circular (*vaṭṭa*), hexagonal (*khaḍ-aṃsa*), octagonal (*aṭṭh-aṃsa*), spiral (*saṃkha-vatta*), and so on. They are calculated according to the known standards (*diṭṭha-pamāṇa*).

Thus the 18 verses for the rules in the procedure for piling are completed.

3.6 Procedure for Sawing

Now he tells an operational rule for the procedure for sawing (*krakaca-vyavahāra*).

3.87. A log of wood (*dāru*) of desired size is made into planks of desired ⟨size⟩. Multiply the thickness (*dala*), the length, and the breadth, these three ⟨with each other⟩, of both ⟨the log and the plank, separately⟩, and divide ⟨the volume of the log⟩ by ⟨the volume of⟩ the plank. ⟨This gives the number of planks.⟩

3.88. A log of wood is eight *karas* long, half a *kara* wide, and one-third *kara* thick. How many planks, each one ⟨*kara*⟩ long, one quarter wide and one-ninth thick, can be obtained from it?

Now the computation of what is cut out of timber by a saw —

3.89. The number of the ⟨cutting⟩ lines (*līha*) made by the saw, multiplied by the length ⟨in *hatthas* of the log⟩, gives ⟨the total length of the cut in⟩ *hatthas*. ⟨From this⟩, according to the width ⟨of the log⟩, the wages for sawing (*cirāvaṇī*) in *koḍīs* ⟨is calculated⟩ by means of the price rate (*aggha-māṇa*), ⟨which is given below⟩.

3.90–92. If the width is one to one and a half *visuvas*, one *koḍī* ⟨is paid⟩ for one hundred *gajas* ⟨of the total length⟩; if two to three ⟨*visuvas*⟩, for eighty *gajas*; if four *visuvas*, for sixty *gajas*; if five to nine ⟨*visuvas*⟩, for forty ⟨*gajas*⟩.// If the width is ten to thirteen *visuvas*, one *koḍī* ⟨is paid⟩ for thirty of them (i.e., *gajas*); if above it up to sixteen ⟨*visuvas*⟩, for twenty *gajas*.// Know that, if above it up to twenty *visuvas*, one *koḍī* ⟨is paid⟩ for ten *gajas*. Above it, the saw does not work. This has been told by carpenters (*suttahārī*).

3.93. A piece of timber, seven *gajas* long and eight or seven *visovagas* wide, is cut (*cīriya*) into eleven planks along ten ⟨sawing⟩ lines. How many *koḍiyas* are there?

[13] 'The circumference at the centre of the wall of a well'?

3.94. Eight *javas* make one *kaṃviy-aṃgula.*[14] A line (*līha*) on a plank made by a saw is one *java* ⟨wide⟩. When one cuts a circular ⟨log of wood⟩, regard its thickness (*piṃḍa*) as the length (*dīha*).

3.95. Mahuva, Vaḍa, Sāla, Sīsama, Niṃva, Sirīsa, etc. are of the same wages (*cirāvaṇiya*) ⟨as those defined above⟩. Multiply by one and a quarter for Khayara, Aṃjaṇa and Kīra, and by one less a quarter for Seṃvala and Suradāru.

Thus the procedure for sawing is completed. 9 verses.

3.7 Procedure for Mounds of Grain

Now he tells the procedure for mounds of grain (*rāśi-vyavahāra*).

3.96. A mound of grain heaped (*anna-rāsī*) on an even ground (*sama-bhū*). The square on one-sixth of its circumference, multiplied by the height, gives ⟨the volume of the grain in⟩ cubic *hatthas*. One cubic *hattha* is a *patta.*[15]

3.97. In the case of ⟨fine⟩ grains like sesamum (*tila*) and Kuddava, the height of the mound is one-ninth of its circumference; in the case of mūṃg pulses (*mugga*) and wheat (*gohuma*), one-tenth; in the case of Vora and horse beans (*kulattha*), one-eleventh.

3.98–99. A circular mound is standing like a peak of a mountain (*sihara*). Its height is four and circumference thirty-six ⟨*karas*⟩. If the mound is piled against the side of a wall, the circumference is half; if against the inside of a corner, it is a quarter;// if against the outside of a corner, the circumference is less by a quarter. Know that the height is the same. Tell what will be the volumes of these mounds separately in ⟨cubic⟩ *karas*.

3.100. Having multiplied the half, the quarter, and the quarter-less circumferences by two, four, and one and one-third, respectively, obtain the volumes as before and then divide ⟨the results⟩ by the respective multipliers.

Thus the rules in the procedure for mounds of grain are completed. 5 verses.

3.8 Procedure for Shadows

Now he tells an operational rule for the procedure for shadows (*chāyā-vyavahāra*).

3.101. Having measured the shadow (*chāyā*) of a pillar (*thaṃbha*), etc., or of a wall (*bhitti*), with a staff (*daṃḍa*), multiply ⟨the number⟩ by the length of the staff ⟨in *aṃgulas*⟩, and divide it by the shadow of the staff ⟨measured in *aṃgulas*⟩. By means of the result, ⟨tell⟩ the height ⟨of the pillar, etc.⟩

3.102. With a staff of twenty-four *aṃgulas*, the shadow of a pillar is three and

[14] See under GSK 1.5–6 in Part IV.

[15] Tr 61 calls the same amount *khārī* and L 7 and 227 '*khārī* (or *khārikā*) of Magadha.'

one-quarter staffs ⟨long⟩, and ⟨the shadow of⟩ the staff is eighteen and one-quarter *aṃgulas* ⟨long⟩.[16] What is the height of the pillar?

Now an operation for the determination ⟨of the cardinal directions by setting up a gnomon⟩ and the calculation ⟨of time⟩.

3.103. On an even ground, ⟨draw⟩ a double-lined circle (*du-reha-vaṭṭa*) two *karas* wide (distant); at the centre, ⟨erect⟩ a gnomon (*saṃka*) of 'sun' (12) ⟨*aṃgulas*⟩. At the middle (*gabbha*) of the first and the last shadows is the south-north ⟨line⟩, ⟨on which the noon shadows lie⟩. By half ⟨the length of daylight from the time of a noon shadow⟩, the rising and the setting (*uday-attha*) ⟨of the sun occurs⟩.[17]

3.104. The constant quantities (*dhuva-rāsī*) (i.e., the lengths of the noon shadows) are four, four, one ⟨*aṃgulas* for the sun⟩ in ⟨the zodiacal signs⟩ beginning with Mayara and five, three, one ⟨*aṃgulas* for the sun⟩ in ⟨the zodiacal signs⟩ beginning with Kakkaḍa. The quotient of ⟨the division of⟩ seven *aṃgulas* by the sum (*juva*) of the shadow (*paha*) and 'sages' (7), decreased or increased by one (*rū-gaya-jutta*[18]), is the elapsed (*gaya*) or the remaining (*sesa*) ⟨length⟩ of daylight (*divasa*).[19]

Thus the rules in the procedure of shadows are completed. 4 verses. In total, 104 verses ⟨in this chapter⟩.

Thus the eight kinds of procedures (*vyavahāra*) in ⟨the book for⟩ the procedural mathematics (*pāṭī*), Gaṇita-sāra-kaumudī (Moonlight of the Essence of Mathematics), composed by the most ⟨devout⟩ Jain, son of the respected Candra, Ṭhakkura Pherū, are completed.

Thus ⟨ends⟩ the third chapter.

[16]PM A2 reads: *daṃḍassa ya addhārasa aṃgula* ('and ⟨the shadow⟩ of the staff is eighteen *aṃgulas* ⟨long⟩').

[17]Tentative translation.

[18]Or, 'subtracted from, or added to, one.'

[19]Tentative translation.

Chapter 4

Four Special Topics

4.1 Region

Now he tells the topic of region (*desa-adhikāra*).

4.1. Pherū, son of Caṃda, speaks of the regional method of writing and counting (*desa-leha-payaḍī*) which is to be observed in transactions for profit in Delhi (*ḍhilliya*) and in Rajasthan (*rāyaṭṭhāṇa*).

4.2. Whatever be allotted and given ⟨as investments⟩ are each multiplied by the produced property in *jīvalaïs* and divided by the *dammas* of all ⟨people⟩. The quotient is the share (*bhāya*) in *jīvalaïs*.

4.3. Attend upon the king (? *seva rāyaha*). Five people went ⟨for business⟩. The ⟨capital in⟩ *jīvalaïs* of them all is thirty thousand in total, and three, thirteen, five, two, and seven thousand by divided numbers (*bhinna-rūva*) (i.e. of each). Having made expenditure (*vaya*), they all got together nine thousand, each obtaining (*kaḍḍha*) one's own *jivalās*. What did each of them get?[1]

4.4. O learned man, that money which becomes *upakkhaï*,[2] after multiplying it by hundred, one should divide by the *caṭṭī*.[3] What is obtained will be the percentage (*saī*).

4.5. In a village, in a town, and in a country, if the *caṭṭī* is nine lakh and fifty thousand and the *upakkhaï* seventy thousand, what will then be its percentage?

4.6. Of that ⟨amount of money⟩ which is to be divided among one hundred ⟨persons⟩ so that each of them may obtain the same amount, wipe off (*phusa*) the last two digits. ⟨The remaining digits indicate *dammas* in the quotient.⟩ One-fifth of these ⟨two digits wiped off⟩ gives the *visuvas* ⟨in the quotient⟩.

4.7. If ⟨a given amount in *dammas* is to be divided⟩ by thousand, wipe off the last three digits. ⟨The remaining digits indicate *dammas* in the quotient.⟩ Write

[1]Lit., 'What out of them comes to whom?'

[2]S. *upakṣaya* (loss).

[3]The word *caṭṭī* occurs with *saī* (percentage) in this verse as well as in verses 5 and 8. Cf. *caṭṭiya* (an account) in GSK 5.17 and 5.33 below.

down twice, four times, and eight times ⟨the three digits which were wiped off⟩, respectively. Know these to be the *visuvas*, etc. (i.e., *vissaṃsas* and *paḍi-vissaṃsas*) ⟨in the quotient⟩. The same process is to be followed for division by ten thousand and by one lakh (*lakkha*) also.

4.8. If the original amount of *caṭṭī*, two lakh nine thousand five hundred and thirty-four, is divided by one hundred (*saī*), how much money will it be?

1. Now on the regional method of accountancy (*deśa-aṃka*).

4.9. Of a given amount of money ⟨expressed in *dammas*⟩, wipe off the last digit. Twice that ⟨digit which was wiped off⟩ represents the *visuvas* in one-tenth part. Wipe off the last two digits. One-fifth of these ⟨two digits which were wiped off⟩ represents the *visuvas* in one-hundredth part.

4.10. Know that, in one-twentieth part of the given amount ⟨in *dammas*⟩, its last digit is the *visuvas*, that the remaining (i.e., fractional) ⟨portion of the original amount expressed in *visuvas*, *visuvaṃsagas*, etc., will become⟩ *visuvaṃsagas*, etc. respectively, and that half of the former digits is the *dammas*.

2. Now he tells the wages (*mukātaya*).

4.11a. Having multiplied whatever is the *mukkātaï* per year (i.e., the annual wages) by the elapsed days, divide by the number of days in a year.

4.11b. If the *mukātaï* is five thousand, how much is it in four months and nine days?

4.12. When the number of *dammas* given as *maseliya* in one month (i.e., the monthly wages) is diminished by its one-third, the remainder becomes the number of *visovās* per day. Thus it should be known.

3. Now on the movements of runners (*dhāvaka-gati*).

4.13. The product of the daily movement (*gaï*) of the slow runner (*lahu-gaï*) and the number of the days ⟨already travelled by him⟩ should be divided by the difference in the daily movements of the slow and the fast runners (*dīha-gaï*). The quotient is the number of days in which the slow runner (*appa-gaï*) and the fast runner meet.

4.14. On the ninth day after ⟨the departure of⟩ one who walks four *joyaṇas* per day, another who walks seven *joyaṇas* per day started in order to kill him. In how many days will he meet ⟨the former⟩?

4.15. A camel (*karaha*) walks five *joyaṇas* on the first ⟨day⟩ and increases ⟨the daily movement⟩ by two *joyaṇas* each day. In how many days will he meet a female camel who walks fourteen *joyaṇas* per day?

4.16. The ⟨three⟩ quantities (*rāsī*) ⟨given in a problem of this type having been put down in three places⟩, first, middle and last, the last term is diminished by the first and divided by the middle. The quotient, multiplied by two and increased by one, is the number of days necessary for the camels ⟨to meet each other⟩.

4. Now he tells how to calculate the years.

4.17. Convert the years from the beginning of the Vikrama era and the months from Caitra into ⟨solar⟩ days. Divide it by 'six, sages, Nandas' (976). The quotient is the number of intercalary months. ⟨Convert⟩ them into years and add ⟨to the given date⟩, and then subtract ⟨from it⟩ 'nine, treasures, tastes' (699) years, two months and two days. ⟨The result is⟩ the Hijrī year (*tājiya-vacchara*), the months starting from Muḥarram.[4] Do the same for the Hijrī date,[5] and then subtract from it the intercalary months and add 'nine, sages, six' (679) years and two days, respectively. O learned man, tell ⟨the result as⟩ the Vikrama date.

Thus the operational rules in ⟨the section for⟩ the regional topics are completed.

4.2 Cloth

Now he tells the topic of cloth (*vastra-adhikāra*).

4.18. *Juja*, *paṭṭolaya*[6], *atalasa*[7], *sāra*, etc., are ⟨varieties of⟩ silk cloth. *Kara*, *vāsaka*, *tāṇa*, etc. are ⟨varieties of⟩ fine ⟨cotton?⟩. *Sāḍa*, etc. are ⟨varieties of⟩ coarse ⟨cotton?⟩.

4.19. All varieties of cloth (*kappaḍa*) diminish by one and a half *karas* out of one hundred *hatthas* (= *karas*) in sewing (*sīvāṇa*), by one in cutting (*kattaraṇa*), and by one, two, and three *kora*[8] in washing (*dhuvaṇa*) silk, etc. respectively (i.e., according to whether it is silk, or fine ⟨cotton?⟩, or coarse ⟨cotton?⟩).

4.20.[9] For all tents (*khīma*) the tape (*navāra*) is equal to ⟨the length of⟩ the cloth, or slightly less or more. All, except the door (*dahalīja*), are one and a quarter of the pole's height.

4.21. Twelve *visuvas* of the height at the bottom of the *kamara* (waist?) and eight at the top for all. With one pole or two poles.[10] From now on we shall speak of the stitched (*siya*, white?) cloth.

4.22. For all pieces of cloth (*paḍa*) take half the sum of the bottom and the top and multiply ⟨the result⟩ by the height. What is obtained is the *kamara*. The breadth of the foundation (? *piṭṭhī*) is multiplied by the length and increased

[4] Or, 'Calculate (*ṇaya*) the months starting from Muḥarram ⟨from the remainder of the year⟩.' The word *māīṇaya* after *muḥarama* is not understood.

[5] That is, convert the years and months into lunar days and from the latter obtain the number of intercalary months.

[6] Cf. 'PATOLA', silk, in Hobson-Jobson.

[7] Cf. 'ATLAS', satin, in Hobson-Jobson.

[8] This is probably identical with *kara* (lengthened metri causa) if it is not another linear measure.

[9] The exact meaning of the rest of this section up to verse 37 is not clear. The following translation is tentative.

[10] Plate XI of Abū al-Faḍl [2004] shows, among other tents and pavilions, a tent with one tent pole and another with two tent poles.

by one-eighth ⟨of itself⟩ and added. ⟨This gives the area of⟩ the cloth (*vaṭṭha*)[11].

4.23. Four times that of the middle beam (*ḍaṃḍa*) is the entrance (*pavesa*) with grass ground (*kaḍa-yala*) ⟨to the *kamara*⟩ of a tent. One and a half of this is the circumference, and less by one-twelfth in a square.[12]

4.24. The central pole is four *karas*, the *kamara* sixteen ⟨*karas*⟩, and the circumference twenty-two ⟨*karas*⟩. O learned man, what is the size of the cloth (*vaṭṭha*)[13] of that tent?

4.25. One-eighth and then in a circle[14] three times is the *kamara*. That increased by half the width is the *daüra*. This multiplied by the *dhara* (base?) is *sīmāṇaya*. Four times the pole is the tent-rope (*tanāva*).

4.26. The *taṃgoṭī* has one pole. Half the sum of the bottom and the top, multiplied by the height, is ⟨the area of⟩ the cloth (*vaṭṭha*)[15]. Five times the pole is the circumference,[16] which, increased by the middle cloth, is ⟨the circumference of the tent⟩ with two poles.

4.27. A folding tent (*khargaha*) and the top of a pavilion (*maṃḍava*). As many *karas* in both directions there are, half of them is the height. Five times of this is the circumference. Half the circumference multiplied by the height is ⟨the area of⟩ the cloth (*vaṭṭha*)[17].

4.28. Cloth for a pair of circular walls and cloth for a pair of doors. The height is multiplied by the length. This is the cloth (*vaṭṭha*)[18], half each ⟨of which⟩ is for a pavilion with one door and one wall.

4.29. The assembly hall (*vārigaha*) is hexagonal or octagonal like an umbrella or a pavilion.[19] Of their *tarakka*, half the sum of the bottom and the top is multiplied by the height.

4.30. ⟨In a tent⟩ with one pole, the circumference is six times. ⟨In a tent⟩ with two poles, the circumference is ⟨further⟩ increased by the middle sheet. One-eighth is added in (for?) the height. The tent rope (*tanāva*) is five times the pole.

4.31. In the assembly hall of the *mīras*, the *cilaṃga* is of square form having one or two poles. The equal heights are four times; the *kamara* is double; the circumference is ⟨further⟩ increased by one-eighth.

4.32. The *vallahali*, mixed with a bunch of leaves and vines, has lace (*jhallariyā*, a cymbal?) in the middle. The cloth of the *puḍa* is again greater than that of these by one-third.

---

[11]Or, 'the surface area.' Cf. *vaṭṭha* (S. *pṛṣṭha*) in AHK.
[12]Or, 'One and a half of this, less by one-twelfth, is the circumference in a square.'
[13]See the footnote under verse 22.
[14]Or, 'In an octagonal and a circular ⟨tent⟩.'
[15]See the footnote under verse 22.
[16]Or, 'The pole is five times the circumference.'
[17]See the footnote under verse 22.
[18]See the footnote under verse 22.
[19]Plate XI of Abū al-Faḍl [2004] shows a large rectangular *vārigaha* (Pe. *bárgáh*).

4.33. The sheets for shadow (*chāyā-paḍa*) are canopy (*caṃdovaya*) and *sarāi*; *cāja-maṇiyā* (decorated with jewels, etc.) are the sheets of walls. The width multiplied by the length, subtracted (or purified?, *sujjhaṃti*), without amusing paintings (*viṇoya-citta*).

4.34. The door (*dahalīja*), *rūi*, *tākā* (= *tāka*, an arch?), *chajjaya*, *kuvvāya*, *carakha* (wheel?), *paḍirūva* (replicas?), *chattālaṃva* (stick of an umbrella?), banners (*nisāṇa*); these should be known to have the measure of *ṭippa*.

4.35. An example: The wages for stitching (*siyāvaṇiya*) of one hundred *gajas* of the tape is sixteen *dammas*. Painting (colour?) for one *gaja* of the back door and *sarāi* is two ⟨*dammas*⟩.

4.36. *Kimisa* per *gaja*, when the colour is fine, is twenty-four ⟨*dammas*⟩ and, when coarse, twenty ⟨*dammas*⟩. Four *ṭaṃkas* of string (*ḍorī*), one thread red or blue.

4.37. A tape is made from *sara*, leather, or blue or red or black cloth. It is evident (*payaḍa*). The thread and tape: ⟨the former⟩ for one hundred *gajas* ⟨weighs⟩ 'kings' (16) less by a quarter ⟨*seras*⟩, and the other half a *sera*.

Thus the 20 verses on the topic of cloth are completed.

4.3 Magic Squares

Now he tells operational rules for the topic of magic squares (*jaṃtra-adhikāra*).

4.38. 'Sun' (12), 'fires' (3), 'tastes' (6), thirteen, fourteen, 'sense organs' (5), 'Yugas' (4), and 'Īśvara' (11). Write down these 'six, one' (16) ⟨numbers⟩, one by one, assembled in ⟨sixteen⟩ cells (*kuṭṭha*): an attractive ⟨figure will be obtained⟩. 'Hands' (2), 'treasures' (9), sixteen, 'seas (*uvahi*)' (7), 'Vasus' (8), 'lunar days' (15), 'directions' (10), and 'moon' (1). Halving ⟨it⟩ optionally, removing one (? *rū harivi*), and putting ⟨them⟩ in order, know other magic squares (*jaṃta*). Whatever be ...[20] in the right order is held (? *dhuya*) in the reverse order in a magic square (*jaṃtara*).[21] Whatever be ⟨the number of⟩ all cells (*geha*) ⟨on each side of the square⟩ is either odd (*visama*) or even (*sama*) or evenly-odd (*sama-visama*). ⟨For example⟩, 'Yugas' (4) is an even number (*sam-aṃka*).

A magic square with six cells ⟨on its side⟩:

4.39–40. ⟨Write down the sequence⟩ beginning with one and increasing by seven in the diagonal (*kanna*) to the right (*dāhiṇa*, south) and ⟨the sequence⟩ beginning with six and increasing by five ⟨in the diagonal⟩ to the left (*vāma*). 'Teeth' (32), thirty-four, 'gods' (33), 'arrows' (5), 'sages' (7), twenty minus one (19), eighteen, twenty-five,// thirty-five, three, four, two, 'sun' (12), thirteen, 'Jinas' (24), thirty, 'lunar mansions' (28), twenty-seven, 'Manus' (14), seventeen, ten, nine, 'nails' (20), and twenty-three: ⟨these are arranged clockwise in the square⟩,

[20] *sunnuvāri* or *sunnuvāritā*.

[21] This line is not understood.

beginning in the east (*puvva*) (i.e., top). A magic square (*jaṃta*) with six cells (*cha-giha*) ⟨on each side is obtained⟩.

4.41. Write down, from the first ⟨component square⟩ on, the ⟨first⟩ quarter sequence (*pā-oli*), and further add ⟨the numbers up to⟩ half, ⟨moving⟩ upwards.[22] This quarter sequence (*pāya-oli*) is ⟨arranged⟩ in such a way that a quarter square (*pā-jaṃta*) is produced. ⟨Put⟩ half ⟨the sequence from its⟩ middle ⟨again⟩ from the beginning (i.e., from the first component square). ⟨Make⟩ the last quarter square also, in the order ⟨of the numbers in the model square of four⟩. These are magic squares beginning with four cells and increasing by four ⟨on each side⟩, for the sequence that increases by one.

4.42ab. 'Lunar days' (15), 'treasures' (9), 'tastes' (6), 'Yugas' (4), 'Vasus' (8), 'hands' (2), thirteen, eleven, 'moon' (1), 'sages' (7), 'sun' (12), 'Manus' (14), 'directions' (10), 'moon's digits' (16), 'properties' (3), 'arrows' (5). ⟨These, when arranged in a square in the natural order, constitute a magic square with four cells on each side⟩.

4.42cd. There are four quarters, up and down, in ⟨a square of⟩ sixty-four cells. Write down ⟨the numbers⟩ sequentially, beginning with one, ⟨according to the given pattern, by choosing the quarters⟩ in the order of one-three-two-four.

Now on calculating magic squares of odd orders:

4.43–44. A desired sequence ⟨of natural numbers⟩ beginning with one ⟨is employed here⟩. The first sequence ⟨of numebrs is obtained by⟩ adding one to the number of cells ⟨on each side of the square⟩ and increasing it by the previous term. With it the centre ⟨column is made⟩. Increase each cell of the centre ⟨column⟩ by the ⟨number of⟩ cells ⟨on each side and put it⟩ by the horse move (*asu-kama*).// If in a cell the number ⟨obtained⟩ happens to be greater than the last number of the first sequence, then subtract from it the number of all the cells and write ⟨the remainder⟩. This is a magic square of the odd number of cells.

4.45. 'Yugas' (4), 'planets' (9), 'eyes' (2), 'Hara's eyes' (3), 'sense organs' (5), 'sages' (7), eight, 'moon' (1), and 'tastes' (6): this is a magic square ⟨of three cells on each side⟩. Write down ⟨numbers⟩ beginning with one in eighty-one cells ⟨so that a magic square with nine cells on each side may be obtained⟩.

Thus the topic of magic squares is completed. 8 verses.

4.4 Miscellaneous

Now he tells the topic of miscellaneous problems (*prakīrṇakādhikāra*).

1. Now he tells how to calculate flowers.

[22]The word order in line 1 suggests the reading: 'Write down, from the first ⟨component square⟩ on, *half of* the ⟨first⟩ quarter sequence, and further add the upper ⟨numbers⟩.' See Kusuba 1993, 179. But, then, the two important concepts related to each other, the 'quarter sequence' (*pā-oli*/*pāya-oli*) and the 'quarter square' (*pā-jaṃta*), mentioned twice each in the present verse, seem to loose their raison d'être. See under GSK 4.41 in Part IV.

4.46. Repeatedly double what remains (*uvvarahi*), adding three each time. If no flower (*kusuma*) remains, three is declared in the beginning ⟨of computation⟩.

4.47. There is a temple (*sura-giha*) with four doors (*duvāra*). At each of them is one Yakṣa (*jakkha*) ⟨image⟩ equal to the door (*vāra-tulla*) ⟨in height⟩. In the middle of them (i.e., at the centre of the temple) is ⟨located the image of⟩ the lord of gods (Indra). A pious man (*dhammiya*) laid all the plentiful flowers by half each time ⟨he paid homage to⟩ the image (*biṃba*), giving one to Yakṣa while passing through each door. If the remainder, twenty, remains with him, how many ⟨flowers⟩ did he possess in total?

2. Now he tells how to calculate mangos.

4.48. Multiply the ⟨number of the fruit⟩ obtained ⟨by each soldier⟩ by the fraction (*aṃsa*), diminish by the first term (*āi*), and divide by the increase (*vuḍḍhi*). The quotient is doubled and increased by one. O learned man, they are said to have gone ⟨to pluck the fruit⟩. By means of ⟨the rule for⟩ the sum of series (*seḍhiya saṃkaliya*), the number of ⟨all⟩ the fruit is ⟨calculated⟩. When it is divided by the ⟨number of the fruit⟩ obtained ⟨by each soldier⟩, the number of ⟨all⟩ people ⟨in the army⟩ will be ⟨obtained⟩.

4.49. One-eighth part from an army (*kaṭaka*) went to a mango-plucking forest (*aṃba-toḍaṇa-vaṇa*) in order to eat them. ...[23] The first ⟨soldier⟩ brought four ⟨mangos to the camp⟩ and all others ⟨did the same⟩ by increasing ⟨the number⟩ by six each. Every ⟨soldier⟩ belonging to the army obtained twenty ⟨mangos⟩. How many people went ⟨to the forest⟩? What was the size of the army? How many mangos were brought by them?

3. Now he tells how to calculate ⟨the number of⟩ *varisolas*[24] ⟨offered⟩ to the sons-in-law.

4.50. Having put down the ⟨given⟩ multipliers in order ⟨from bottom to top⟩, repeatedly multiply ⟨them with each other from the bottom⟩ upwards (*uvar-uppari*) beginning with unity, adding one each time. One should tell ⟨the result as⟩ the first total number of *varisolas*. The same numbers (*aṃka*), without the unity, are multiplied one by one with each other. The result is ⟨the number of *varisolas*⟩ each of them eats. Thus state the learned men.

4.51. Five sons-in-law (*jamāiya*) went to their father-in-law's ⟨house⟩ (*sāsu-raya*).[25] The mother-in-law (*sāsu*), having filled the plate (? *taṭṭhiya*[26]), gave ⟨them⟩ one after the other *varisolas*, increasing twice, three times, four times, and five times of what remained ⟨on the plate⟩ after they had eaten ⟨severally⟩. The last one ate all ⟨the *varisolas* offered to him⟩. She said: 'Many were eaten

[23] *āesi rāṇaya* (*āpasa rāyaṇa* in PM A10).

[24] *Varisola* here is apparently different from *varisolaga* in GSK 5.12 below. The latter is some byproduct in the process of sugar manufacture, while the former is a sweetmeat described in the MU (vol. 2, p. 121, verses 15b–17a) under the name *varṣolaka*. It is made of thick sugar syrup (*kharapāka*) to which rice flour, milk, cardamom, saffron, and camphor are added and rolled into small balls (*golaka*).

[25] PM A12 reads: *ke pi paṃca jāmātaraḥ śvaśrūgṛhe gatāḥ*.

[26] S. *tat-sthita* 'that which was there'?

by this one.'[27] ⟨But, actually⟩ all have eaten one ⟨and the same number⟩. How many were there ⟨originally⟩? How many were eaten ⟨by each son-in-law⟩?

4. Now he tells how to calculate the area of cloth (*vastra-phala*).

4.52. The number of persons who hold one *hattha* is multiplied by four and by the ⟨square⟩ *karas* (= *hatthas*) of the piece of cloth (*vattha*) obtained ⟨by each person⟩. The result is the length and the width of the ⟨original square piece of⟩ cloth. ⟨The same⟩, when multiplied by the number of persons per *kara* and by four, is the number of all persons.

4.53. An excellent ⟨piece of square⟩ cloth: On its four sides (*caü-ddisi*),[28] each ⟨length of⟩ one *kara* was held by three persons. ⟨But when the cloth was divided equally among all the persons holding the cloth⟩, nine ⟨square⟩ *karas* were obtained by each person. How many persons ⟨were there⟩? How many ⟨square⟩ *hatthas* (= *karas*) ⟨did⟩ the excellent cloth ⟨measure⟩?

5. Now he tells about the movements of camels.

4.54. = GSK 4.16.

4.55. A camel walks four *joyaṇas* on the first ⟨day⟩ and increases ⟨his daily movement⟩ by three ⟨*joyaṇas*⟩ in every ⟨succeeding day⟩. In how many days will a female camel going sixteen *joyaṇas* per day meet ⟨him⟩?

6. Now he tells the inverse problem (*viparīta-uddeśaka*).

4.56. Add the subtrahend to the remainder (*sesa*) and square it. Subtract the increase (*ahiya*), draw its square root (*mūla*), multiply by the divisor (*bhāya*), and then divide by the multiplier (*guṇa-yāra*). The result should be known as the unknown quantity (*amuṇiya-rāsi*).

4.57. ⟨Tell⟩ the original quantity of the example for the inverse operation (*viviriya uddesago*), which, when multiplied by five, then divided by nine, then squared, increased by nine, then reduced to its square root, and then diminished by two, leaves the remainder three.

7. Now he tells how to know the other's mind (*para-cintā*).

4.58. Seventy is multiplied by the ⟨unknown number⟩ diminished by three ⟨repeatedly⟩; twenty-one by ⟨the same number diminished⟩ by five ⟨repeatedly⟩; fifteen by ⟨the same number diminished⟩ by seven ⟨repeatedly⟩. Give one hundred and five to the sum (*piṃḍa*) and divide ⟨the total by one hundred and five⟩. Know ⟨that the remainder is⟩ the ⟨number⟩ thought of by the other person (*para-citta*).

4.59. The 'son' (*suya*) that was thought of (*ciṃtiya*) is increased by two, multiplied by two, increased by one, multiplied by five, increased by the 'daughter' (*suyā*), multiplied by ten, and decreased by 'sky, five, hands' (250). Know that

[27] She joked thus because the fifth son-in-law ate all the *varisolas* offered to him and did not leave some on the plate as the other four had done previously.

[28] Lit., 'In its four directions.'

the remaining ⟨digits⟩, without the zero (*sunna*) ⟨in the units' place, are the 'son' and the 'daughter'⟩, respectively.

8. Now he tells how to know the erased digits (*marddita-aṃka*).

4.60. (Not understood.)

9. Now he tells how to calculate the same digits (*sadṛśa-aṃka*) (i.e., produce a number containing the same digit repeatedly).

4.61. The ⟨number consisting of the eight digits⟩ beginning with one and ending with nine, without eight, when multiplied by any optional digit (*icchiy-aṃka*) and by nine, becomes ⟨a number consisting of nine⟩ same digits ⟨each of which is⟩ equal to the number beginning with one ⟨expressed⟩ by the previous ⟨optional⟩ digit.

10. Now he tells how to calculate the number of cows (*go-saṃkhyā*).

4.62. Those ⟨given numbers⟩ should be put down one after another from top to bottom.[29] Multiply all ⟨terms from bottom⟩ upwards (*uvaruppari*). Thus the number of cows should be known.

4.63. The cows ⟨of a certain town⟩, going out of four gates, go to ⟨drink⟩ water in five rivers,[30] sitting down at the feet of seven trees, come to nine watering spots,[31] are made to sit in six orchards,[32] and are protected by eight cowherds. They are everywhere the same ⟨in number in each grouping⟩. O learned man, how many cows are there in total in that town (*nayara*)?

11. Now he tells the distribution of cow's milk (*go-dugudha-vaṃṭana*).

4.64. (Not understood.)

Thus the four topics (*adhikāra*) beginning with the region in the Gaṇitasāra (Essence of Mathematics), composed by the son of the respected Candra, Ṭhakkura Pherū, are completed. 64 verses.

[29] The word order of the first line suggests the reading: 'Those ⟨given numbers⟩ which are ⟨placed⟩ from top (*uparāo*) to bottom (*hiṭṭhi*) should be put down (*ṭhavijja*) one after another (*aṇukkamihi*).'

[30] *gaya pāṇī paṃca sari* (S. *gatāḥ pānīyāya pañcasu saritsu*). PM A16 reads: *paṃca-sarassu pivaṃti* ('They drink ⟨water⟩ at five lakes').

[31] *āvaṃti vārihi navihi païsi* (S. *āyānti vāriṣu navasu pradeśeṣu*). PM A16 reads: *9 pratolīṣu praviśaṃti* ('They enter nine streets').

[32] *chac ca vāḍihi niviṭṭhiya* (S. *ṣaṭsu ca vāṭeṣu niveṣṭitāḥ*). PM A16 reads: *6 vaneṣu caraṃti* ('They roam in six orchards').

Chapter 5

Quintet of Topics

Now he tells the rules for a quintet of topics (*uddeśa-paṃcaga*).

5.1 Yield of Grains

5.1. Having saluted the Creator (*siṭṭhi-kara*), I state the five topics (*uddesa*) dealing with the yield (*nippatti*) of grains, sugarcane (*ikkhu*) and oil (*cuppaḍa*), of regional tax (*desa-kara*) and price (*aggha*), and of measures (*māṇa*).

5.2. Grain (*anna*) grows everywhere, but because of the quality of the soil (*bhūmi*), there is much difference ⟨in the yield⟩. Delhi (*ḍhilliya*), Hansi (*āsiya*), and Narhad (*narahaḍa*): Know that these are irrigated regions (*varuṇa-paesa*).

5.3. The length and the width of an area, when multiplied ⟨with each other, give⟩ the measure of the land in *viggahayas*.[1] Twenty *kamas* or sixty *kaṃviyas* in length and breadth make one *vīgahaya*.[2]

5.4–8. The yield (*phala*) of food-grains is obtained at harvest (*nippanna*) from ⟨an area of⟩ one *vīgaha* of twenty *visuvas* as follows. Know sixty *maṇas* of kodrava grains (*kuddava*), twenty four of kidney beans (*maüṭṭha*),// twenty-two *maṇas* of chaula beans (*caüla*), sixteen of sesamum (*tila*), eighteen of *mūṃg* pulses (*mugga-māsa*), twenty of Italian millet (*kaṃguṇiya*), fifteen of millet (*cīṇaya*), and 'horses' (7) of rice king (*kūrīsa*),// sixteen *maṇas* of cotton (*kappāsa*), forty of Indian millet (*juvāri*), ten of flax (*saṇa*), and so also for sugarcane (*ikkhu*). ⟨These are⟩ the autumn harvest (*savāṇiya-sāhā*). Know, from now on, the spring harvest (*āsāḍhiya*).// Forty-five ⟨*maṇas*⟩ of wheat (*gohuva*), thirty-two of rice (*kalāva*), lentil (*massūra*), and chickpeas (*caṇaya*), fifty-six *maṇas* of barley (*java*), ten of mustard (*sarisama*), linseed (*alasī*) and safflower (*karaḍa*),// fourteen *maṇas* of val pulses (? *vaṭulā*), Indian rape (*tori*), and horse grains (*kulattha*). All ⟨these three kinds of⟩ seeds (*kaṇa*) are the same ⟨in weight⟩. Ten *maṇas* of cumin (*jīra*) and coriander (*dhaṇiya*). Others are

[1] The word order here is rather clumsy.

[2] The second half of this verse, with *kama* for *kami* and *kaṃvī* for *kaṃviya*, occurs also in GSK 1.4b.

included in *sikkaya*[3].

5.9. All the spices (*vesavāra*), garden cresses (*hālima*), fenugreek (*metthī*), leafy vegetables (*saggavattī*), *kora* grains, etc. are *seṃkkaya*[4], which is for a tax of one hundred *dammas* per *viggahaya*.

Thus the yield of the harvest (*dhānya-utpatti-phala*).

5.2 Yield of Sugarcane Juice and Oil

5.10–12. Sugarcane juice weighing fifty *maṇas* ⟨is obtained from⟩ nine *khāris* ⟨of sugarcane⟩. Its one-fifth is jaggery (*gula*), one-sixth brown sugar (*sakkara*), and one-sixteenth sugar candy (*khaṃḍa*).// One and a half times this is thickened juice (*ravva*), a little more or less than this is watery marrow (*nīravasa*), and again the same amount is *navi*. Whatever is said is gained through experience (*diṭṭha-patta*).// Less than sugar candy by one-third is *nivāta* and less by one-quarter is *varisolaga*. Very pure (*aï-cukkha*), remaining *sīra*,[5] equal ⟨in weight⟩ to the sugar candy, is obtained only once.

Thus the yield of sugarcane juice (*ikṣu-rasa-phala*).

5.13. From one *maṇa* of sesamum, sarson and safflower come nine, seven and five *visuvas* respectively of oil. One-ninth or one-eighth of milk is butter (*lūṇiya*) and a quarter less than that is ghee.

Thus the yield of oil (*sneha-phala*).

5.3 Yield of Regional Tax

5.14. Ten for a she-goat, the same for a cow, twice that for a she-buffalo, and four *vayallas* for a plough. For fire of hearth, the hearth tax (*kuḍhiyā*), excepting things related to barbers (? *nāviya*), necklaces (? *valahāra*) and dowry (? *mahara*).[6]

5.15. *devaï*(?), *kannacalā* (ones who move ones' ears?), *nīlī* (blue ones), *kavilīya* (tawny ones), cows, *adaṃtīya* (those without teeth), *brāhmaṇas* (*vippa*) with houses (or clothes?, *vāsaṇa*): no movable (? *cara*) tax is levied on these.[7]

5.16. For a plough of thirty-two *ṭaṃkas*, the hearth tax is of three kinds: one, one and a half, and two *ṭaṃkas*. One ⟨*ṭaṃka*⟩ for an ⟨old⟩ she-buffalo, a half ⟨*ṭaṃka*⟩ for an ⟨old⟩ cow, and ⟨one⟩ *ṭaṃka* for an old bull.[8]

5.17. The example given here may increase or decrease (i.e., change) according to the account (*caṭṭiya*). One-half, one-third, one-quarter, and other ...

[3] Cf. *seṃkkaya* in verse 9.
[4] Cf. *sikkaya* in verse 8.
[5] Cf. S. *sitā*, candied sugar.
[6] Tentative translation.
[7] Tentative translation.
[8] Tentative translation.

Thus the yield of regional tax (*deśa-kara-phala*).

5.4 Yield of Price

5.18. If a certain number of *pāīs* are ⟨obtained⟩ for one *damma*, then three times that of *seīs* and twice that of *pāīs* are, one should know, ⟨obtained⟩ for one *ṭaṃka*.

5.19. If a certain number of *seras* are told for one *damma*, then one and a quarter times that of *maṇas* are for one *ṭaṃka*. ⟨If a certain number of *maṇas* are told for one *ṭaṃka*⟩, throw off one-fifth part of the *maṇas*. Know that the remainder is the *seras* for one *damma*.

5.20. O learned man, if a one-*maṇa* container ⟨of grain is obtained⟩ for a certain number of *dammas*, know that one *sera* ⟨of the same is obtained⟩ for half that number of *visovas*.

5.21. If one score (*koḍiyā*) of things which are sold by number is for a certain number of *dammas*, one of those things is obtained for the same number of *visovas*.

Thus the yield of price (*argha-phala*).

5.5 Measurement

Now measurement:

5.22. The diameter of a circle, multiplied by three and increased by its one-sixth, gives the circumference. By multiplying it by a quarter of the diameter, whatever is obtained is the area.

[9]Showing ⟨the figure⟩ graphically (*darśana*): 

The circumference is 19. The area of the figure is 28⟨||⟩.[10]

Thus the circle (*vṛtta*).

5.23. ⟨When one calculates the area of⟩ a square from ⟨that of the circumscribing⟩ circle, it is twelve *visuvas* and slightly more (lit., with difference). ⟨When one calculates the area of⟩ a circle from ⟨that of the circumscribing⟩ square, the circle is less by one-fifth.

5.24. ⟨When one calculates the area of⟩ a circle from ⟨that of the circumscribing⟩ triangle, the area is twelve and a half *visovas*. ⟨When one calculates the area of⟩ a triangle from ⟨that of the circumscribing⟩ circle, seven and a half *visovagas*.

[11]Showing these ⟨figures⟩ in particular graphically:

[9]One prose line.

[10]Here and hereafter, one vertical line, |, denotes a quarter. The same sign is used in other works of Pherū also. See the Nahatas' edition of *Ratnaparīkṣā*, pp. 13-14; *Dravyaparīkṣā*, pp. 22-31, etc.; *Jyotiṣasāra*, pp. 8 and 10; and *Vāstusāra*, p. 98.

[11]One prose line.

[12]

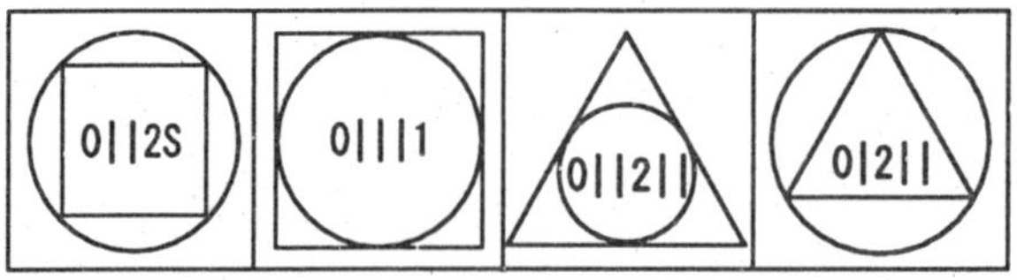

Thus the measurement of the field (*kṣetra-māna*).

5.25. The cube of the diameter (*udaya*, lit. height) of a sphere, decreased by ⟨its own⟩ quarter and again decreased by ⟨its own⟩ quarter, gives the volume (*pāhāṇa*, lit. stone). The circumference, multiplied by a quarter of the circumference and increased by ⟨its own⟩ one-ninth, gives the surface area.

[13]Setting-down (*nyāsa*): (6) Answer (*labdha*): The volume of the sphere is 121||,[14] and the surface area 100SS6.[15] ⟨Calculation:⟩ The cube ⟨of the diameter⟩, 216, decreased by a quarter, 162, again decreased by a quarter, 121||.[16] ⟨A quarter of⟩ the circumference, 4|||, is multiplied by ⟨the circumference⟩, 19; the product is 90⟨|⟩. One-ninth part of this is 10⟨$\frac{1}{36}$⟩. Thus the surface area is 100⟨SS6⟩.

5.6 Masonry

5.26. A cubic *kaṃviya* of all varieties of stone originating in Delhi (*ḍhilliya*) weighs fifty *maṇas* and twenty-four hundred *tullas* (i.e., *tolas*).[17]

5.27. *Vaṃsī* stone weighs forty-eight ⟨*maṇas*⟩, marble (*mamāṇīya*) sixty, black stone (black granite) sixty-two, *jajjāvara* and of Kannāṇā forty-nine, and *kuḍukkaḍa* sixty.[18]

5.28. ⟨A cubic *kaṃviya* of⟩ clay (*maṭṭī*) ⟨weighs⟩ twenty-five *maṇas*, kodrava grains (*tusaṃnna*) eight *maṇas*, wild grains (*vaṇ-aṃnna*) twelve, oil and ghee ten *maṇas* ⟨each⟩, and salt sixteen *maṇas*. This is an example.

5.29. One mason (*rāja*) along with three men gather, for a wall whose volume (*pāhaṇa*, lit. stone) is twelve ⟨cubic⟩ *gajas*, fourteen hundred bricks and thirty water pots. This is an example.

5.30. Twenty-seven *maṇas* of *hakka*, nine of lime (*cunna*), twice that of mortar

[12]In the Nahatas' text:

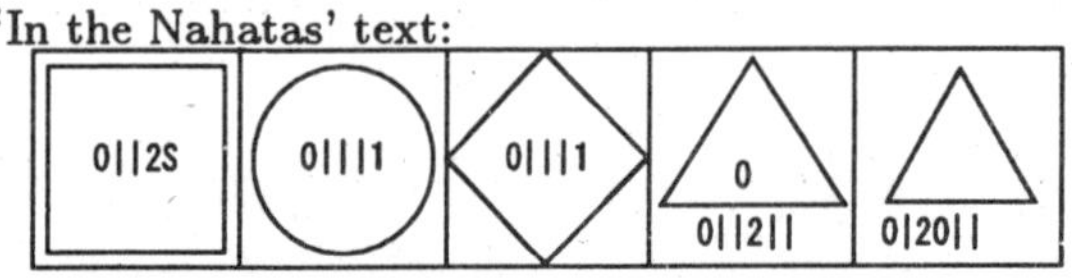

The third figure of the Nahatas' text is superfluous. The symbol, 'S', used at the end of the number in the first figure seems to mean 'and slightly more'. See verse 23.

[13]Three prose lines.

[14]120 in the Nahatas' text.

[15]The symbol, 'SS', seems to separate the units' place and the twentieths' place.

[16]120 in the Nahatas' text.

[17]This is almost the same as GSK 3.67 but inserts 'hundred' after 'twenty-four'.

[18]This is the same as GSK 3.68 except for minor phonetic variants (*mamāṇīya* for *mammāṇī* and *jajjāvara* for *jajjāvaya*).

(*khora*), and nine *maṇas* of *cijja*. Know these exactly for a wall whose volume (*pāhāṇa*) is one ⟨cubic⟩ *gaja*.

5.31–32. In plastering, one less by a quarter *maṇa* of pure lime (*kevaṇa cunna*) is accompanied by a quarter *sera* of flax (*saṇa*), one-third of mortar (*khora*) for dry land (*talavaṭṭa*), and a half ⟨of mortar⟩ for watery place (*jala-ṭhāṇa*),// ⟨with which⟩ forty *maṇas* of cow-dung (*chāṇaya*), sixty of limestone (*kakkara*) roasted and powdered, sixty of carried-with-protection (*rakkha-pavāhiya*) ⟨sand?⟩ and forty of ⟨carried-⟩without-protection (*arakkha-*) ⟨sand?⟩ are mixed.

5.33. This quintet of topics has been told by Pherū, son of Caṃḍā, such as (including) the regional tax yields. One should know ⟨these⟩ on the occasion of ⟨making⟩ an account (*caṭṭiya*).

Thus the quintet of topics is completed.

Thus the rules in ⟨the book for⟩ the procedural mathematics (*pāṭī*), Gaṇita-sāra-kaumudī (Moonlight of the Essence of Mathematics), composed by the most ⟨devout⟩ Jain, son of the respected Candra, Ṭhakkura Pherū, are completed.

All the contents combined and verses added together 311.

Written down (i.e., copied) on the 5th day of the white half of Caitra month, Vikrama year 1404 (= Saturday, 17 March 1347)[19].

[19] For the date see §0.7 of Part II.

Chapter 6

List of Rules

6.0. The number (*saṃ°* [1]) of the rules (*sūtra*), the verses (*gāthā*), and the contents (*vastu*) combined, counted 311.[2]

6.1. 15 verses (1.1–15): Establishing the basic plan (*mūla-prabandha*[3]).

6.2. 78 verses (1.16–93): 25 fundamental operations or procedures.

6.3. 5 verses (1.16–20): Rules for producing the sum, 1.

6.4. 6 verses (1.21–26): Computation (*gaṇanā*) of difference, 2.

6.5. 6 verses (1.27–32): 2 kinds of computation for multiplication, 3.

6.6. 1 verse (1.33): Producing the computation for division, 4.

6.7. 3 verses (1.34–36): Computation for producing the square number, 5.

6.8. 2 verses (1.37–38): Computation for producing the square root number, 6.

6.9. 4 verses (1.39–42): Computation for producing the cube, 7.

6.10. 3 verses (1.43–45): Computation for producing the cube root, 8.

6.11. 4 verses (1.46–49): Computation of the fundamental operations of fractions, 9.[4]

6.12. 3 verses (1.50–52): Computation of the sum of fractions, 10.[5]

6.13. 2 verses (1.53–54): Computation of multiplication of fractions, 11.

6.14. 2 verses (1.55–56): Computation of division of fractions, 12.

---

[1]An abbreviation of *saṃkhyā*. The same abbreviation is used in 6.7 and 6.8, too. Abbreviations used in this List of Rules: *ga°* = *gaṇanā*, *gā°* = *gāha/gāhā*, *jā* = *jāti*, *dutī* = *dutīka*, *pratha.* = *prathama*, *vi°* = *vidhi*, *vyava°* = *vyavahāra*, *saṃ/saṃ°* = *saṃkhyā*.

[2]Actually, 311 is the total number of the verses in the GSK (93 + 17 + 104 + 64 + 33 = 311), although this List of Rules covers the first three chapters only. In the lines that follow, we have supplied the serial numbers of the verses in parentheses.

[3]This word occurs also in 6.75 below but it is uncommon in mathematics. It probably has the same meaning as the word *paribhāṣā* in other *pāṭī* works.

[4]The word *abhinna* (integer) in N seems to be a simple corruption of *bhinna* (fraction) and the serial number 9 should go with the next line, although these occur in the erroneous introduction to 1.46 also in N; see the fn. in the Text. The word *parikrama* is a wrong Sanskritization of *parikamma* (S. *parikarman*), which occurs in 1.15.

[5]This line should be replaced by the following two lines:

'1 verse (1.50): Computation of the sum of fractions, 9.'

'2 verses (1.51–52): Computation of the difference of fractions, 10.'

6.15. 2 verses (1.57–58): Computation of the squares of fractions, 13.
6.16. 1 verse (1.59): Computation of the square roots of fractions, 14.
6.17. 2 verses (1.60–61): Computation of the cubes of fractions, 15.
6.18. 1 verse (1.62): Computation of the cube roots of fractions, 16.
6.19. 9 verses (1.63–71): Computation of the rule of three, 17.
6.20. 6 verses (1.72–77): Computation of the rule of five, 18.
6.21. 1 verse (1.78): Computation of the rule of seven, 19.
6.22. 1 verse (1.79): Computation of the rule of nine, 20.
6.23. 1 verse (1.80): Computation of the rule of eleven, 21.
6.24. 5 verses (1.81–85): Computation of the inverse rule of three, 22.
6.25. 4 verses (1.86–89): Computation of a variety called purchase and sale, 23.
6.26. 2 verses (1.90–91): Computation of barter, 24.
6.27. 2 verses (1.92–93): Computation of the sale of living beings, 25.
6.28. Thus 78 verses for 25 fundamental operations as the seed of the rules. Let it be auspicious.
6.29. Another ⟨chapter⟩ for 8 classes of the reduction of fractions with eight names, 17 verses for rules (2.1–17).
6.30. 1. Homogenization of fractions, 1 verse (2.1).
6.31. 2. Multi-part class, 1 verse (2.2).
6.32. 3. Part-part class, 3 verses (2.3–5).
6.33. 4. Part-addition class, 2 verses (2.6–7).
6.34. 5. Computation of part-subtraction ⟨class⟩, 2 verses (2.8–9).[6]
6.35. 6. Part-mother class, 2 verses (2.10–11).
6.36. 7. Chain-reduction ⟨class⟩, 2 verses (2.12–13).
6.37. 8. Pillar-problem class, 4 verses (2.14–17).[7]
6.38. Another ⟨chapter⟩ for the computations in the 8 procedures, 104 verses for rules (3.1–104).
6.39. 1. The first procedure, for mixture, 25 verses (3.1–25).
6.40. 2 verses (3.1–2): Computation in mixture as the first item, 1.
6.41. 2 verses (3.3–4): Computation of the commission of surety as the second item, 2.
6.42. 2 verses (3.5–6): Rule for the conversion of several bonds into one, 3.
6.43. 4 verses (3.7–10): Investments (proportionate distribution), 4.
6.44. 4 verses (3.11–14): Equal or different (purchase in proportion), 5.
6.45. 2 verses (3.15–16): Procedure for gold, 6.
6.46. 4 verses (3.17–20): Fractions for gold, 7.[8]

---

[6] The word *bhāgapravāha* seems to be a wrong Sanskritization of *bhāgapavāha* (S. *bhāga-apavāha*), which occurs in the introduction to 2.8.

[7] The class-name, 'pillar-problem' (*staṃbhoddesa*, S. *stambha-uddeśa*), is not Pherū's but Śrīdhara's. See §2.2.16 of Part I and the first item of Appendix B.

[8] This line should be replaced by the following two lines:
'1 verse (3.17): Fractions for gold, 7.'
'3 verses (3.18–20): Refinement of gold, 8.'

6.47. 5 verses (3.21–25): Lost (unknown) purities ⟨and weights⟩ of gold pieces as the eighth item, 8.[9]

6.48. 2. The second procedure for series, 9 verses (3.26–34).

6.49. 2 verses (3.26–27): Computation (*gaṇita*) of the procedure for series, 1.

6.50. 1 verse (3.28): Calculation (*ānayana*) of the lost (unknown) first term, 2.

6.51. 1 verse (3.29): Calculation of the lost (unknown) common difference, 3.

6.52. 1 verse (3.30): Calculation of the lost (unknown) number of terms, 4.

6.53. 2 verses (3.31–32): Calculation of the sum of the sums, 5.

6.54. 1 verse (3.33): Calculation of the sum of the square series and the sum of the cubic series, 6.

6.55. 1 verse (3.34): Computation of the sum of the sum, the square, and the cube, 7.

6.56. 3. Procedure for plane figures, 19 verses for rules (3.35–53).

6.57. 1. Square figures.[10]

6.58. 2. Rectangular figures.

6.59. 3. One or more equi-perpendicular figures.[11]

6.60. 4. Triangular figures.

6.61. 5. Pentagonal figures.[12]

6.62. 6. Right-angled triangles.[13]

6.63. 7. Circles and disks.

6.64. 8. Bow-like figures.

6.65. 9. Elephant's tusk figures.

6.66. 10. Thunderbolt figures.

6.67. 11. Drum-like figures.

6.68. 12. Various rules.

6.69. 4. Procedure for excavations, 15 verses (3.54–68).

6.70. 4 verses (3.54–57): Computation (*gaṇana*) in various operations for excavation, 1.

6.71. 3 verses (3.58–60): Calculation (*ānayana*) of the capacity of a circular well, 2.

6.72. 6 verses (3.61–66): Calculation of the volumes of stones, 3.

6.73. 2 verses (3.67–68): Calculation of the weights of stones, 4.

6.74. 5. Procedure for piling, 19 verses (3.69–86).

6.75. 1 verse (3.69): Basic plan, 1.[14]

6.76. 4 verses (3.70–73): Bricks and stones of the wall, 2.

6.77. 3 verses (3.74–76): Computation of the piling of the dome, 3.

6.78. 2 verses (3.77–78): The wall of the spiral stairway, 4.[15]

---

[9] The serial number should be 9.

[10] In this section, the number of verses for each topic is not given.

[11] This is a tentative interpretation of *ekādi sālaṃba*.

[12] Figures of this category are not treated in the GSK.

[13] *trikoṇa-vikaṭa*, lit. 'beautiful triangular ⟨figures⟩.' This expression is uncommon. A right-angled triangle is usually called 'a well-born trilateral' (*jātya-tryasra*).

[14] Read 1 for *gā°* 1 at the end of the line.

[15] Actually, this topic is treated in 3 verses (3.77–79) and the next one in 1 verse (3.80).

6.79. 2 verses (3.79–80): The number computed for the minaret, 5.
6.80. 2 verses (3.81–82): Correction in the computation of the arch, 6.
6.81. 1 verse (3.83): Computation for the staircase, 7.
6.82. 1 verse (3.84): Computation for the bridge structure, 8.
6.83. 1 verse (3.85): Computation of the number (or volume) of the cylindrical well, 9.
6.84. 1 verse (3.86): Computation of the number (or volume) of the stepwell, 10.
6.85. 6. Procedure for sawing, 9 verses (3.87–95).
6.86. Rule for the length of timber.
6.87. Cutting by a saw.
6.88. Various kinds of timber.
6.89. These verses are 9 ⟨in number⟩.
6.90. 7. Procedure for mounds of grain, 5 verses (3.96–100).
6.91. Mound of grain heaped.
6.92. Length and height.
6.93. Width.
6.94. Essence of mathematics (? *gaṇitasāri*).
6.95. 8. Procedure for shadows, 4 verses (3.101–104).
6.96. Determination of shadows.
6.97. Determination of the cardinal directions.
6.98. Thus the rules for the eight types of procedure, 104 verses.

Part IV

Mathematical Commentary

Chapter 1

Twenty-five Fundamental Operations

1.1 Weights and Measures

GSK 1.1–2. Introduction.

Pherū refers to three kinds of sources of this book: (1) the ⟨books on⟩ algorithms of computations (*gaṇaṇa-pāḍī*, S. *gaṇanā-pāṭī*) told (*vuttā*, S. *uktā*) by his predecessors, (2) direct experience (*aṇubhūya*, S. *anubhūya*, lit. having experienced), and (3) hearing (*suṇiūṇa*, S. *śrutvā*, lit. having heard from contemporaries). For the compound, *gaṇaṇa-pāḍī*, see the expressions, *gaṇitasya pāṭīm* (GT 1) and *pāṭīṃ sad-gaṇitasya* (L 1). Cf. also *sva-viracita-pāṭyā gaṇitasya sāram uddhṛtya* (Tr intro. 1), *vakṣye golaṃ pāṭīṃ kuṭṭaṃ* (MS 14.1), and *vakṣye sugamāṃ pāṭīṃ* (MS 15.1).

GSK 1.3–4a. Monetary units.

20 *paḍikāiṇis* = 1 *kāiṇi*,
20 *kāiṇis* = 1 *paḍivissaṃsa*,
20 *paḍivissaṃsas* = 1 *vissaṃsa*,
20 *vissaṃsas* = 1 *visova*,
20 *visovas* = 1 *damma*,
50 *dammas* = 1 *ṭaṃkaya*.

The Sanskrit terms corresponding to these units are: *pratikākiṇī*, *kākiṇī*, *prativiṃśāṃśaka*, *viṃśāṃśaka*, *viṃśopaka*, *dramma*, *ṭaṅka(ka)*.

The word *visova* (also spelled *visovaga*, *visuva*, etc.) can be used for 'one-twentieth part' of any standard unit. Thus, it denotes one-twentieth part of *guṃja* (weight, GSK 2.13), of *maṇa* (weight, GSK 5.13), of *gaja* (length, GSK 3.89–93), of *vīgahaya* (area, GSK 5.4), and of an area unit without denomination (GSK 5.23–24). In the BAB, one *viṃśopaka* is 'one-twentieth part' of one *rūpaka* (see Shukla's introduction, p. lxxx). Cf. also EI 11, 34 and EI 30, 166 for *viṃśopaka*.

For a similar vigesimal subdivision of the area unit *bigha* see GSK 5.4–8 below.

The *koḍī* used for wages of sawing (GSK 3.89–93) also is a monetary unit; it is the cowry shell, known as *varāṭaka* in Sanskrit. Pherū, however, does not define it anywhere. Tr paribhāṣā 4 and L 2 record the relationship, 20 *varāṭakas* = 1 *kākiṇī*. Note, however, that this *kākiṇī* is not equivalent to the *kāiṇi* of the above list because, according to this list, 160000 *kāiṇis* = 1 *damma*, while L 2 gives the relationship of 64 *kākiṇīs* = 1 *dramma*.

For *jīvalaï* or *jivalā*, which seems to be another monetary unit, see GSK 4.2–3.

GSK 1.4b. Area measures.

20 *kamas* × 20 *kamas* = 1 *vīgahaya*,
60 *kaṃvīs* × 60 *kaṃvīs* = 1 *vīgahaya*.

Sanskrit terms: *kama* from S. *krama* (step?), *kaṃvī* from S. *kambi/kambī* (joints of a bamboo), *vīgahaya* from S. *vigrahaka* (streching out).

The linear measure *kama* does not occur elsewhere in this work but the relationship, 20 × 20 sq. *kamas* = 1 *vīgahaya*, seems to be supported by the measures, *gatha* and *kāthi*, recorded by the translator of the *Ā'īn-i-Akbarī* [Abū al-Faḍl 2004, 666, note 97]: '... and the *jarib* is equal to 5 chains of 11 yards each, or to 60 *gaz* or 20 *gathas* or knots. ... The Maratha *bigha* is called 20 *pānds* or 400 sq. *kāthis* or rods of (each) 5 cubits and 5 hand-breadths. ...' If this interpretation is correct, 1 *kama* = 3 *kaṃvīs*, which is about 2 meters (see GSK 5.28). This may be too long for 'a step' (*krama*).

The linear measure *kaṃvī* (also called *kaṃviya* or *kaṃbiya*) is equivalent to *gaja* or *hattha* (S. *hasta*). See GSK 1.5–6 below. The Gujarātī word *kaṃbā* means 'a chip of a bamboo' or 'a measure of 24 inches used by carpenters in measuring timber' (MGED).

Vīgaha(ya) > H. *bīghā*. Abū al-Faḍl [2004, 585: Book III, Ā'īn 10] says: 'The *bigha* is a name applied to the *jarib*. It is a quantity of land 60 *gaz* long by 60 broad.' According to Prinsep [1858, 127], 'standard *bīghā* of western provinces = 60 × 60 *gaz* = 3600 *gaz*.'

Professor Irfan Habib suggests the following method of determining the length of the *gaz* in Delhi Sultanate, ca. 1370, which could be valid also for Pherū's time.

The anonymous *Sīrat-i Fīrūzshāhī* composed at the court of Fīrūz Shāh Tughluq at Delhi in AH 772 (AD 1370–71) states (p. 201) as follows about the Aśokan pillar which was brought to Delhi and erected at the Fīrūz Shāh Koṭlā under Fīrūz's orders: 'The pillar is 22 *gaz* in height, out of which two *gaz* are covered [by masonry] while the rest is visible. Thus $21\frac{1}{2}$ *gaz* is the height of the building [up to the base of the visible part of the pillar] and twenty *gaz* of stone is the visible height of the stone [pillar] ... ' According to Zafar Hasan [1915–20, II, p. 79], the visible part of the Aśokan pillar measured 42′7″ in height.

Thus 20 *gaz* = 42′7″ = 12.979 m and therefore 1 *gaz* = 0.64897 m. If 1 *kaṃvī* is equal to 1 *gaz*, a *vīgahaya* or *bīghā* would be 0.64897 × 60 × 60 sq. m = 2336.292 sq. m.

GSK 1.4b = GSK 5.3b.

GSK 1.5–6. Linear measures.

24 *pavvaṃgulas* = 32 *karaṃgulas*
(i.e. 3/4 *pavvaṃgula* = 1 *karaṃgula*).

8 *javas* = 1 *pavvaṃgula*,
24 *aṃgulas* = 1 *hattha*,
4 *hatthas* = 1 *ḍaṃḍa*,
2000 *ḍaṃḍas* = 1 *kosa*,
4 *kosas* = 1 *joyaṇa*.

Sanskrit terms: *parva-aṅgula*, *kara-aṅgula*, *yava*, *hasta*, *daṇḍa*, *krośa*, *yojana*.

Judging from the continuity of the definitions, '*aṃgula*' of the third equation must be the same as '*pavv-aṃgula*' of the second equation. These relationships, except the first, constitute the standard table of linear measures employed in popular mathematical textbooks like Tr paribhāṣā 7 (except 8 *javas* = 1 *pavv-aṃgula*) and L 5–6.

Besides the units given above, the linear measures, *kama*, *kaṃviya* (also spelled *kaṃbiya* or *kaṃvī*), *kaṃviy-aṃgula*, *gaja*, and *visuva* (or *visovaga*), occur without definitions in our text.

We estimate that a *kaṃviya* was about 63 ~ 72 cm or 24.8 ~ 28.3 in (see GSK 5.28 below). The *kaṃviya* is, therefore, very close to the unit called *gaz* (Prinsep 1858, 127–28), which is expressed as *gaja* in our text. Moreover, in the section for sawing (GSK 3.87–95), *gaja* is used side by side with *hattha* and 8 *javas* are said to make a *kaṃviy-aṃgula*, which is therefore equivalent to a *pavv-aṃgula*. These two words, *kaṃviy-aṃgula* and *pavv-aṃgula*, seem to be used synonymously in the sense of 'a joint digit' or 'a digit for the interval of two consecutive joints of a bamboo'. Hence follows the following system of linear measures:

8 *javas* = 1 *kaṃviy-aṃgula* (or *pavv-aṃgula*),
24 *kaṃviy-aṃgulas* (or *pavv-aṃgulas*)
= 1 *kaṃviya* (or *hattha* or *gaja*) = 63 ~ 72 cm,
20 *visuvas* = 1 *gaja*,
3 *kaṃviyas* = 1 *kama* (? see GSK 1.4b above).

A similar system is used in Śambhudāsa's commentary (AD 1428/29) on the PV (Hayashi 1991, 447):

24 *jyeṣṭha-aṅgulas* = 1 *gaja*,
20 *visās* = 1 *gaja*.

From the relation, 24 *pavv-aṃgulas* = 32 *kar-aṃgulas*, given in GSK 1.5, we can reconstruct another system for linear measures:

6 *javas* = 1 *kar-aṃgula*,
24 *kar-aṃgulas* = 1 *kara* = 47 ~ 54 cm.

This *kara* is therefore equivalent to the ordinary *hasta* ('an arm' or cubit, i.e. the length from the elbow to the tip of the middle finger) but otherwise Pherū does not distinguish *kara* and *hattha*. He uses them side by side synonymously in GSK 1.64, 1.79, 3.56, 3.59, 3.62, 3.64, 3.82, 4.19, 4.52, and 4.53. In the last two, for example, '⟨the number of⟩ the people who seize one *hattha*' (*je jaṇa gahaṃti hatthaṃ te*) is also called '⟨the number of⟩ the people per *kara*' (*kara-jaṇa*).

GSK 1.7. Definition of three kinds of *hattha*.

sara-hattha = *hattha* defined above,
paḍa-hattha = width (*vikkhaṃba*) × length (*āyāma*),
ghaṇa-hattha = area (*vitthāra*) × height (*udaya*).

Sanskrit terms: *śara-hasta* ('arrow' or linear *hasta*), *paṭa-hasta* ('cloth' or square *hasta*), *ghana-hasta* ('solid' or cubic *hasta*).

Note that *vitthāra* (*vistāra*) is used here not in the sense of 'width,' but of 'area.'

Cf. *Aṇuogaddāra* [AD] 337: *se samāsao tivihe paṇṇatte / taṃ jahā — sūti^aṃgule 1 payar-aṃgule 2 ghaṇ-aṃgule 3 / aṃgula-āyatā ega-padesiyā seḍhī sūi^aṃgule 1 sūyī sūyīe guṇiyā payar-aṃgule 2 payaraṃ sūīe guṇitaṃ ghaṇ-aṃgule 3 /* That is to say, *sūti^aṃgula* (or *sūi-* or *sūyī-*) = a line of spacial points whose length is one *aṃgula* and whose width and thickness are one spacial point (*ega-padesiyā* = *eka-pradeśikā*), *payar-aṃgula* = *sūyī* × *sūyī*, and *ghaṇ-aṃgula* = *payara* × *sūyī*.

Sanskrit terms: *sūcī-aṅgula* (or *sūci-*), *pratara-aṅgula*, *ghana-aṅgula*.

GSK 1.8. Volume measures.

4 *karapuḍas* = 1 *pāī*,
4 *pāīs* = 1 *māṇa(a)*,
4 *māṇas* = 1 *seī*,
16 *seīs* = 1 *pattha*.

Besides these, Pherū defines a unit of capacity called *patta* in GSK 3.96,

1 cubic *hattha* = 1 *patta*,

but he never uses it. Instead, he uses the term *khāri* without a definition in GSK 5.10; it is presumably equivalent to *patta*. See under GSK 5.10–12. Pherū uses another conversion ratio between *seī* and *pattha*, 10 *seīs* = 1 *pattha*, in an example for the inverse three-quantity operation (GSK 1.82).

Sanskrit terms: *kara-puṭa*, *pāli*? *pādi(kā)*?, *māna(ka)*, *seti(kā)*, *prastha*, *pātra(?)*, *khārī*/*khāri(kā)*.

The BAB uses a similar system except for *patta* and *khāri* (see Shukla's introduction, p. lxxx):

4 *mānakas* = 1 *setikā* (or *setika*),
4 *setikās* = 1 *kuḍava*,
4 *kuḍavas* = 1 *prastha* (hence 16 *setikās* = 1 *prastha* as in the GSK).

For similar measures found in an Old-Gujarātī commentary on the *paribhāṣā* section of the Tr, see Srinivasan 1979, 77–78.

GSK 1.9–10a. Weight measures (2 systems).

System 1:

6 *guṃjas* = 1 *māsa(ya)*,
4 *māsayas* = 1 *ṭaṃka*,
10 *ṭaṃkas* = 1 *pala*,
6 *palas* = 1 *sera* (hence 60 *ṭaṃkas* = 1 *sera*),
40 *seras* = 1 *maṇa*. (It follows that 2400 *ṭaṃkas* = 1 *maṇa*.)

Sanskrit terms: *guñjā*, *māṣa(ka)*, *ṭaṅka*.

Pherū uses the two conversion ratios, 66 *ṭaṃkas* = 1 *sera* and 72 *ṭaṃkas* = 1 *sera*, in an example for the inverse three-quantity operation (GSK 1.83). The latter of the two ratios is actually defined in L 8[a] (an additional verse that appears after L 8 in Mahīdhara's text):

$\frac{3}{4}$ *gadyāṇaka* = 1 *ṭaṅka*,
72 *ṭaṅkas* = 1 *sera*,
40 *seras* = 1 *maṇa*. (It follows that 2880 *ṭaṃkas* = 1 *maṇa*.)

System 2 (for gold):

16 *javas* = 1 *māsa(ya)*,
4 *māsayas* = 1 *ṭaṃka*,
3 *ṭaṃkas* = 1 *tolaya*.

Sanskrit terms: *yava*, *māṣa(ka)*, *ṭaṅka*, *tola(ka)*.

GSK 1.10b. Purity of gold.

16 *javas* = 1 *vannī*,
12 *vannīs* = *mahā-kaṇaya* (great gold, i.e., pure gold).

Sanskrit terms: *yava*, *varṇa/varṇi(kā)*, *mahā-kanaka*.

Note that Pherū does not use the 16-*varṇa* but the 12-*varṇa* system of purity of gold. It is based on his weight system, where 12 *māsas* = 1 *tola* (see System 2 under GSK 1.9–10a above). Cf. S. R. Sarma 1983.

GSK 1.11. Time units.

60 *palas* = 1 *ghaḍiyā*,
60 *ghaḍiyās* = 1 *diṇa-rayaṇī* (day and night or nychthemeron),
30 *diṇa-rayaṇīs* = 1 *māsa* (month),
12 *māsas* = 1 *varisa* (year).

Sanskrit terms: *ghaṭikā*, *dina-rajanī* (or *-rajani*), *varṣa*.

Pherū uses *divasa* and *diṇa* (*dina*) also for 'one day' (see GSK 1.70 and 71).

Cf. Tr paribhāṣā 8.

Table 1.1: Names of decimal places by Jain authors with Hindi terms

	AD 204/326	GSS 1.63–68	GSK 1.12–14	Hindi*2
10^0	*ekka*	*eka*	*ega*	*eka*
10^1	*dasa(ga)*	*daśa*	*daha*	*daśa/dasa*
10^2	*sata/saya*	*śata*	*saya*	*sau*
10^3	*sahassa*	*sahasra*	*sahasa*	*hajāra*
10^4	*dasasahassa*	*daśasahasra*	*dasahasa*	*dasa hajāra*
10^5	*satasahassa*	*lakṣa*	*lakkha*	*lākha*
10^6	*dasasatasahassa*	*daśalakṣa*	*dasalakkha*	*dasa lākha*
10^7	*koḍi*	*koṭi*	*koḍi*	*karoḍa*
10^8	*dasakoḍi*	*daśakoṭi*	*dasakoḍī*	*dasa karoḍa*
10^9	*koḍisata*	*śatakoṭi*	*avva**1	*araba*
10^{10}	*dasakoḍisata*	*arbuda*	*dasaavva*	*dasa araba*
10^{11}	–	*nyarbuda*	*khavva*	*kharaba*
10^{12}	–	*kharva*	*dasakhavva*	*dasa kharaba*
10^{13}	–	*mahākharva*	*saṃkha*	*padma*
10^{14}	–	*padma*	*dasasaṃkha*	*dasa padma*
10^{15}	–	*mahāpadma*	*paüma*	*nīla*
10^{16}	–	*kṣoṇi*	*dasapaüma*	*dasa nīla*
10^{17}	–	*mahākṣoṇi*	*nīla*	*śaṃkha*
10^{18}	–	*śaṅkha*	*dasanīla*	*dasa śaṃkha*
10^{19}	–	*mahāśaṅkha*	*nīlasaya*	
10^{20}	–	*kṣiti*	*nīlasahasa*	
10^{21}	–	*mahākṣiti*	*dasasahasanīla*	
10^{22}	–	*kṣobha*	*nīlalakkha*	
10^{23}	–	*mahākṣobha*	*nīladasalakkha*	
10^{24}	–	–	*koḍinīla*	
	–	–	etc.	

*1 S. *arbuda*. Cf. Ap. *avvua* (AHK) for S. *arbuda*.
*2 See Gupta 2001b, 88. According to Agar Chand Nahata cited by Gupta [2001b, note 27], this list is extended to a list of 97 places in an anonymous *Amalasiddhi*.

GSK 1.12–14. Names of 25 decimal places.

Pherū gives the names of the first twenty-five decimal places. Compare them with those of other Jain authors (Table 1.1). Pherū uses the centesimal scale for the numbers/places greater than *sahasa* or thousand as in, and more strictly than, the GSS (see *śata-koṭi* of the latter) and introduces a new term, *nīla*, for 10^{17}, which made it possible for him to discard *kṣoṇi*, *śaṅkha*, *kṣiti*, and *kṣobha* of the GSS. The word *iccāi* (S. *ity-ādi*, 'and so on', lit. 'beginning with these') at the end of the list shows that Pherū had a longer list in mind. The list, in fact, can be easily and systematically extended as follows: *dasa-koḍi-nīla* (10^{25}), ..., *dasa-paüma-nīla* (10^{33}), *nīla-nīla* (10^{34}), *dasa-nīla-nīla* (10^{35}), etc. Pherū's list is very close to the traditional Hindi list of 19 terms (Gupta 2001b, 88), which is obtained by shifting the pair, *saṃkha* and *dasa-saṃkha*, from the fourteenth-fifteenth places to the eighteenth-ninteenth places.

GSK 1.15. Forty-five gateways to mathematics.

45 *dvāras*	25 *parikarmans*	(treated in Adhyāya 1, 93 verses)
	8 *jātis*	(treated in Adhyāya 2, 17 verses)
	8 *vyavahāras*	(treated in Adhyāya 3, 104 verses)
	4 *adhikāras*	(treated in Adhikāras 4, 45+19 verses)

In addition to these 45 'gateways' to mathematics, the GSK has a supplementary section (33 verses) called Uddesa-paṃcaga (Uddeśa-pañcaka) or 'Quintet of Topics'. We will refer to the verses of that section as GSK 5.–. See GSK 5.1 below. For the structure of the GSK see §2.1 of Part I and §0.1 of Part II.

1.2 Eight Fundamental Operations of Integers

Fundamental operation 1: Sum.

GSK 1.16a. Rule 1 (sum of natural series).

$$S(n) = \frac{n+1}{2} \cdot n.$$

Note that the n is here called *icchā* (the requisite).

GSK 1.16b. Missing.

GSK 1.17. Rule 2 (sum of natural series).

$$S(n) = \begin{cases} \frac{n}{2} \cdot (n+1) & \text{if } n \text{ is even} \\ n \cdot \frac{n+1}{2} & \text{if } n \text{ is odd.} \end{cases}$$

Note that the n is here called *diṇa* (S. *dina*) or '⟨the number of⟩ days.'

Cf. PG 14a = Tr 1a.

GSK 1.18. Rule 3 (sum of natural series).

$$S(n) = \frac{(nx + x) \cdot n}{2x}.$$

The superfluous x is called *paṇh-akkhara* (*praśna-akṣara*) or '⟨the number of⟩ the letters in question'.

Exactly the same formula is given in an Āryā verse cited by Abhayadeva in his commentary on *Ṭhāṇaṃga* [SA] 747, where x is called *vāñcha* or 'optional ⟨number⟩.'

> *gaccho vāñchābhyasto vāñchayuto gacchasaṅguṇaḥ kāryaḥ /*
> *dviguṇīkṛtavāñchahṛte vadanti saṅkalitam ācāryāḥ //* (in ASA 747)
> 'The number of terms should be multiplied by an optional ⟨number⟩, increased by the optional, and multiplied by the number of terms. ⟨The result⟩, when devided by twice the optional, the teachers say, is the sum.'

This Sanskrit verse is cited also in PM B6. Moreover, PM B8, too, gives the same formula, in which x is called *śabda* or 'word.'

> *śabdena guṇayed icchāṃ śabdaṃ tu prakṣipet tataḥ/*
> *icchayā tāḍitaṃ kuryād dvighnaśabdena bhājayet//* PM B8 //
> 'One should multiply the requisite by the 'word' and then add the 'word' ⟨to the result⟩. One should make ⟨the sum⟩ multiplied by the requisite and divide ⟨the product⟩ by twice the 'word'.'

GSK 1.19. Ex.: Purely numerical.

Calculate the following:
(1) the sum $S(n)$ when $n = 10, 20, 30, 40, 50, 60, 70, 80, 90, 100$.
(2) the number of terms from the sums obtained.
Answer: (1) $S(10) = (10/2) \times (10 + 1) = 55$, etc. (2) See the next verse.

Note that the word *mūla* (root) is used here and in the next verse in the sense of 'the number of terms.' This seems to be influenced by *pada*, which means the number of terms as well as the square root. Pherū uses *pada* and *gaccha* for the number of terms throughout in the section on *śreḍhī* (GSK 3.26–34).

Cf. PG (1) = Tr (1).

GSK 1.20. Rule 4 (number of terms, two rules).

Rule 1:

$$n = \frac{\sqrt{8S(n)+1} - 1}{2}.$$

Cf. PG 15 = Tr 2.

Rule 2:

$$\begin{aligned} n &= [\text{the integer part of } \sqrt{2S(n)}] \\ &= [\text{the remainder of the integer root}]. \end{aligned}$$

Cf. PG 14b = Tr 1b.

Ex. (given in GSK 1.19 above).
Answer: When $S(n) = 55$, we have: $\sqrt{2 \cdot 55} = \sqrt{110} = \sqrt{10^2 + 10}$, from which we have $n = 10$; etc.

Fundamental operation 2: Difference.

GSK 1.21. Definition.

For two integers, n and m ($n > m$), let $S(n, m)$ be defined as:

$$S(n, m) = S(n) - S(m).$$

Then, either $S(m)$ or $S(n, m)$ is called *vimakaliya* (S. *vyavakalita*, that which is subtracted) and in either case the number of terms left in the summation $S(n)$

decreases one by one from n, which is called the root-quantity (*mūla-rāsī*, S. *mūla-rāśi*), as the *vimakaliya* increases.

GSK 1.22. Rule 1 (difference of natural series).

$$S(n,m) = \frac{(m+1+n)(n-m)}{2},$$

where m = *vimakaliya-paya* and $S(n,m)$ = *vimakaliya-sesa*.

$$\overbrace{\underbrace{1+2+\cdots+(m-1)+m}_{\textit{vimakaliya: } S(m)}+\underbrace{(m+1)+(m+2)+\cdots+(n-1)+n}_{\textit{vimakaliya-sesa: } S(n,m)}}^{\textit{saṃkaliya: } S(n)}.$$

Cf. PG 16 = Tr 3.

GSK 1.23. Ex.: Purely numerical.

Calculate $S(n,m)$ when $n = 1000$ and $m = 10, 20, 30, \ldots$
Answer: $S(1000,10) = (10+1+1000) \times (1000-10)/2 = 50445$, etc.

GSK 1.24. Rule 2 (number of terms of the difference).

$$\begin{aligned} m &= [\text{the integer part of } \sqrt{2\{S(n)-S(n,m)\}}] \\ &= [\text{the remainder of the integer root}]. \end{aligned}$$

See Rule 2 of GSK 1.20 above.

Cf. PG 17 = Tr 4.

GSK 1.25. Rule 3 (difference of natural series).

The second line of the verse seems to be corrupt. It contains two *vimakaliya*: the former is an element of the calculation and the latter the result. Therefore, we have to distinguish them by supplying some words. The most natural way seems to add *paya* to the former and *sesa* to the latter. Moreover, the result of the calculation, being a linear expression of the number of terms, cannot be the sum of the terms. Therefore, we have to read *guṇa* instead of *juya*. Thus emended, the verse means:

$$S(m) = \frac{2n+1-k}{2} \cdot k,$$

where $k = n - m$ = *vimakaliya-paya* and $S(m)$ = *vimakaliya-sesa*. Note that here the *vimakaliya* (that which is subtracted) and the *vimakaliya-sesa* (the remainder of the subtraction) exchange their positions.

$$\overbrace{\underbrace{1+2+\cdots+(m-1)+m}_{\textit{vimakaliya-sesa: } S(m)}+\underbrace{(m+1)+(m+2)+\cdots+(n-1)+n}_{\textit{vimakaliya: } S(n,m)}}^{\textit{saṃkaliya: } S(n)}.$$

The type of this formula, where $S(m)$ is asked for when n and k $(= n - m)$ are given, agrees with the example in the next verse, but it is still wrong because the correct formula is:

$$S(m) = \frac{(n-k)(n-k+1)}{2}.$$

GSK 1.26. Ex.: Purely numerical.

Calculate $S(100) - (98 + 99 + 100)$.

Fundamental operation 3: Multiplication

GSK 1.27. Rule 1 (*kavāḍa-saṃdhi*).

An algorithm for the multiplication by notational places. See Datta & Singh 2001, Part 1, pp. 136–43.

Cf. PG 18–19a = Tr 5–6a.

GSK 1.28. Ex.: Purely numerical.

Calculate the following: 2000×32, 964×27, 108×60.

GSK 1.29. Rule 2 (multiplication 'by parts').

This verse seems to prescribe either of the following two methods or both, although the text is not clear.

(1) The method called *sthāna-vibhāga-khaṇḍa-guṇana* or 'the multiplication by parts after the division ⟨of the multiplier (or the multiplicand)⟩ into notational places'.

(2) The method called *rūpa-vibhāga-khaṇḍa-guṇana* or 'the multiplication by parts after the division ⟨of the multiplier (or the multiplicand)⟩ into integers'.

$$nm = n(m_1 + m_2 + \cdots + m_k) = nm_1 + nm_2 + \cdots + nm_k.$$

Cf. PG 20a = Tr 7a.

GSK 1.30. Verification of multiplication.

Rule 1: Take the sum of the digits in n. If the result has more than one notational place, take again the sum of the digits in it. Repeat the same procedure until the sum of the digits has only one notational place and let the result be $SD(n)$. Obtain $SD(m)$ and $SD(nm)$, too. Then,

$$SD(SD(n) \times SD(m)) = SD(nm).$$

Rule 2: Let the remainders of the divisions of n and m by nine be $R_9(n)$ and $R_9(m)$, respectively. Then,

$$R_9(R_9(n) \times R_9(m)) = R_9(nm).$$

This is the so-called 'verification by nine.'

According to the introductory phrase, these rules are meant for the results obtained by means of 'the second ⟨rule of⟩ multiplication,' which seems to refer to verse 29.

Cf. MS 18.67–70 for Rule 1, and GK 11.11 for Rule 2.

GSK 1.31. Arithmetical operations with zero.

$$0 + n = n, \quad n \pm 0 = n, \quad 0 \times n = 0, \text{ etc.}, \quad n \times 0 = 0.$$

Judging from the example in the next verse, the expression, 'etc.' (*āi*), seems to imply:

$$0 \div n = 0, \quad 0^2 = 0, \quad \sqrt{0} = 0, \quad 0^3 = 0, \quad \sqrt[3]{0} = 0.$$

Note that this rule does not include $0 - a$ and $a \div 0$. The same is true with the PG, the Tr, and the L. The L includes $a \div 0 \times 0$.

Cf. PG 21 = Tr 8.

GSK 1.32. Ex.: Purely numerical.

Calculate the following: $0 \times n$, $0 \div n$, 0^2, $\sqrt{0}$, 0^3, etc.
Answer: 0, 0, 0, 0, 0, etc.

Fundamental operation 4: Division.

GSK 1.33. Rule.

An algorithm for the division by notational places. See Datta & Singh 2001, Part 1, pp. 151–53.

Cf. PG 22 = Tr 9.

Fundamental operation 5: Square.

GSK 1.34. Rule 1.

An algorithm for calculating the square by notational places. See Datta & Singh 2001, Part 1, pp. 156–59.

Cf. PG 23 = Tr 10.

GSK 1.35. Rule 2 (two formulas).

Rule 1:

$$n^2 = n \times n.$$

Rule 2.

$$n^2 = (n - a)(n + a) + a^2.$$

Cf. PG 24 = Tr 11 (which contains another rule: $n^2 = 1+3+5+\cdots+(2n-1)$).

GSK 1.36. Ex.: Purely numerical.

Table 1.2: Extraction of the square root of 186624

Mark odd and	e	o	e	o	e	o	
even places:	1	8	6	6	2	4	Given number
Subtract 4^2 from		**2**	6	6	2	4	Remainder
the first o-place:							
Put the root, 4,		2	6	6	2	4	Remainder
below that o-place:		**4**					Line of root
Double the 4 and move		2	6	6	2	4	Remainder
the result to the next:			**8**				Line of root
Divide the above and put			**2**	6	2	4	Remainder
the quotient in the line:			8	**3**			Line of root
Subtract its square from			**1**	**7**	2	4	Remainder
the above and double it:			8	**6**			Line of root
Move the line of root			1	7	2	4	Remainder
to the next place:				**8**	**6**		Line of root
Divide the above and put						4	Remainder
the quotient in the line:				8	6	**2**	Line of root
Subtract its square from						**0**	Remainder
the above and double it:				8	6	**4**	Line of root
Halve the line of root:						0	Remainder
				4	**3**	**2**	Line of root

Calculate the following: 1^2 to 9^2, 16^2, 24^2, 28^2.
Answer: 1, 4, 9, ... 81, 256, 576, 784.

Cf. PG (4) = Tr (5).

Fundamental operation 6: Square root.

GSK 1.37–38. Rule.

An algorithm for calculating the square root by notational places. We illustrate it by Shukla's example (PG, tr., p. 10). See Table 1.2. Cf. Datta & Singh 2001, Part 1, pp. 170–72.

Cf. PG 25–26 = Tr 12–13.

Fundamental operation 7: Cube.

Table 1.3: Calculation of the cube of 652

Put $6^3 = 216$:	**2**	**1**	**6**								Line of cube
Add $3 \cdot 6^2 \cdot 5 =$		5	4	0							
at the next place:	2	**7**	0	0							Line of cube
Add $3 \cdot 6 \cdot 5^2 =$			4	5	0						
at the next place:	2	7	**4**	5	0						Line of cube
Add $5^3 =$				1	2	5					
at the next place:	2	7	4	**6**	**2**	**5**				($= 65^3$)	Line of cube
Add $3 \cdot 65^2 \cdot 2 =$			2	5	3	5	0				
at the next place:	2	7	**7**	1	6	0	0				Line of cube
Add $3 \cdot 65 \cdot 2^2 =$						7	8	0			
at the next place:	2	7	7	1	6	**7**	**8**	0			Line of cube
Add $2^3 =$									8		
at the next place:	2	7	7	1	6	7	8	0	**8**		Line of cube

GSK 1.39–40. Rule 1.

An algorithm for calculating the cube by notational places. For an illustration of the algorithm, see Table 1.3. Cf. Datta & Singh 2001, Part 1, pp. 163–66. The additive term at each step, 540, etc., seems to have been only memorized and directly added to the line of cube because Pherū does not use the verb *ṭhā* (S. *sthā*, to put down) for them.

Cf. PG 27–28 (1st quarter) = Tr 14–15 (1st quarter).

GSK 1.41. Rule 2 (two formulas).

Rule 1:

$$n^3 = n \times n \times n,$$

where the three n's are aligned vertically.

Rule 2:

$$n^3 = n(n-1) \times 3 + (n-1)^3 + 1.$$

Cf. PG 28 ≈ Tr 15 (except 1st quarter):

niryuktarāśir antyaṃ ⟨*tasya*⟩ *ghano 'sau samatrirāśihatiḥ* /
ekādicaye cāntye tryādihate pūrvaghanayutiḥ saike // PG 28 //
niryuktarāśir antyas tathā ghano 'sau samatrirāśihatiḥ /
khaikādicayenāntye tryādihate vā yutiḥ saike // Tr 15 //

The beginning of the second line of Tr 15 is corrupt. See GSK 4.43, where a similar corruption, *khaïgāi* for *egāi*, occurs.

Table 1.4: Extraction of the cube root of 277167808

Mark cube and	n	n	c	n	n	c	n	n	c	
non-cube places:	2	7	7	1	6	7	8	0	8	Given number
Diminish the first cube place by 6^3:		**6**	**1**	1	6	7	8	0	8	Remainder
Put the root, 6, below		6	1	1	6	7	8	0	8	Remainder
the third from c-place:					**6**					Line of root
Put $3 \cdot 6^2$ behind it:		6	1	1	6	7	8	0	8	Remainder
		1	**0**	**8**	6					Line of root
Divide the above (611)			**7**	**1**	6	7	8	0	8	Remainder
by it (108):		1	0	8	6					Line of root
Put the quotient in line			7	1	6	7	8	0	8	Remainder
and remove the divisor:					6	**5**				Line of root
Subtract $3 \cdot 6 \cdot 5^2$ ⟨from			**2**	**6**	**6**	7	8	0	8	Remainder
the third from c-place⟩:					6	5				Line of root
Subtract 5^3 ⟨from the			**2**	**5**	**4**	**2**	8	0	8	Remainder
above or next c-place⟩:					6	5				Line of root
Put the root obtained			2	5	4	2	8	0	8	Remainder
below the third:							**6**	**5**		Line of root
And so on:										

GSK 1.42. Ex.: Purely numerical.

Calculate the following: 1^3 to 9^3, 16^3, 225^3, 309^3.

Answer: 1, 8, 27, ..., 729, 4096, 11390625, 29503629.

Cf. PG (5) = Tr (6).

Fundamental operation 8: Cube root.

GSK 1.43–45. Rule.

An algorithm for calculating the cube root by notational places. We explain the algorithm by Shukla's example (PG, tr., p. 13). See Table 1.4.

Cf. PG 29–31 = Tr 16–18.

1.3 Eight Fundamental Operations of Fractions

GSK 1.46. Expression of fractions.

Pherū mentions the following three rules.

(1) A mixed number, $n + \frac{b}{a}$, is expressed as $\begin{bmatrix} n \\ b \\ a \end{bmatrix}$.

(2) When the fractional part is negative, $n - \frac{b}{a}$, a dot is put above the numerator: $\begin{bmatrix} n \\ \dot{b} \\ a \end{bmatrix}$.

(3) When a number has no denominator (that is, the number is an integer), 'one' is written down as the denominator: $\begin{bmatrix} n \\ 1 \end{bmatrix}$.

Concerning the arrangement of the numerator and the denominator of a fraction, Gaṇeśa (on L 30) says:

> *ūrdhvam aṃśas tadadho haraḥ sthāpya iti jyotirvidāṃ pāramparyam/*
> 'The numerator should be placed above and the denominator below it. This is the tradition of those who know the luminaries.'

and Siṃhatilaka Sūri (SGT p. 15):

> *rūpaśabdenaikas tadadho dvyādyaṅkena chedāḥ/ nyāsaḥ* $\begin{array}{c|c|c|c|} 1 & 1 & 1 & 1 \\ 2 & 3 & 9 & 18 \end{array}$
> 'With the word *rūpa* ('unity') ⟨used in the preceding sentence, *rūpasya pūrṇasyārdhaṃ* ...⟩, 'one' ⟨is written down⟩ and with the digits, two, etc., its divisors ⟨are written down⟩ below it. Setting-down: $\begin{array}{c|c|c|c|} 1 & 1 & 1 & 1 \\ 2 & 3 & 9 & 18 \end{array}$,

Concerning the three terms that constitute a mixed number, MU 2.2.118a says:

> *rūpam ūrdhvam adhaś cāṃśas tasyādhaḥ cheda iṣyate/* MU 2.2.118a /
> 'It is desired that *rūpa* ('units') is ⟨put⟩ above with the numerator below and the denominator ⟨further⟩ below it.'

Concerning the sign of a negative term (or a subtrahend), Bhāskara II in the auto-commentary on an example of his BG says:

> *tathā yāny ūnagatāni tāny ūrdhvabindūni ca/* BG 5 /
> 'Likewise, those (digits) which are in the state of being a subtrahend have over dots.'

and Siṃhatilaka Sūri (SGT p. 37):

> *teṣāṃ bhāgānāṃ ca viyojyatvopalakṣaṇāya paścāt teṣāṃ śūnyaṃ deyam/*
> *prathamodāharaṇanyāsaḥ* $\begin{array}{r|r|r|} 10 & 1 & 2 \\ \circ\, 1 & \circ\, 2 & \circ\, 2 \\ 4 & 2 & 3 \end{array}$

'In order to indicate that those parts are to be subtracted, zero ⟨symbol⟩ should be put behind them. Setting-down for the first example:

10	1	2
∘ 1	∘ 2	∘ 2
4	2	3

'

For an earlier negative sign see Hayashi 1995a, 88-90.

Cf. PG 32 (2nd quarter) = Tr 19 (2nd quarter), GT 35 for 'the denominator of an integer = 1'.

GSK 1.47a. Reduction to the same colour (*savaṃnaṇa*).

A mixed number is transformed into a fraction before an operation is applied to it.

$$\begin{bmatrix} n \\ b \\ a \end{bmatrix} \to \begin{bmatrix} na+b \\ a \end{bmatrix}, \quad \begin{bmatrix} n \\ \dot{b} \\ a \end{bmatrix} \to \begin{bmatrix} na-b \\ a \end{bmatrix}$$

The result is an improper fraction except when $n = 1$ and the sign is negative.

These two operations are usually included in the *bhāga-anubandha-* and *bhāga-apavāha-jātis* belonging to the so-called *jāti* (class) operations or *kalā-savarṇana* (the homogenization of fractions, i.e., the reduction of composite fractions to simple ones). Pherū treats eight kinds of *jāti* operations in Chapter 2 but what he treats there under these headings are actually *sva-bhāga-anubandha* and *sva-bhāga-apavāha* only.

Cf. PG 39a+40a = a verse cited under Tr (7), Tr 24a (= PG 39a) (in the quotation and in verse 24a, the Tr replaces *rūpagaṇaś* of PG with *rūpaguṇac*, which is wrong).

GSK 1.47b. Reduction to the same denominator (*sadisa-ccheya*).

Two fractions are reduced to the same denominator before the sum or the difference is calculated.

$$\begin{bmatrix} b_1 \\ a_1 \end{bmatrix}\begin{bmatrix} b_2 \\ a_2 \end{bmatrix} \longrightarrow \begin{bmatrix} a_2b_1 \\ a_1a_2 \end{bmatrix}\begin{bmatrix} a_1b_2 \\ a_1a_2 \end{bmatrix}$$

This is a preliminary to the summation and subtraction of fractions, while the 'part class' (GSK 2.1a) is a class of mixed (or composite) fractions of the type $\left(\frac{b_1}{a_1} + \frac{b_2}{a_2}\right)$, which is to be reduced to the simple fraction, $\frac{a_2b_1+a_1b_2}{a_1a_2}$.

Cf. PG 36, Tr 23a.

GSK 1.48–49. Sum and difference of fractions.

For the sum or the difference of two fractions, the sum or the difference, respectively, of their numerators is calculated after they have been reduced to the same denominator.

$$\begin{bmatrix} b_1 \\ a_1 \end{bmatrix} \pm \begin{bmatrix} b_2 \\ a_2 \end{bmatrix} \to \begin{bmatrix} a_2b_1 \\ a_1a_2 \end{bmatrix} \pm \begin{bmatrix} a_1b_2 \\ a_1a_2 \end{bmatrix} \to \begin{bmatrix} a_2b_1 \pm a_1b_2 \\ a_1a_2 \end{bmatrix}$$

In verses 48b–49a Pherū tells the reader not to add up the two denominators because otherwise the denominator of the result is doubled and therefore the 'upper quantity' or the numerator becomes half of the correct value.

$$\begin{bmatrix} a_2b_1 \pm a_1b_2 \\ 2a_1a_2 \end{bmatrix} = \begin{bmatrix} (a_2b_1 \pm a_1b_2)/2 \\ a_1a_2 \end{bmatrix}$$

Fundamental operation 9: Sum of fractions.

GSK 1.50a. Rule.

In this rule, the sum of the numerators is divided by the denominator.

$$\begin{bmatrix} b_1 \\ a_1 \end{bmatrix} + \begin{bmatrix} b_2 \\ a_2 \end{bmatrix} \rightarrow \begin{bmatrix} a_2b_1 \\ a_1a_2 \end{bmatrix} + \begin{bmatrix} a_1b_2 \\ a_1a_2 \end{bmatrix} \rightarrow (a_2b_1 + a_1b_2) \div a_1a_2.$$

Cf. PG 32a = Tr 19a.

GSK 1.50b. Ex.: Purely numerical.

Calculate $\frac{1}{3} + \frac{1}{6} + \frac{1}{5} + \frac{1}{9}$ and $(3\frac{\dot{1}}{4}) + (1\frac{1}{2}) + \frac{1}{3}$.

Here and hereafter $a\frac{\dot{c}}{b}$ means $a - \frac{c}{b}$.

Cf. PG (6) = Tr (7).

Fundamental operation 10: Difference of fractions.

GSK 1.51. Rule.

In this rule, positive and negative mixed numbers are separately reduced to improper fractions (or proper fractions in some cases); the results are separately reduced to the same denominators; the numerators obtained are added together separately; and the difference of the two sums is calculated.

$$\sum_i \begin{bmatrix} n_{1,i} \\ b_{1,i} \\ a_{1,i} \end{bmatrix} - \sum_i \begin{bmatrix} n_{2,i} \\ b_{2,i} \\ a_{2,i} \end{bmatrix}$$

$$\rightarrow \sum_i \begin{bmatrix} a_{1,i}n_{1,i} + b_{1,i} \\ a_{1,i} \end{bmatrix} - \sum_i \begin{bmatrix} a_{2,i}n_{2,i} + b_{2,i} \\ a_{2,i} \end{bmatrix}$$

$$\rightarrow \begin{bmatrix} b_1 \\ a_1 \end{bmatrix} - \begin{bmatrix} b_2 \\ a_2 \end{bmatrix} \rightarrow \begin{bmatrix} a_2b_1 - a_1b_2 \\ a_1a_2 \end{bmatrix}$$

Cf. PG 32b = Tr 19b (Tr reads *ādyānya* instead of *āyavyaya* of PG).

GSK 1.52. Ex.: Purely numerical.

Calculate (1) $8 - \frac{1}{2} - \frac{1}{3} - \frac{1}{6} - \frac{1}{9}$; or $8 - \frac{1}{2}$, $8 - \frac{1}{3}$, $8 - \frac{1}{6}$, and $8 - \frac{1}{9}$, and (2) $\frac{1}{6} - (3\frac{1}{2}) - (5\frac{1}{3}) - \frac{1}{9}$(?); or $(5\frac{1}{3}) - (3\frac{1}{2})$ and $\frac{1}{6} - \frac{1}{9}$.

Cf. PG (8) = Tr (8).

Fundamental operation 11: Multiplication of fractions.

GSK 1.53. Rule.

$$\begin{bmatrix} b_1 \\ a_1 \end{bmatrix} \times \begin{bmatrix} b_2 \\ a_2 \end{bmatrix} \rightarrow b_1 b_2 \div a_1 a_2.$$

Cf. PG 33a = Tr 20a.

GSK 1.54. Ex.: *damma*.

Calculate $(5\frac{1}{4})$ *dammas* $\times$ $(8\frac{1}{3})$ *dammas*(?) and $\frac{1}{2} \times \frac{1}{6}$.

It is strange that both the multiplier and the multiplicand of the first problem have the denomination of currency, *damma*. As we mentioned in the footnote to the translation, the problems intended by Pherū were probably these: $(5\frac{1}{4})$ *dammas* $\times$ $(8\frac{1}{3})$ and $\frac{1}{2}$ *damma* $\times \frac{1}{6}$.

Cf. PG (10), Tr (9).

Fundamental operation 12: Division of fractions.

GSK 1.55. Rule.

$$\begin{bmatrix} b_1 \\ a_1 \end{bmatrix} \div \begin{bmatrix} b_2 \\ a_2 \end{bmatrix} \rightarrow \begin{bmatrix} b_1 \\ a_1 \end{bmatrix} \times \begin{bmatrix} a_2 \\ b_2 \end{bmatrix}$$

Cf. PG 33b = Tr 20b.

GSK 1.56. Ex.: *dramma*.

Calculate $(7\frac{1}{4}) \div (1\frac{1}{2})$ *damma* and $(6\frac{1}{4}) \div (4\frac{1}{3})$.

It is strange that both the multiplicand and the multiplier of the first problem have the instrumental plural forms.

Cf. PG (11) = Tr (10) (Tr adds *tathā* before *tribhiḥ* in the first line and reads *bhāgāptam* instead of *pṛthagāptam* of PG at the end of the verse).

Fundamental operation 13: Square of a fraction.

GSK 1.57. Rule.

$$\begin{bmatrix} b \\ a \end{bmatrix}^2 \rightarrow b^2 \div a^2.$$

Cf. PG 34a = Tr 21a.

GSK 1.58. Ex.: Purely numerical.

Calculate $(1\frac{1}{2})^2$, $(5\frac{1}{3})^2$, $(7\frac{1}{4})^2$, $(\frac{1}{2})^2$, and $(\frac{1}{3})^2$, or $(\frac{1}{2}\cdot\frac{1}{3})^2$ instead of the last two.

Cf. PG (12) = Tr (11).

Fundamental operation 14: Square root of a fraction.

GSK 1.59. Rule.

$$\begin{bmatrix} b \\ a \end{bmatrix}^{\frac{1}{2}} \rightarrow \sqrt{b} \div \sqrt{a}.$$

In this rule Pherū mentions the algorithm for the square root of an integer (GSK 1.37–38) after the division, but it must be applied before it.

Cf. PG 34b = Tr 21b.

Fundamental operation 15: Cube of a fraction.

GSK 1.60. Rule.

$$\begin{bmatrix} b \\ a \end{bmatrix}^{3} \rightarrow b^3 \div a^3.$$

Cf. PG 35a = Tr 22a (Tr reads *phalaṃ* instead of *padaṃ* of PG: the former is better).

GSK 1.61. Ex.: Purely numerical.

Calculate: $(7\frac{1}{2})^3$, $(15\frac{1}{4})^3$, $(\frac{1}{4})^3$, and $(\frac{1}{3})^3$.

Cf. PG (13) = Tr (12).

Fundamental operation 16: Cube root of a fraction.

GSK 1.62. Rule.

$$\begin{bmatrix} b \\ a \end{bmatrix}^{\frac{1}{3}} \rightarrow \sqrt[3]{b} \div \sqrt[3]{a}.$$

In this rule, as in GSK 1.59, Pherū mentions the algorithm for the cubic root of an integer (GSK 1.43–45) after the division, but it must be applied before it.

Cf. PG 35b = Tr 22b (Tr reads *chedanamūloddhṛte mūlam* instead of *ghanamūlaṃ chedamūlahṛte* of PG).

1.4 Proportion

Fundamental operation 17: Three-quantity operation.

GSK 1.63. Rule.

The three quantities given in a problem were presumably aligned horizontally by the author of our text also, although the text does not specify the direction,

$$[a, b, c],$$

whose first and last terms should be of the same denomination. In addition, the two categories to which a and b belong ought to have a proportional relationship such as the quantity-price relationship of a commodity, distance-time relationship of a runner, etc., although the text does not refer to this point. Then, the middle term is multiplied by the last and divided by the first. The quantity obtained is to c just as b is to a.

$$(b \times c) \div a.$$

DP 60 = GSK 1.63. Cf. PG 43 = Tr 29, MS 15.24–25.
Cf. Sarma 2002.

GSK 1.64. Ex. 1: Price–length *trairāśika* for a piece of cloth (*kappaḍa*, S. *karpara*).

[11 *dammas*, 7 *karas*, 24 *dammas*] = $15\frac{3}{11}$ *hatthas* = 15 *hatthas* 6 *aṃgulas* $4\frac{1}{11}$ *javas*.

GSK 1.65. Ex. 2: Number–price *trairāśika* for coins (*muṃda*, S. *mudrā*).

[9, 25 *dammas*, 16] = $44\frac{4}{9}$ *dammas* = 44 *dammas* 8 *visovas* 17 *vissaṃsas* 15 *paḍivissaṃsas* 11 *kāiṇis* $2\frac{2}{9}$ *paḍikāiṇis*.

In DP 49–50 Pherū gives a rule intended specifically for the exchange of coins:

> *bhaṇisu hava nāṇavaṭṭaṃ dammittihi jāma ittiyaṃ muddaṃ/*
> *iya aggha-pamāṇeṇaṃ ittiya muṃdāṇa kaï mullaṃ//* DP 49 //
> *rāsiṃ tihāi guṇiyaṃ majjhima hariūṇa bhāu jaṃ laddhaṃ/*
> *taṃ tāṇa muṃda mullaṃ na saṃsayaṃ bhaṇaï pheru tti//* DP 50 //
> 'I will tell now the exchange of coins. By these many of *dammas* (b), so many coins (a) ⟨of a certain kind are obtained⟩. By this price standard, what is the price (x) of so many coins (c) ⟨of the same kind⟩?// Multiply the ⟨given⟩ quantity (c) ⟨of coins⟩ by the first ⟨term⟩ (b) and divide ⟨the product⟩ by the middle ⟨term⟩ (a). What is obtained is the price of those many of coins. There is no doubt ⟨in this calculation⟩. Pherū says so.'

That is to say,

$$x = (b \times c) \div a.$$

GSK 1.66. Ex. 3: Weight–price *trairāśika* for sandalwood (*caṃdaṇa*, S. *candana*).

[$1\frac{1}{4}$ *palas*, $9\frac{1}{3}$ *dammas*, $6\frac{1}{6}$ *palas*] = $43\frac{5}{9}$ *dammas* = 43 *dammas* 11 *visovas* 2 *vissaṃsas* 4 *paḍivissaṃsas* 8 *kāiṇis* $17\frac{7}{9}$ *paḍikāiṇis*.

Cf. PG (25) = Tr (30).

GSK 1.67. Ex. 4: Price–weight *trairāśika* for long pepper (*pippali*).

[$7\frac{1}{4}$ *dammas*, $2\frac{1}{6}$ *seras*, $9\frac{1}{3}$ *dammas*] = $2\frac{154}{261}$ *seras* = 2 *seras* 3 *palas* 5 *ṭaṃkas* 1 *māsayas* $3\frac{171}{261}$ *guṃjas*.

Cf. PG (26) = Tr (31) (for *marica* or pepper).

GSK 1.68. Ex. 5: Price–volume *trairāśika* for rice (*taṃdula*, S. *taṇḍula*), etc.

[$120\frac{1}{4}$ *dammas*, $5\frac{1}{3}$ *patthas*, 1 *dammas*] = $\frac{64}{1437}$ *patthas* = 2 *māṇas* 3 *pāīs* $1\frac{871}{1437}$ *karapuḍas*.

Cf. PG (27) = Tr (33) (for *dhānya* or grain), PG (28) ≈ Tr (32) (differ in the 3rd terms, for *dhānya*).

GSK 1.69. Ex. 6: Weight–price *trairāśika* for gold pieces of purity 12 (*bārahavannī kaṇaya*, S. *dvādaśa-varṇa-kanaka*).

[1 *tola*, $100\frac{1}{3}$ *dammas*, $1\frac{1}{10}$ *māsayas*] = $7\frac{21}{40}$ *dammas* = 7 *dammas* 1 *visovā* 1 *vissaṃsā*.

Purity 12 indicates pure gold. In the DP Pherū says that one *tola* of pure gold costs 478 *dammas*.

Cf. PG (29) = Tr (34).

GSK 1.70. Ex. 7: Distance–time *trairāśika* for a lame man (*paṃgula*).

[$\frac{1}{6}$ *joyaṇa*, 7 days, 60 *joyaṇas* (S. *yojanas*)] = 2520 days = 7 years.

Cf. PG (30) ≈ Tr (35) (differ in the 2nd term).

GSK 1.71. Ex. 8: Distance–time *trairāśika* for a worm (*kīḍaya*, S. *kīṭaka*).

[$\frac{1}{7}$ *aṃgula*, $\frac{1}{6}$ day, 8×2 *joyaṇas*] = 14336000 days = 39822 years + 2 months + 20 days.

The worm has to go the distance of eight *joyaṇas* and come back along the same way.

Cf. PG (31) = Tr (36).

Fundamental operations 18–21: Five-, seven-, nine-, and eleven-quantity operations.

GSK 1.72. Rule for $(2n+1)$-quantity operation.

When n pairs of quantities plus another quantity ($a_1, \ldots, a_n; b_1, \ldots, b_n$ and p) concerning one substance are given, they are arranged vertically in two columns (see below), and then 'the fruit number' (*phal-aṃka*, S. *phala-aṅka*), p, which is put at the bottom (*hiṭṭhima*) of one of the two columns, is transposed to the opposite side, and finally the product of the terms of the longer side is divided by the product of those of the shorter side (*thov-aṃka-rāsi*, S. *sthoka-aṅka-rāśi*, lit. '⟨the product of⟩ the quantities of ⟨the side which contains⟩ the smaller number of digits') .

$$\begin{bmatrix} a_1 & b_1 \\ a_2 & b_2 \\ a_3 & b_3 \\ \vdots & \vdots \\ a_n & b_n \\ p & \end{bmatrix} \rightarrow \begin{bmatrix} a_1 & b_1 \\ a_2 & b_2 \\ a_3 & b_3 \\ \vdots & \vdots \\ a_n & b_n \\ & p \end{bmatrix} \rightarrow \frac{b_1 \cdot b_2 \cdots b_n \cdot p}{a_1 \cdot a_2 \cdots a_n}.$$

The $(2n+1)$-quantity operation can be regarded as a compound computation that consists of n three-quantity operations. That is, when the $(2n+1)$ quantities satisfy the condition:

$$\begin{aligned} a_1 : p &= b_1 : x_1, \\ a_2 : x_1 &= b_2 : x_2, \\ a_3 : x_2 &= b_3 : x_3, \\ \vdots \quad & \quad \vdots \quad \vdots \\ a_n : x_{n-1} &= b_n : x, \end{aligned}$$

an algorithm devised for obtaining x is 'the $(2n+1)$-quantity operation'.

Brahmagupta, Śrīdhara, etc. add special proviso for fractions (transposition of the denominators) but Pherū does not. See BSS 12.11b–12, PG 45 = Tr 31.

GSK 1.73. Ex. 1 for *pañca-rāśika*: Interest (*phala*), time (*kalā*) and capital (*mūla* or *pamāṇa-dhaṇa*, S. *pramāṇa-dhana*).

time	1	1	month, year
capital	100	60	
interest	5	·	

Answer: $x = 36$.

time	1	·	month, year
capital	100	60	
interest	5	36	

Answer: $x = 1$ year.

time	1	1	month, year
capital	100	·	
interest	5	36	

Answer: $x = 60$.

Cf. PG (39) = Tr (44) (numerically the same as the GSK). Cf. also L 83.

GSK 1.74. Ex. 2 for *pañca-rāśika*: Interest (*vavahāra*, S. *vyavahāra*).

time	$1\frac{1}{3}$	$9\frac{1}{4}$	months
capital	$100\frac{1}{2}$	$17\frac{1}{4}$	*dammas*
interest	$1\frac{1}{2}$	·	*dammas*

Answer: $x = 3\frac{15}{32}$ *dammas* = 3 *dammas* 9 *visovas* 7 *vissaṃsas* 10 *paḍi-vissaṃsas*.

Cf. PG (40) ≈ Tr (45) (differ in one term; GSK is closer to Tr).

GSK 1.75. Ex. 3 for *pañca-rāśika*: Wages for a porter (*bhāḍaya*, S. *bhāṭaka*).

weight	$8\frac{1}{2}$	$9\frac{1}{4}$	*maṇas*
distance	$1\frac{1}{3}$	$10\frac{1}{4}$	*joyaṇas*
wages	$2\frac{1}{4}$	·	*dammas*

Answer: $x = 17\frac{1969}{2176}$ *dammas* =17 *dammas* 18 *visovas* 1 *vissaṃsas* 18 *paḍi-vissaṃsas* 19 *kāiṇis* $8\frac{4}{17}$ *paḍikāiṇis*.

Cf. PG (43) = Tr (49).

GSK 1.76. Ex. 4 for *pañca-rāśika*: Wages for workers (*kammayara*, S. *karma-kara*).

workers	12	8	people
time	4	45	days
wages	30	·	*dammas*

Answer: $x = 225$ *dammas*.

Cf. PG (44) = Tr (50).

GSK 1.77. Ex. 5 for *pañca-rāśika*: Price of a piece of gold (*suvanna*, S. *suvarṇa*).

purity	12	$10\frac{1}{2}$	*vannīs*
weight	$3m\dot{1}g$	$2m1g$	$m =$ *māsas*, $g =$ *guṃja*
price	$20\frac{1}{4}$	·	

Answer: $x = 13\frac{351}{1632}$.

The denomination of the price is not given. If it is *damma*, one *tola* of pure gold in this problem costs $83\frac{11}{17}$ *dammas* as the following *trairāśika* shows.

$$\left[3 \text{ } m\bar{a}sas \text{ } \dot{1} \text{ } gu\dot{m}ja, 20\frac{\dot{1}}{4}, 1 \text{ } tola\right] = 83\frac{11}{17}.$$

See GSK 1.69 above, where one *tola* of pure gold costs $100\frac{1}{3}$ *dammas*.

Cf. PG (41) = Tr (46), Tr (47), PG (42) = Tr (48).

GSK 1.78. Ex. for *sapta-rāśika*: Price of blankets (*kaṃvala*, S. *kambala*).

length	6	9	⟨*karas*⟩
width	3	5	*karas*
number	2	7	
price	90	·	*dammas*

Answer: $x = 787\frac{1}{2}$ *dammas* = 787 *dammas* 10 *visovās*.

Cf. PG (45) = Tr (51).

GSK 1.79. Ex. for *nava-rāśika*: Price of garments (*cīra*).

number	12	9	
colours	5	4	
length	7	8	*karas*
width	3	5	*hatthas*
price	600	·	*dammas*

Answer: $x = 685\frac{5}{7}$ *dammas.*

For *nava-rāśika*, Śrīdhara gives two examples, one for the price of rectangular stones in PG (46) = Tr (52) and the other for the feed of elephants in PG (47) = Tr (53).

GSK 1.80. Ex. for *ekādaśa-rāśika*: Price of small pulses (*mumga*, S. *mudga*).

volume	2	9	*patthas*
volume	6	3	*seīs*
volume	3	2	*māṇas*
volume	2	1	*pāīs*
volume	1	3	*karapuḍas*
price	60	·	*dammas*

Answer: $x = 135$ *dammas.*

It is impossible to imagine a real situation where this solution is valid. In fact, the product of the five volumes has fifteen dimensions and does not make sense in this real world. The correct answer is, of course, obtained by means of the three-quantity operation: [2489 *karapuḍas*, 60 *dammas*, 9447 *karapuḍas*] = $227\frac{1817}{2489}$ *dammas.*

Cf. L 87 for an example of the rule of eleven. Śrīdhara gives no examples for it.

Fundamental operation 22: Inverse three-quantity operation.

GSK 1.81. Rule.

The three quantities given in a problem are aligned just as in the case of the three-quantity operation,

$$[a, b, c].$$

Here also, the first and the last terms should be of the same denomination as mentioned in the text, but the two categories to which a and b belong ought to have an inversely proportional relationship like the purity-weight relationship of gold alloys that contain the same amount of gold. Then, the middle term is multiplied by the first and divided by the last.

$$(b \times a) \div c.$$

Note that the multiplier (the third term, c) and the divisor (the first term, a) of the middle term (b) in the case of the three-quantity operation are here operated inversely as the divisor and the multiplier, respectively.

Cf. PG 44b = Tr 30a.

GSK 1.82. Ex. 1: *seī-pattha* relation for the volume of a certain thing.

[10 *seīs*, 27 *patthas*, 16 *seīs*] = $16\frac{7}{8}$ *patthas* = 16 *patthas* 14 *seīs.*

The second conversion ratio, 16 *seīs* = 1 *pattha*, is the one that has been adopted in our text (GSK 1.8), but the first one, 10 *seīs* = 1 *pattha*, has not been found anywhere else.

Cf. PG (34) = Tr (38) = Tr (40) (ex. for *hāra* or necklaces).

GSK 1.83. Ex. 2: *ṭaṃka–sera* relation for the weight of grain in a container (*vakkhara*).

[66 *ṭaṃkas*, 20 *maṇas*, 72 *ṭaṃkas*] = $18\frac{1}{3}$ *maṇas* = 18 *maṇas* 13 *seras* 2 *palas*.

Both conversion ratios used here, 66 *ṭaṃkas* = 1 *sera* and 72 *ṭaṃkas* = 1 *sera*, are not included in the weight systems defined in GSK 1.9–10a, where 60 *ṭaṃkas* = 1 *sera*, although the latter one, 72 *ṭaṃkas* = 1 *sera*, actually occurs in L 8^a (see under GSK 1.9–10a above).

Cf. PG (35) ≈ Tr (41) (differ in the 3rd term; ex. for *suvarṇa* or gold).

GSK 1.84. Ex. 3: Purity–weight relation of gold alloys that contain a certain amount of gold.

[$11\frac{1}{2}$ *vannīs*, $40\frac{1}{2}$ *tolas*, $10\frac{1}{4}$ *vannīs*] = $45\frac{18}{41}$ *tolas* = 45 *tolas* 1 *ṭaṃka* 1 *māsa* $4\frac{12}{41}$ *javas*.

Cf. PG (38) = Tr (42).

GSK 1.85. Ex. 4: Size–number relation of blankets made from a certain amount of raw material.

[9 × 3, 220, 5 × 2] = 594.

The linear measure is not specified but it must be *hattha* (S. *hasta*). Some other writers treat this as a case of the *vyasta-pañca-rāśika* (inverse five-quantity operation).

Cf. PG (37) = Tr (43).

Fundamental operation 23: Purchase and sale.

A merchant buys a certain amount of a certain commodity and sells the same for profit. Let us use the following notation.

$$\text{buying rate } (\textit{kaya}) = k = \frac{c_k}{d_k} = \frac{\text{quantity } (\textit{vatthu})}{\text{price } (\textit{damma})},$$

$$\text{selling rate } (\textit{vikaya}) = v = \frac{c_v}{d_v} = \frac{\text{quantity } (\textit{vatthu})}{\text{price } (\textit{damma})},$$

$$\text{capital } (\textit{mūla}) = m, \quad \text{profit } (\textit{lāha}) = \ell,$$

$$\text{total product or gain } (\textit{savva\^{}uppattī}) = M.$$

Sanskrit terms: *kraya*, *vikraya*, *vastu* (⟨the number, or quantity, of⟩ a thing), *dramma*, *lābha*, *sarva-utpatti*.

Concerning these quantities, we have two basic relationships:

$$\ell + m = M, \qquad km = vM.$$

Āryabhaṭa II gives various formulas obtained from these two relationships.

$$m = \frac{v\ell}{k-v}, \qquad m = \frac{\ell}{c_k d_v - c_v d_k} \cdot c_v d_k \quad \text{(MS 15.28b–29)},$$

$$m = \frac{vM}{k}, \qquad \ell = \frac{(k-v)M}{k} \quad \text{(MS 15.34)},$$

$$k = \frac{vM}{m}, \qquad v = \frac{mk}{M}. \quad \text{(MS 15.35)},$$

$$m = \frac{c_v d_k M}{c_k d_v}, \qquad \ell = \frac{(c_k d_v - c_v d_k)M}{c_k d_v} \quad \text{(MS 15.44–45a)}.$$

The *Bakhshālī Manuscript* has five formulas (algorithms) for the buying and selling problems. None of those algorithms is the same as Āryabhaṭa II's, although they are of course equivalent. They are (BM 54–57; see Hayashi 1995a, 386)

$$m = \frac{\ell}{k/v - 1}, \quad \ell = m\left(\frac{k}{v} - 1\right), \quad m = \frac{M}{k/v}, \quad \ell = M - m, \quad m = \frac{-\ell}{1 - k/v}.$$

An anonymous Sanskrit mathematical anthology (Hayashi 2006, 202–204) contains five formulas; four of them are the same as the second and the third pairs of Āryabhaṭa II and the remaining one is obtained by substituting m/v for M/k in the latter of the second pair. GSS 6.167 gives a formula obtained by dividing the numerator and the denominator of the latter of the first pair by $c_v d_k$.

Pherū devotes four verses to the topic of purchase and sale (*kraya-vikraya*). He gives three formulas (one of them is implicit) in the first verse and one in the last verse, and two examples in between them. Two of those formulas are the same as Āryabhaṭa II's first pair.

GSK 1.86. Rule 1: Capital and total product.

When the buying rate (k), the selling rate (v), and the profit (ℓ) are known, write them down in that order,

$$k \quad v \quad l.$$

Multiply the middle term by the last, and the last term by the first, and divide the two products by the difference, $k - v$, separately. Then, the capital and the total gain are obtained:

$$m = \frac{v\ell}{k-v}, \qquad M = \frac{k\ell}{k-v}.$$

When m, instead of ℓ, is known, the formula intended by Pherū is presumably this:

$$\ell = \frac{(k-v)m}{v}.$$

GSK 1.87. Ex. 1: No commodity mentioned.

Given: $k = 17/1$ *maṇas/ṭaṃka*, $v = 15/1$ *maṇas/ṭaṃka*, $\ell = 10$ *ṭaṃkas*.
Answer: $m = 75$ *ṭaṃkas*.

GSK 1.88. Ex. 2: A certain thing (*vatthu*, S. *vastu*).

Given: $k = 5/3$ pieces/*damma*, $v = 7/9$ pieces/*damma*, $\ell = 12$ *dammas*.
Answer: $m = 10\frac{1}{2}$ *dammas* = 10 *dammas* 10 *visovās*.

GSK 1.89. Rule 2: Capital.

Pherū asks the reader first to write down the rate prices above and the rate quantities below and then to perform two cross-multiplications to obtain two products, $c_k d_v$ and $c_v d_k$:

$$\begin{bmatrix} d_k & d_v \\ c_k & c_v \end{bmatrix} \quad \rightarrow \quad [c_v d_k, c_k d_v]$$

The formula given is this:

$$m = \frac{c_v d_k \ell}{c_k d_v - c_v d_k}.$$

Cf. MS 15.28b–29 (see above).

Fundamental operation 24: Barter.

A certain amount (p) of a certain commodity (*bhaṃḍa*, S. *bhāṇḍa*) is exchanged for a certain amount (x) of a counter-commodity (*paḍibhaṃḍa*, S. *pratibhāṇḍa*) on the assumption that both have the same commercial value. Let their buying (or selling) rates be c_1/d_1 and c_2/d_2.

GSK 1.90. Rule.

As in the case of the *pañca-rāśika*, etc., the given terms are arranged in two vertical columns (this is not explicitly stated) and the 'prices' and the 'fruits' are transposed to the opposite sides. Then, the product of the longer side is divided by that of the shorter side. Here, the 'fruits' are the quantities to be exchanged.

commodity	I	II
quantity	c_1	c_2
price	d_1	d_2
fruit	p	x

$\rightarrow$

I	II
c_1	c_2
d_2	d_1
x	p

$$\rightarrow \quad x = \frac{c_2 d_1 p}{c_1 d_2}.$$

Cf. PG 46a = Tr 32a.

GSK 1.91. Ex.: long pepper (*pippali*) and ginger (*suṃṭhī*, S. *śuṇṭhī*).

commodity	pepper	ginger	
quantity	2	5	*maṇas*
price	200	300	*dammas*
fruit	100		*maṇas*

Answer: $x = 166\frac{2}{3}$ *maṇas* = 166 *maṇas* 26 *seras* 48 *ṭaṃkas*.

Cf. PG (48) = Tr (54) (Tr reads *paṇaiḥ* for *palam* of PG; the latter is better). Cf. also PG (49) = Tr (55) (for *āmra* or mango and *kapittha* or wood-apple).

Fundamental operation 25: Selling of living beings.

The commercial value of a group of the same kind of living beings is inversely proportional to their age and proportional to their number.

GSK 1.92. Rule.

As in the case of the *pañca-rāśika*, etc., the given terms are arranged in two vertical columns (this is not explicitly stated) and the 'ages' and the 'fruits' are transposed to the mutually opposite sides. The rest of the rule is the same as the previous one, that is, the product of the longer side is divided by that of the shorter side. Here, the 'fruits' are the prices and the price of a living thing is regarded as inversely proportional to the age.

$$\begin{array}{r|cc|} \text{age} & y_1 & y_2 \\ \text{number} & n_1 & n_2 \\ \text{fruit} & p & x \end{array} \quad \rightarrow \quad \begin{array}{|cc|} y_2 & y_1 \\ n_1 & n_2 \\ x & p \end{array} \quad \rightarrow \quad x = \frac{y_1 n_2 p}{y_2 n_1}.$$

Cf. PG 46b = Tr 32b (Tr reads *vikrayaḥ* for *vikraye* of PG; the latter is better).

GSK 1.93. Ex.: Camels (*karaha*, S. *karabha*).

age	10	9	years
number	3	5	
fruit	108	·	*ṭaṃkas*

Answer: $x = 200$ *ṭaṃkas*.

Cf. PG (50) = Tr (56) (for *nārī* or women, which word occurs in PG but not in Tr), PG (51) (for camels, numerically the same as GSK), GT 117 (for camels, numerically the same as GSK except for one term).

It is remarkable that, while Mahāvīra (GSS 5.40), Śrīdhara (PG (50) = Tr (56)), Śrīpati (GT 116), Bhāskara II (L 79), Nārāyaṇa (GK I, p. 53), and Gaṇeśa (GM 99) give examples of sale of human beings, Pherū refrains from doing so.

Chapter 2

Eight Classes of Reduction of Fractions

Class 1: Part class.

GSK 2.1a. Rule.

$$\begin{bmatrix} b_1 & b_2 \\ a_1 & a_2 \end{bmatrix}^{bh} \rightarrow \begin{bmatrix} a_2b_1 & a_1b_2 \\ a_1a_2 & a_1a_2 \end{bmatrix} \rightarrow \begin{bmatrix} a_2b_1 + a_1b_2 \\ a_1a_2 \end{bmatrix}$$

Cf. GSK 1.48. Cf. PG 36 ≈ Tr 23a (the PG first removes the common factor from the denominators).

PG 37 gives another rule:

$$\begin{bmatrix} b_1 \\ a_1 \\ b_2 \\ a_2 \end{bmatrix}^{bh} \rightarrow \begin{bmatrix} a_2b_1 \\ a_1 \\ b_2 \\ a_1a_2 \end{bmatrix} \rightarrow \begin{bmatrix} a_2b_1 + a_1b_2 \\ a_1a_2 \end{bmatrix}$$

GSK 2.1b. = GSK 2.2b, which is an example for the rule of GSK 2.2a.

Class 2: Multi-part class.

GSK 2.2a. Rule.

$$\begin{bmatrix} b_1 & b_2 \\ a_1 & a_2 \end{bmatrix}^{pb} \rightarrow \begin{bmatrix} b_1 \times b_2 \\ a_1 \times a_2 \end{bmatrix}$$

Cf. GSK 1.53 (multiplication of fractions). Cf. PG 38a = Tr 23b.

GSK 2.2b. Ex.: Purely numerical.

Three interpretations are possible: (1) $\left[\frac{1}{2}, \frac{1}{2}\right]^{pb}$, $\left[\frac{1}{2}, \frac{1}{5}\right]^{pb}$, and $\left[\frac{1}{2}, \frac{1}{6}\right]^{pb}$, or (2) $\left[\frac{1}{2}, \frac{1}{2}\right]^{pb}$ and $\left[\frac{1}{2}, \frac{1}{5}, \frac{1}{6}\right]^{pb}$, or (3) $\left[\frac{1}{2}, \frac{1}{2}, \frac{1}{5}, \frac{1}{6}\right]^{pb}$.

Answer: (1) $\frac{1}{4}$, $\frac{1}{10}$, and $\frac{1}{12}$. (2) $\frac{1}{4}$ and $\frac{1}{60}$. (3) $\frac{1}{120}$.

Cf. PG (15), Tr (14), (15).

Class 3: Part-part class.

GSK 2.3. Rule.

The first line of the verse prescribes:

$$\begin{bmatrix} 1 \\ b \\ a \end{bmatrix}^{bb} \rightarrow \begin{bmatrix} 1 \times a \\ b \end{bmatrix} \left\langle \rightarrow \begin{bmatrix} a \\ b \end{bmatrix} \right\rangle$$

which means:

$$1 \div \frac{b}{a} = \frac{1 \times a}{b} \left\langle = \frac{a}{b} \right\rangle.$$

This is the rule for 'the part-part class' according to Śrīdhara (PG 38b = Tr 25a) and Āryabhaṭa II (MS 15.19a), but in the second line of the verse Pherū adds further operations:

$$p_i = \frac{\frac{a_i}{b_i} \cdot M}{\sum_{j=1}^{n} (a_j/b_j)},$$

where a_i/b_i is obtained by means of the above rule. M and p_i are simply called *dhana* (the value ⟨of the total sum⟩) and *phala* (the fruit or result), respectively. This is the so-called 'investments' or the proportionate distribution. Cf. GSK 3.7 below.

This verse is cited and paraphrased in Sanskrit prose in PM A33 as follows.

> *chedena rūpe guṇite cchedagame kṛte saty aṃśānāṃ yutiṃ vidhāya tayā svayutyā pūrvaṃ pṛthag aṃśān dhanena guṇayitvā bhāgo hāryaḥ/ eṣa bhāgabhāgavidhiḥ//* prose part of PM A33 //
>
> When unity is multiplied by the denominator and the elimination of the denominator is made, having made the sum of the parts, one should first multiply each part by the value ⟨of the total sum⟩ and then divide ⟨the product⟩ by that sum ⟨of the parts⟩. This is the rule for the part-part ⟨class⟩.

GSK 2.4. Ex. 1: Proportionate distribution of 100 *dammas* into three parts

Divide the sum (M) of 100 *dammas* into three in the ratio, (2 parts among 3 parts) : (7 parts among 9 parts) : (3 parts among 6 parts).

$$\begin{bmatrix} 1 & 1 & 1 \\ 3 & 9 & 6 \\ 2 & 7 & 3 \end{bmatrix}^{bb} \rightarrow \begin{bmatrix} 2 & 7 & 3 \\ 3 & 9 & 6 \end{bmatrix}, \quad \frac{2}{3} + \frac{7}{9} + \frac{3}{6} = \frac{35}{18}.$$

Since $M = 100$ *dammas*, each part is,

$$p_1 = \left(\frac{2}{3} \cdot 100\right) / \frac{35}{18} = 34\frac{2}{7} \ \textit{dammas},$$

$$p_2 = \left(\frac{7}{8} \cdot 100\right) / \frac{35}{18} = 40 \ \textit{dammas},$$

$$p_3 = \left(\frac{3}{6} \cdot 100\right) / \frac{35}{18} = 25\frac{5}{7} \ \textit{dammas}.$$

In Pherū's terminology, 'igi *b*-bhāya-*a*-bhāya' seems to mean '*a* parts among *b* parts of unity,' that is, a/b. Śrīdhara expresses it as '*a*-chedaka-*b*-aṃśa-rāśi-saṃbhakta-rūpa' or 'unity divided by the quantity which has the denominator *a* and the numerator *b*'. See PG (17) = Tr (19).

This verse is cited and paraphrased in Sanskrit prose in PM A34 and the problem is solved also in Sanskrit prose.

ādyasya puruṣasya bhāgatrayāṇāṃ bhāgadvayam/ anyasya navabhāgānāṃ saptabhāgaḥ/ anyasya ṣaḍbhāgānāṃ bhāgatrayam/ drammaśate rājñā prasāde datte trayo janāḥ pṛthak pṛthag dhanaṃ kiṃ prāpnuvanti// nyāsaḥ/

1	*1*	*1*	*100*
3	*9*	*6*	*1*
2	*7*	*3*	

chedena rūpe guṇite chedagame ca jātaṃ

2	*7*	*3*
3	*9*	*6*

chedābhyām anyonyaṃ kṛte jātaṃ chedasādṛśyam

12	*14*	*9*
18	*18*	*18*

aṃśayutau 35/ aṃśānāṃ pṛthak pṛthak phalena guṇanārthaṃ nyāsaḥ/ 12/ 14/ 19/ śatena dvādaśake guṇite/ 1200/ aṃśayutyā/ 35/ bhāge labdham

34
10
35

ayam ādyasya bhāgaḥ/ śatena caturdaśake guṇite/ 1400/ aṃśayutyā/ 35/ bhāge labdham 40/ ayaṃ dvitīyasya bhāgaḥ/ śatena navake guṇite/ 900/ aṃśayutyā/ 35/ bhāge labdham

25
25
35

ayaṃ tṛtīyasya// prose part of PM A34 //

'Two parts among three parts are for the first person; seven parts among nine parts are for another ⟨person⟩; three parts among six parts are for ⟨yet⟩ another ⟨person⟩. ⟨These are the shares of the three persons.⟩ When one hundred *drammas* are given ⟨to them⟩ as a favour by a king, how much does each of the three persons obtain? Setting-down:

1	1	1	100
3	9	6	1
2	7	3	

When unity is multiplied by the denominator and the elimination of the denominator ⟨is made⟩, what is produced is

2	7	3
3	9	6

When mutual ⟨multiplication of the denominator and the numerator of a fraction⟩ by the ⟨other⟩ two denominators is made, the sameness of the denominators is produced.

12	14	9
18	18	18

When the sum of the numerators is taken: 35. Setting-down for multiplying each numerator by the fruit: 12, 14, 9.

When twelve is multiplied by one hundred: 1200. When this is divided by the sum of the numerators, what is obtained is

34
10
35

This is the share of the first ⟨person⟩. When fourteen is multiplied by one hundred: 1400. When this is divided by the sum of the numerators, what is obtained is 40. This is the share of the second. When nine is multiplied by one hundred: 900. When this is divided by the sum of the numerators, what is obtained is

25
25
35

This is of the third.'

GSK 2.5. Ex. 2: A cistern filled by four pipes.

$$\begin{bmatrix} 1 & 1 & 1 & 1 \\ 1 & 1 & 1 & 1 \\ 1 & 2 & 3 & 4 \end{bmatrix}^{bb} \rightarrow \begin{bmatrix} 1 & 2 & 3 & 4 \\ 1 & 1 & 1 & 1 \end{bmatrix} \text{(cisterns/day)}.$$

$$\frac{1}{1}+\frac{2}{1}+\frac{3}{1}+\frac{4}{1} = 10 \text{ (cisterns/day)}.$$

Therefore, the time required is:

$$t = \frac{1}{10} \text{ day.}$$

Since $M = 6$ cubic *karas*, the water supplied by each pipe is:

$$p_1 = \left(\frac{1}{1}\cdot 6\right)/10 = \frac{3}{5} \text{ cubic } kara, \quad p_2 = \left(\frac{2}{1}\cdot 6\right)/10 = \frac{6}{5} \text{ cubic } karas,$$

$$p_3 = \left(\frac{3}{1}\cdot 6\right)/10 = \frac{9}{5} \text{ cubic } karas, \quad p_4 = \left(\frac{4}{1}\cdot 6\right)/10 = \frac{12}{5} \text{ cubic } karas.$$

This verse is cited and the problem is solved in Sanskrit prose in PM A40.

nyāsaḥ

1	*1*	*1*	*1*
1	*1*	*1*	*1*
1	*2*	*3*	*4*

rūpe cchedena guṇite cchedagame ca

1	*2*	*3*	*4*
1	*1*	*1*	*1*

chedasādṛśyād yuktau/ 10/ dinasya daśāṃśena pūrayanti/ pūrvavat kṛte labdhaṃ yathākramam

0	*1*	*1*	*2*
6	*2*	*8*	*4*
10	*10*	*10*	*10*

ime pratinālakaṃ hastā aṃśāś ca/ eṣā sarvāpi bhāgabhāgajātiḥ//
prose part of PM A40 //

'Setting-down:

1	1	1	1
1	1	1	1
1	2	3	4

When unity is multiplied by the denominator and the elimination of the denominator ⟨is made⟩,

1	2	3	4
1	1	1	1

Since the denominators are the same, the sum is

taken: 10. ⟨Therefore, the four pipes⟩ fill ⟨the cistern⟩ in one-tenth of a day. When ⟨the calculation is⟩ made as before, what is obtained is, in the same order,

0	1	1	2
6	2	8	4
10	10	10	10

These are the ⟨cubic⟩ *hastas* and ⟨its⟩ part for each pipe. All this is the part-part class.'

Cf. PM A35 and A37. Cf. also PG 69, (16), (91), Tr (18) (= PG (16)), GSS 8.32b–34, MS 15.43, L 96–97, GK I, p. 94.

Class 4: Part-addition class.

GSK 2.6. Rule.

$$\begin{bmatrix} b_1 \\ a_1 \\ b_2 \\ a_2 \end{bmatrix}^{ba} \rightarrow \begin{bmatrix} b_1(a_2+b_2) \\ a_1a_2 \end{bmatrix}$$

This is called *sva-bhāga-anubandha-jāti* or 'one's own part addition class' by the anonymous commentator of the PG. In the example that follows, we express this as $\frac{b_1}{a_1}\left(1+\frac{b_2}{a_2}\right)$ for convenience.

Cf. PG 39b = Tr 24b.

GSK 2.7. Ex.: Purely numerical.

Calculate (1) $3\frac{1}{2}\left(1+\frac{1}{4}\right)\left(1+\frac{1}{6}\right)\left(1+\frac{1}{2}\right)$ and (2) $\frac{1}{2}\left(1+\frac{1}{3}\right)\left(1+\frac{1}{4}\right)$.
Answer: (1) $7\frac{21}{32}$, (2) $\frac{5}{6}$.

Cf. PG (19) = Tr (17).

Class 5: Part-subtraction class.

GSK 2.8. Rule.

$$\begin{bmatrix} b_1 \\ a_1 \\ -b_2 \\ a_2 \end{bmatrix}^{ba} \rightarrow \begin{bmatrix} b_1(a_2-b_2) \\ a_1a_2 \end{bmatrix}$$

This is called *sva-bhāga-apavāha-jāti* or 'one's own part subtraction class' by the anonymous commentator of the PG. In the following example, we express this as $\frac{b_1}{a_1}\left(1-\frac{b_2}{a_2}\right)$ for convenience.

Cf. PG 40b (the Tr does not have this rule).

GSK 2.9. Ex.: Purely numerical.

Calculate (1) $3\left(1-\frac{1}{2}\right)\left(1-\frac{1}{4}\right)\left(1-\frac{1}{6}\right)$ and (2) $\frac{1}{2}\left(1-\frac{1}{3}\right)\left(1-\frac{1}{4}\right)$.
Answer: (1) $\frac{15}{16}$, (2) $\frac{1}{4}$.

Cf. PG (21).

Class 6: Part-mother class.

GSK 2.10. Rule.

This is a combination of the five preceding 'classes'. Pherū tells the reader first to perform the operation of each 'class' and then to add up the results after reducing them to the same denominator.

$$\left[\begin{matrix} b_1 \\ a_1 \end{matrix} \begin{bmatrix} b_2 & b_3 \\ a_2 & a_3 \end{bmatrix}^{pb} \begin{bmatrix} n \\ b_4 \\ a_4 \end{bmatrix}^{bb} \begin{bmatrix} b_5 \\ a_5 \\ b_6 \\ a_6 \end{bmatrix}^{ba} \begin{bmatrix} b_7 \\ a_7 \\ -b_8 \\ a_8 \end{bmatrix}^{ba}\right]^{bh}$$

Cf. PG 42 ≐ Tr 25b–26a.

GSK 2.11. Ex.: Purely numerical.

Calculate the following:

$$\left[\begin{matrix} 1 \\ 2 \end{matrix} \begin{bmatrix} 1 & 1 \\ 4 & 4 \end{bmatrix}^{pb} \begin{bmatrix} 1 \\ 1 \\ 3 \end{bmatrix}^{bb} \begin{bmatrix} 1 \\ 2 \\ 1 \\ 2 \end{bmatrix}^{ba} \begin{bmatrix} 1 \\ 3 \\ -1 \\ 2 \end{bmatrix}^{ba}\right]^{bh}$$

Answer:

$$\rightarrow \begin{bmatrix} 1 & 1 & 3 & 3 & 1 \\ 2 & 16 & 1 & 4 & 6 \end{bmatrix}^{bh} \rightarrow \begin{bmatrix} 215 \\ 48 \end{bmatrix} \rightarrow \begin{bmatrix} 4 \\ 23 \\ 48 \end{bmatrix}$$

Cf. PG (23) = Tr (20) (numerically the same as GSK 2.11).

Class 7: Chain-reduction class.

GSK 2.12. Rule.

When the $(i+1)$-th unit is one p_{i+1}-th part of the i-th unit, a quantity expressed in the two units can be converted to the upper unit by the following operation:

$$\begin{bmatrix} a_i \\ p_i \\ \pm a_{i+1} \\ p_{i+1} \end{bmatrix}^{vs} \rightarrow \begin{bmatrix} a_i p_{i+1} \pm a_{i+1} \\ p_i p_{i+1} \end{bmatrix}$$

This operation can be applied to any part of a long 'chain' of measures.

Cf. PG 41 = Tr 26b–27a. The commentator of the PG (p. 35) remarks that this is a combination of the part-addition (*bhāga-anubandha*) and the part-subtraction (*bhāga-apavāha*) classes.

GSK 2.13. Ex.: Weight expressed in four units.

Reduce the following chain of measures to *tola*.

$$\begin{bmatrix} 2\left(1-\frac{1}{7}\right) & \textit{tolas} \\ 3\left(1-\frac{1}{7}\right) & \textit{māsayas} \\ 4\left(1-\frac{1}{7}\right) & \textit{guṃjas} \\ 5\left(1-\frac{1}{7}\right) & \textit{visuvās} \end{bmatrix}$$

Answer:

$$\begin{bmatrix} \frac{12}{7} \\ \frac{18}{7} \\ 12 \\ \frac{24}{7} \\ 6 \\ \frac{30}{7} \\ 20 \end{bmatrix}^{vs} \to \begin{bmatrix} \frac{12}{7} \\ \frac{18}{7} \\ 12 \\ \frac{510}{7} \\ 120 \end{bmatrix}^{vs} \to \begin{bmatrix} \frac{12}{7} \\ \frac{2670}{7} \\ 1440 \end{bmatrix}^{vs} \to \begin{bmatrix} \frac{19950}{7} \\ 1440 \end{bmatrix} \to 1\frac{987}{1008} \ \textit{tolas}.$$

The first column in the answer is obtained by applying the part-subtraction class to each term of the original column. For the conversion ratios, 12 and 6, see GSK 1.9–10a above. Naturally, the conversion ratio for the *visuvā* (= *visovā*, 'one-twentieth') is 20.

Doing the part-subtraction after the chain-reduction will make the computation easier.

$$\begin{bmatrix} 2 \\ 3 \\ 12 \\ 4 \\ 6 \\ 5 \\ 20 \end{bmatrix}^{vs} \to \begin{bmatrix} 2 \\ 3 \\ 12 \\ 85 \\ 120 \end{bmatrix}^{vs} \to \begin{bmatrix} 2 \\ 445 \\ 1440 \end{bmatrix}^{vs} \to \begin{bmatrix} 3325 \\ 1440 \end{bmatrix} \to \frac{665}{288}.$$

$$\begin{bmatrix} 665 \\ 288 \\ -1 \\ 7 \end{bmatrix}^{ba} \to \begin{bmatrix} 3990 \\ 2016 \end{bmatrix} \to 1\frac{987}{1008} \ \textit{tolas}.$$

Cf. PG (22) = Tr (22).

Class 8: Pillar-part class.

GSK 2.14. Rule.

When an unknown quantity x is decreased by its own parts $(b_i/a_i)x$ and the remainder p is seen (*paccakkha*, S. *pratyakṣa*), the present rule prescribes

$$x = \frac{p}{1 - \sum \frac{b_i}{a_i}},$$

where $\sum \frac{b_i}{a_i}$ is obtained through the reduction to the same denominator and the summation of the numerators.

Cf. PG 74a, Tr (27b), GSS 4.4, GT 64, MS 15.20, L 53.

The PG treats this topic in the *miśraka-vyavahāra*. The GSS (4.4) includes this in the *prakīrṇaka-vyavahāra*, calling it *bhāga-jāti*. The L treats it under *iṣṭakarman*. The GT (64) calls it *dṛśya-jāti* and places it between the *vallī-savarṇana-jāti* and the *śeṣa-jāti*. The GT does not make a clear distinction between the two categories, the mixed fractions and the algebraic equations, calling both *jāti*, while the SS of the same author Śrīpati includes only the former category in which the *dṛśya-jāti* does not occur. Śrīpati, therefore, seems to have put the *dṛśya-jāti* in the latter category. For the two categories called *jāti* see §2.2.16 of Part I.

GSK 2.15. Ex. 1: Pillar (*thaṃbha*, S. *stambha*).

Calculate the height of a pillar when $p = 3$ *kaṃviyas*, and $b_i/a_i = 1/2$, $1/6$, $1/12$.
Answer: 12 *kaṃviyas*.

For the linear measure *kaṃviya*, see GSK 1.4b above.

Cf. PG (96), Tr (23), (24), (25).

GSK 2.16. Ex. 2: Herd of cows (*thaṇa-vāla*, S. *stana-pāla*).

Calculate the number (x) of cows in a herd when $p = 141$ and $b_i/a_i = 1/5$, $1/8$, $1/16$, $1/4$, $1/6$.
Answer: 720.

The Nahatas' text reads 145 (*saü paṇayālu*) instead of 141 but the number 145 will bring us the fractional solution, $145 \times \frac{240}{47}$, which is impossible because we are calculating the number of cows. We propose to read *igayālu* (forty-one) instead of *paṇayālu* (forty-five). With this emendation, which does not affect the meter of the verse, we have the integer solution, $141 \times \frac{240}{47} = 720$. Cf. 'AMg. *igayāla*' in Pischel, §445.

Cf. Tr (26) (for a pearl-necklace), (28) (for a flock of bees), (29) (purely numerical).

GSK 2.17. Ex. 3: Troop (*jūha*, S. *yūtha*) of probably elephants.

Calculate the number (x) of elephants in a troop when

$$x - \frac{x}{2}\left(1+\frac{1}{3}\right) - \frac{x}{6}\left(1+\frac{1}{7}\right) - \frac{x}{8}\left(1+\frac{1}{9}\right) = 4.$$

Answer: $x = 1008$.

Cf. Tr (27) (for elephants; numerically the same).

Chapter 3

Eight Types of Procedures

3.1 Procedure for Mixture

GSK 3.1. Separation of capital and interest.

$$\begin{array}{r|cc|} \text{time} & t & T \\ \text{capital} & c & C \\ \text{interest} & f & F \end{array} \qquad M = C + F.$$

We have the relationship:

$$\frac{f}{tc} = \frac{F}{TC}.$$

When M is given instead of C and F, the latter are calculated by the computational rules:

$$C = \frac{tcM}{tc + fT}, \qquad F = \frac{fTM}{tc + fT},$$

which are obained from the two relationships, $M = C + F$ and $C : F = tc : fT$, by means of the proportionate distribution. Cf. GSK 3.7 below.

Cf. PG 47 = Tr 33.

GSK 3.2. Ex.

$$\begin{array}{|cc|} 1 & 4 \\ 100 & \cdot \\ 5 & \cdot \end{array} \quad M = 520 \ \textit{dammas}. \qquad \text{Answer: } C = 433\tfrac{1}{3},\ F = 86\tfrac{2}{3}\ \textit{dammas}.$$

Cf. PG (52) = Tr (57), PG (53) = Tr (58).

GSK 3.3. Separation of capital, interest, commission of surety (*bhāvyaka*), etc.

$$\begin{array}{r|cc|} \text{time} & t & T \\ \text{capital} & c & C \\ \text{fruit} & f_i & F_i \end{array} \qquad F = \sum F_i, \qquad M = C + F,$$

where F_i are the interest (*phala*), the commission of the surety, the fee (*vitti*, S. *vṛtti*) of the calculator, the charges of the scribe (*lehaga*, S. *lekhaka*), etc. We have the relationship:

$$\frac{f_i}{tc} = \frac{F_i}{TC},$$

from which the following two computational rules are obtained:

$$C = \frac{tc}{tc + \sum(f_i T)} \cdot M, \quad F_i = \frac{f_i T}{tc + \sum(f_i T)} \cdot M.$$

Cf. PG 48 = Tr 34.

GSK 3.4. Ex.

1	12
100	·
f_i	·

$f_1 = 5, \quad f_2 = 1, \quad f_3 = \frac{1}{2}, \quad f_4 = \frac{1}{4}, \quad M = 905.$

Answer: $C = 500$, $F_1 = 300$, $F_2 = 60$, $F_3 = 30$, $F_4 = 15$.

Cf. PG (54) = Tr (59), GT 119.

GSK 3.5. Conversion of several bonds into one (*eka-patrī-karaṇa*).

time	t	T_i
capital	c	C_i
interest	f_i	F_i

We have the relationship:

$$\frac{f_i}{tc} = \frac{F_i}{T_i C_i}.$$

Let T be the average elapsed time to be assumed for the single bond. Then, the sum of the interests can be expressed in two ways:

$$\sum F_i = \sum \frac{f_i T_i C_i}{tc} = \sum \frac{f_i T C_i}{tc},$$

from which the following computational rule is obtained:

$$T = \left(\sum \frac{f_i T_i C_i}{tc}\right) \div \left(\sum \frac{f_i C_i}{tc}\right).$$

Let F be the average interest to be assumed for the single bond. Then, the sum of the interests for the unit of time can be expressed in two ways:

$$\sum \frac{F_i}{T_i} = \sum \frac{f_i C_i}{tc} = \sum \frac{F C_i}{tc},$$

from which the following computational rule is obtained:

$$F = \left(\sum \frac{f_i C_i}{tc}\right) \times tc \div \sum C_i.$$

This formula utilizes the same $\sum \frac{f_i C_i}{tc}$ as employed in the computation of T above. Pherū does not cancel tc. Instead, assuming that $t = 1$ and $c = 100$, he rewrites this as

$$F = \left(\sum \frac{f_i C_i}{tc}\right) \times 100 \div \sum C_i.$$

Cf. PG 51 = Tr 35.

GSK 3.6. Ex.

Calculate the average time elapsed and the average interest for the single bond when $t = 1$, $c = 100$, $f_i = 2, 3, 4, 5$, $T_i = 4, 6, 10, 8$, $C_i = 100, 200, 300, 600$, respectively.

Answer: $T = 8\frac{2}{25}$ months, $F = 4\frac{1}{6}$.

Cf. PG (57)–(58) = Tr (60)–(61) (numerically the same as the GSK except for $C_4 = 400$), PG (59) = Tr (62).

GSK 3.7. Proportionate distribution of the gain according to the investments (*prakṣepaka*).

Let a_i be the investments, M the total gain, and p_i the share of each. Then,

$$p_i = \frac{M a_i}{\sum_{j=1}^{n} a_j}.$$

If a_i are fractions, Pherū remarks, the sum is taken only after they are reduced to the same denominator. For various applications of this rule see GSK 2.3, 3.1, 3.11, 3.24, and 4.2.

Cf. PG 59a = Tr 38a.

GSK 3.8. Ex. 1: Seeds (*bīya*, S. *bīja*)

Calculate each share (*bhinna-phala*) of the harvest (*halahari dinnāṇa*, S. *haladhareṇa dattānāṃ*), $M = 210$ *maṇas*, when the investments of seeds are $a_i = 2, 3, 5, 4$ *maṇas*.

Answer: $p_1 = 30$, $p_2 = 45$, $p_3 = 75$, $p_4 = 60$ *maṇas*.

Cf. PG (71) = Tr (67) (numerically the same as GSK 3.8), PG (72).

GSK 3.9. Ex. 2: Silk (*paṭṭa*).

Problem (tentative interpretation): A cloth of silk is woven (*vuṇa* < $\sqrt{ve}$, to weave) on the capital(?),

8 *ṭaṃkas* + 6 *dammas* (= 406 *dammas*) invested by a man (?) for *vuṇijjahi/vuṇiya* and

10 *ṭaṃkas* + 5 *dammas* (= 505 *dammas*) invested by another man (?) for *mulla/vuṇāvaviya*,

and 86 *ṭaṃkas* are obtained by selling the cloth. What is the share of each? That is to say, $a_1 = 406$, $a_2 = 505$ *dammas*, and $M = 86$ *ṭaṃkas*.

Answer: $p_1 = \frac{86}{406+505} \cdot 406 = 38\frac{298}{911}$ *ṭaṃkas*. $p_2 = \frac{86}{406+505} \cdot 505 = 47\frac{613}{911}$ *ṭaṃkas*.

Cf. Tr (68)–(69) (for interests).

GSK 3.10. Ex. 3: sandalwood (*caṃdaṇa*, S. *candana*).

Problem (tentative interpretation): There is a plank of sandalwood shaped like a trapezium (*kaṃcolu?*) of a uniform thickness, whose height, top width, and bottom width are seven, seven, and one *aṃgulas*, respectively. It costs ten *dammas*. Calculate the price of each of the seven small pieces cut out of the sandalwood with six straight lines one *aṃgula* distant from each other and parallel to the top and the bottom.

Solution: The seven small pieces also have trapezium-like shapes. The top width of the i-th piece from the bottom, which is equal to the bottom width of the $(i+1)$-th piece, can be calculated by

$$d_i = \frac{6}{7} \cdot i + 1 \quad \textit{aṃgulas} \quad \text{(cf. GK kṣetra 74a)},$$

and the surface area of the i-th piece by

$$A_i = \frac{d_{i-1} + d_i}{2} \quad \text{sq. } \textit{aṃgulas},$$

since the height of every trapezium is one *aṃgula*.

Now, the price of a piece of sandalwood must be proportional to its volume, which is in turn proportional to its surface area (A_i) because the thickness is the same everywhere. The price of a piece is therefore proportional to its surface area ($p_i \propto A_i$).

$$A_i = \frac{10}{7}, \frac{16}{7}, \frac{22}{7}, \frac{28}{7}, \frac{34}{7}, \frac{40}{7}, \frac{46}{7} \quad \text{sq. } \textit{aṃgulas}.$$

Hence we may assume:

$$a_i = 5,\ 8,\ 11,\ 14,\ 17,\ 20,\ 23, \quad \text{and } M = 10 \ \textit{dammas}.$$

Answer: $p_i = \frac{25}{49}, \frac{40}{49}, \frac{55}{49}, \frac{70}{49}, \frac{85}{49}, \frac{100}{49}, \frac{115}{49}$ *dammas*.

We can consider a similar problem with a frustum-like sandalwood, but the computation of the volume of each piece, which is also a frustum, seems to be too much for an example of the proportionate distribution.

GSK 3.11. Purchase in proportion (*sama-viṣama-kraya*, lit. 'purchase of equal or different ⟨quantities⟩').

One buys n kinds of commodities, whose buying rates are c_i/d_i (quantity/price), in the ratio (*aṃsa*, S. *aṃśa*, lit. 'parts'), $a_1 : a_2 : \ldots : a_n$, for a total money (*davva*, S. *dravya*) M. Then, the price of each kind is calculated by:

$$p_i = \frac{\left(\frac{d_i}{c_i} \cdot a_i\right)}{\sum_{j=1}^{n} \left(\frac{d_j}{c_j} \cdot a_j\right)} \cdot M,$$

and the quantity of each kind by means of the three-quantity operation:

$$q_i = [d_i, c_i, p_i].$$

Cf. PG 59b = Tr 38b, MS 15.37–38.

GSK 3.12. Ex. 1: Yellow myrobalan (*haraḍaï*, S. *harītakī*), Terminalia bellerica (*baheḍa*, S. *vibhītikā*, a kind of myrobalan), and emblic myrobalan (*āmalaya*, S. *āmalaka*).

Calculate the price and the quantity of each kind when $c_i/d_i = 1/1$, $3/1$, $6/1$ *seras/damma*, $a_1 : a_2 : a_3 = 1 : 1 : 1$, and $M = 1$ *damma*.
Answer: $p_1 = \frac{3}{2}$, $p_2 = \frac{1}{2}$, $p_3 = \frac{1}{4}$ *damma*; $q_1 = q_2 = q_3 = \frac{3}{2}$ *seras*.

GSK 3.13. Ex. 2: Long pepper (*pippali*), pepper (*miriya*, S. *marica*), and dry ginger (*suṃṭhī*, S. *śuṇṭhī*).

Calculate the price and the quantity of each kind when $c_i/d_i = \frac{1}{2}/3$, $1/9\frac{1}{3}$, $1\frac{1}{4}/4$ *seras/damma*, $a_1 : a_2 : a_3 = 1 : 1 : 1$, and $M = 1$ *damma*.
Answer: $p_1 = \frac{9}{31}$, $p_2 = \frac{14}{31}$, $p_3 = \frac{8}{31}$ *damma*; $q_1 = q_2 = q_3 = \frac{3}{62}$ *sera*.

Cf. PG (75) = Tr (70).

GSK 3.14. Ex. 3: Rice (*taṃdula*, S. *taṇḍula*), *mūṃg* pulses (*muṃga*, S. *mudga*), and clarified butter (*ghia*, S. *ghṛta*).

Calculate the price and the quantity of each kind when $c_i/d_i = 9/1$, $11/1$, $1/1$ *seras/damma*, $a_1 : a_2 : a_3 = 3 : 2 : 1$, and $M = 1\frac{1}{4}$ *dammas*.
Answer: $p_1 = \frac{11}{40}$, $p_2 = \frac{3}{20}$, $p_3 = \frac{33}{40}$ *damma*; $q_1 = 2\frac{19}{40}$, $q_2 = 1\frac{13}{20}$, $q_3 = \frac{33}{40}$ *sera*.

Cf. PG (73)–(74) = Tr (71)–(72).

GSK 3.15. Procedures for gold (*suvarṇa-vyavahāra*).

There are n gold pieces, whose weight and purity are w_i and v_i, respectively. They are smelted into a single alloy, whose weight and purity are W and V, respectively. Since the total weight of pure gold is preserved in the mixture, we have the relationship:

$$\sum v_i w_i = VW,$$

from which the two computational rules given in the present verse are obtained:

$$V = \frac{\sum v_i w_i}{W}, \qquad W = \frac{\sum v_i w_i}{V}.$$

Obviously, in the case of a simple mixture of gold pieces, we have the relationship:

$$W = \sum w_i.$$

Cf. PG 52b, Tr 36.

GSK 3.16. Ex. 1: Purity of the gold alloy.

Calculate the purity of the gold alloy made from four gold pieces when $v_i = 9$, 10, 8, 11 *vannīs*, and $w_i = 3$, 6, 5, 2 *tolas*, respectively.
Answer: $V = 9$ *vannīs* 5 *javas*.

Cf. PG (61) = Tr (63).

GSK 3.17. Ex. 2: Purity of the gold alloy.

Calculate the purity of the gold alloy made from three gold pieces when $v_i = 8,\ 9\frac{1}{4},\ 6\frac{1}{4}$ *vannīs*, and $w_i = 3\frac{1}{3},\ 5\frac{1}{6},\ 2\frac{1}{5}$ *māsayas*, respectively.
Answer: $V = 8$ *vannīs* $2\frac{82}{321}$ *javas*.

Cf. PG (62).

GSK 3.18. Refined gold (*pakva-suvarṇa*).

The same two rules as above hold true in the case of the refinement of gold pieces:

$$V = \frac{\sum v_i w_i}{W}, \qquad W = \frac{\sum v_i w_i}{V}.$$

In this case, however, the total weight of the gold pieces is not preserved because dross is removed in the process of refinement.

$$W < \sum w_i.$$

Cf. PG 53 = Tr 37.

GSK 3.19. Ex. 1: Purity of the gold alloy obtained by refinement.

Calculate the purity of the gold alloy made from four gold pieces by refinement when $v_i = 9,\ 7,\ 10,\ 8$ *vannīs*, $w_i = 6,\ 5,\ 8,\ 7$ *tolas* respectively, and $W = 20$ *tolas*.
Answer: $V = 9$ *vannīs*.

Cf. PG (63) = Tr (65).

GSK 3.20. Ex. 2: Weight of the gold alloy obtained by refinement.

Calculate the weight of the gold alloy made from four gold pieces by refinement when $v_i = 7,\ 8,\ 9,\ 6$ *vannīs*, $w_i = 4,\ 5,\ 5,\ 7$ *tolas* respectively, and $V = 11$ *vannīs*.
Answer: $W = 14$ *tolas* $1\frac{1}{11}$ *māsayas*.

Cf. PG (64), Tr (66).

GSK 3.21. Unknown (*naṣṭa*, lit. 'lost') purity of a component gold piece in a simple mixture.

When the purity (v_j) of a gold piece smelted with other gold pieces into a single alloy is unknown,

$$v_j = \frac{V \sum w_i - \sum_{i \neq j} v_i w_i}{w_j},$$

which is obtained from the basic relationship mentioned above.

Cf. PG 54.

GSK 3.22. Ex.: Purity of a component piece.

Calculate the unknown purity (v_4) of a gold piece when $v_i = 9, 8, 10$ *vannīs*, $w_i = 3, 5, 7, 8$ *māsayas* respectively, and $V = 10$ *vannīs*.
Answer: $v_4 = 11$ *vannīs* 10 *javas*.

Cf. PG (65).

GSK 3.23. Unknown weight of a component piece in a simple mixture.

When the weight (w_j) of a gold piece smelted with other pieces into a single alloy is unknown,

$$w_j = \frac{V \sum_{i \neq j} w_i - \sum_{i \neq j} v_i w_i}{v_j - V},$$

which too is obtained from the basic relationship mentioned above.

Cf. PG 55.

GSK 3.24. Unknown weights of all component pieces in a simple mixture.

When the weights (w_i) of all n component gold pieces for a simple mixture are unknown,

$$w_i = \frac{a_i W}{\sum_{j=1}^{n} a_j}, \quad \text{where} \quad a_i = \frac{1}{|V - v_i|}.$$

For the proportionate distribution see GSK 3.7 above.

Mahāvīra (GSS 6.182–183) gives two rules for the case of a simple mixture of only two gold alloys. Let $v_1 < V < v_2$. Then,

$$w_1 = \frac{\frac{1}{V-v_1}}{\frac{1}{V-v_1} + \frac{1}{v_2-V}} \cdot W, \quad w_2 = \frac{\frac{1}{v_2-V}}{\frac{1}{V-v_1} + \frac{1}{v_2-V}} \cdot W.$$

$$w_1 = \frac{v_2 - V}{(v_2 - V) + (V - v_1)} \cdot W, \quad w_2 = \frac{V - v_1}{(v_2 - V) + (V - v_1)} \cdot W.$$

Mahāvīra explicitly states that these rules are obtained by means of the rule of proportionate distribution (*prakṣepataḥ*) and remarks: 'In this manner many more ⟨weights⟩ can be determined' (*evaṃ bahuśo 'pi vā sādhyam*).

Mahāvīra (GSS 6.185) also gives a rule for the same purpose in the case of the refinement of gold pieces. Assume any optional number (a_i, which have to satisfy the condition, $\sum a_i < W$, although it is not explicitly stated) for every w_i except one, say w_j. Then,

$$w_j = \frac{VW - \sum_{i \neq j} v_i a_i}{v_j}.$$

For an example see GSS 6.186, where $v_i = 5, 6, 7, 8, 11, 13$, $V = 9$, and $W = 60$. The English translator, Raṅgācārya, of the GSS missed the word *pakva* or 'refined' used in the statement of this example and gave a wrong solution: $w_i = 20, 4, 4, 4, 4, 24$, the sum of which is 60 ($= W$). See p. 315 of his translation accompanying his edition of the GSS.

GSK 3.25. Ex.: Weights of all component gold pieces in a simple mixture.

Calculate the weights (w_i) of gold pieces when $v_i = 5, 7, 9, 11$ *vannīs*, $V = 10$ *vannīs*, and $W = 1$ *tola*.

Answer: $a_1 = \frac{1}{5}$, $a_2 = \frac{1}{3}$, $a_3 = a_4 = 1$, $\sum a_i = \frac{38}{15}$. $w_1 = \frac{3}{38}$, $w_2 = \frac{5}{38}$, $w_3 = w_4 = \frac{15}{38}$.

Cf. GSS 6.184.

3.2 Procedure for Series

We use the following notation.

$$A = A(n) = \sum_{k=1}^{n} a_k, \quad \text{where} \quad a_k = a + (k-1)d.$$

$$S(n) = \sum_{k=1}^{n} k, \quad S^2(n) = \sum_{k=1}^{n} k^2, \quad S^3(n) = \sum_{k=1}^{n} k^3.$$

The sum and the difference of the natural series, $S(n)$ and $S(n,m)$, have already been treated in GSK 1.16–26 above.

GSK 3.26. Arithmetical progression.

$$a_n = a + (n-1)d, \qquad \bar{m}_n = \frac{a + a_n}{2}, \qquad A(n) = \bar{m}_n \cdot n.$$

Cf. Tr 39.

GSK 3.27. Ex.: Yellow myrobalan (*haraḍaī*, S. *harītakī*).

Calculate the amount of yellow myrobalam for a horse (*turiya*, S. *turaga*) when $a = 20$, $d = 5$, $n = 7$ days; also calculate a, d, and n, in order, from the rest.

Answer: $a_7 = 50$, $\bar{m}_7 = 35$, $A(7) = 245$.

Another Skt. word that might be proposed for *turiya* is *tvarita* (see Pischel, §152), but Tr (73) supports *turaga*.

Cf. Tr (73) (the same problem as the first half of GSK 3.27), Tr (74) (donation of money).

GSK 3.28. First term of an arithmetical progression.

$$a = \frac{A}{n} - \frac{(n-1)d}{2}.$$

Cf. PG 86a = Tr 40a.

GSK 3.29. Common difference of an arithmetical progression.

$$d = \frac{A/n - a}{(n-1)/2}.$$

Cf. 86b = Tr 40b.

GSK 3.30. Number of terms of an arithmetical progression.

$$n = \frac{\sqrt{8dA + (2a-d)^2} - 2a + d}{2d}.$$

Note: The first term a occurs twice in this formula. It is called 'the first' (*āī*, S. *ādi*) in the first place but simply 'the value' (*dhaṇa*, S. *dhana*) in the second.

Cf. PG 87 = Tr 41.

GSK 3.31. Sum of the sums of the natural series.

$$\sum_{k=1}^{n} S(k) = \frac{S(n)\cdot(n+2)}{3}.$$

PG 103b.

GSK 3.32. Ex.: Purely numerical.

Calculate $\sum_{k=1}^{n} S(k)$, $S^2(n)$, and $S^3(n)$ when $n=5$.
Answer: $\sum_{k=1}^{5} S(k) = 35$, $S^2(5) = 55$, $S^3(5) = 225$.

Rules for the last two problems are given in the next verse.

Cf. PG (116), (117).

GSK 3.33. Sums of the square series and of the cubic series.

$$S^2(n) = \frac{2n+1}{3}\cdot S(n), \qquad S^3(n) = \{S(n)\}^2.$$

Cf. PG 102b–103a.

GSK 3.34. Sum of the natural series and the square and the cube of the number of terms.

$$S(n) + n^2 + n^3 = (2n+1)\cdot\frac{n(n+1)}{2}.$$

Cf. PG 102a.

3.3 Procedure for Plane Figures

GSK 3.35a. Areas of a square and an oblong.

When a and b denote the width (*vikkhaṃbha*, S. *viṣkambha*) and the length (*āyāma*) respectively of a square (*caüraṃsa*, S. *caturasra*), in which case $a = b$, or of an oblong (*dīha-caüraṃsa*, S. *dīrgha-*), the area (*khetta*, S. *kṣetra*) is:

$$A = ab.$$

Cf. Tr 42a.

GSK 3.35b. Ex.

Calculate the area of a square when $a = b = 6$ *karas*, and that of an oblong when $a = 3$ and $b = 5$ *karas*.

Answer: $A = 36$ sq. *karas*. $A = 15$ sq. *karas*.

Cf. PG (122), (123) (both given as exs. of 'equi-perpendicular' figures under PG 115), Tr (75), (76).

GSK 3.36. Areas of a trilateral and a quadrilateral 1.

When a, b, c, and d denote the four (or three) sides of a quadrilateral (*caü-bbhuya*, S. *catur-bhuja*) (or of a trilateral, *ti-bhuya*, S. *tri-bhuja*, where $d = 0$), the area is:

$$A = \sqrt{(s-a)(s-b)(s-c)(s-d)}, \quad \text{where} \quad s = \frac{a+b+c+d}{2}.$$

The same formula is given also in BSS 12.21, PG 117, Tr 43a, GSS 7.50, MS 15.69–70, SS 13.28, L 169, and GK kṣetra 33–34 (part 2, pp. 39–40). According to the BSS and GSS, it gives the accurate (*sūkṣma*) areas. The PG prescribes it for inequi-perpendicular quadrilaterals or quadrilaterals whose perpendiculars from both ends of the top are unequal, that is, for quadrilaterals excepting trapeziums; for trapeziums the PG prescribes the same rule as GSK 3.39 below. The MS, L, and GK point out that it gives only approximate areas. The PG, Tr, SS, and GSK do not refer to the applicability problem.

GSK 3.37. Ex. 1: A quadrilateral.

Calculate the area of a quadrilateral when $a = 25$, $b = 52$, $c = 60$, $d = 39$ *karas*.

Answer: $A = 1764$ sq. *karas*.

Cf. Tr (80).

GSK 3.38. Ex. 2: A trilateral.

Calculate the area of a trilateral when $a = 14$ (which is the base), $b = 13$, $c = 15$ *hatthas*.

Answer: $A = 84$ sq. *hatthas*.

Cf. PG (124), (125), (126) (all given as exs. of 'equi-perpendicular' figures under PG 115), Tr (81).

GSK 3.39. Areas of a trilateral and a quadrilateral 2.

When a, b, and h denote the base (*bhū*), the top (*muha*, S. *mukha*), and the perpendicular (*laṃba*) of a quadrilateral, or of a trilateral (*taṃsa*, S. *tryasra*) where $b = 0$, their respective areas are

$$A = \frac{a+b}{2} \cdot h, \qquad A = \frac{a}{2} \cdot h.$$

Pherū states expressly that the first formula is meant for 'all quadrilaterals' (*sayalāṇa caürasāṇaṃ*), but it gives the correct areas only for trapeziums. The 'perpendicular' seems to be assumed 'at the midpoint' (see *majjha* in GSK

3.40) of the top side of the quadrilateral. Śrīdhara in PG 115 restricts its use to 'equi-perpendicular quadrilaterals' (*sama-lambaka-caturaśra*), i.e., trapeziums, and to trilaterals. In Tr 42b he prescribes the same formula for 'other quadrilaterals ⟨than squares and oblongs within the category of equi-perpendicular quadrilaterals⟩' (*caturasreṣv anyeṣu*).

Cf. PG 115, Tr 42b+43b.

GSK 3.40. Ex.

Calculate the area of a quadrilateral when $a = 21$, $b = 5$, $h = 12$ *hatthas*, and the two flank sides (*bhuva-juva*, S. *bhuja-yuga*) are 13 and 15 *hatthas*.
Answer: $A = 156$ sq. *hatthas*.

Note that this is not a trapezium. If one assumes the figure made of the given four sides to be a trapezium, the 'perpendicular' would be $\sqrt{2074}/4$, which falls in between 11 and 12.

PG (127), (128), (129)–(130), Tr (77), (78), (79) (= PG (128)): all these treat 'equi-perpendicular' quadrilaterals.

GSK 3.41. Perpendicular and two segments of the base of a trilateral.

The perpendicular from the vertex of a trilateral can be calculated by

$$h = \frac{2A}{a}.$$

When a_i $(i = 1, 2)$ denote the two segments of the base (*ahavā*, S. *avadhā/āvādhā*) corresponding to the two flank sides, b_i,

$$a_i = \sqrt{b_i^2 - h^2}.$$

Cf. Tr 49–51a.

GSK 3.42. Hypotenuse of a right-angled triangle (Pythagorean Theorem).

When a, b, and c denote the side (*bhuva*, S. *bhuja/bhujā*, lit. an arm), the upright (*laṃba*, usually called *koṭi*), and the hypotenuse (*kanna*, S. *karṇa*, lit. an ear) respectively of a right-angled triangle,

$$c = \sqrt{a^2 + b^2}.$$

Cf. Tr 51b.

GSK 3.43. Circumference and area of a circle (Rule 1).

When d, C, and A denote the diameter, the circumference, and the area of a cricle respectively,

$$C = \sqrt{10d^2}, \qquad A = \frac{d}{4} \cdot C.$$

Cf. Tr 45.

GSK 3.44. Ex.

Calculate the circumference and the area of a circle whose diameter is 10. Answer: $C = \sqrt{1000}$, $A = \sqrt{6250}$.

Cf. Tr (85) (numerically the same). For the calculation of $\sqrt{1000}$ and $\sqrt{6250}$ Śrīdhara supplies an approximation formula in Tr 46 (= PG 118), namely,

$$\sqrt{K} \approx \frac{[\sqrt{Ka^2}]}{a},$$

where K and a are positive integers and $[\sqrt{Ka^2}]$ is the integer part of $\sqrt{Ka^2}$, which is calculated by means of the popular algorithm based on the decimal notation (cf. GSK 1.37–38). Assuming $a = 100$, Śrīdhara obtains $C = \sqrt{1000} \approx 31\frac{31}{50}$ and $A = \sqrt{6250} \approx 79\frac{1}{20}$. Pherū, too, must have treated the square roots in the same way, although the GSK does not contain the approximation formula.

GSK 3.45. Circumference and area of a circle (Rule 2).

$$C = 3d + \frac{d}{6}, \qquad A = \frac{d}{2} \cdot \frac{C}{2}.$$

The formula for the area derived from the Nahatas' text is

$$A = \frac{\frac{d}{2} \cdot C}{2},$$

but this is rather clumsy and is not likely to be Pherū's intention. We can obtain the above simpler formula by reading *guṇiyaṃ* instead of *guṇiyā* so that it may modify not *parihi* but *dalaṃ*. Therefore we emended the text.

The formula for C is repeated verbatim in GSK 5.22a.

The same two formulas are given also by Bhuvanadeva in his *Aparājitapṛcchā* [AP] 47.22 and in the anonymous *(Gaṇita-)Pañcaviṃśatikā* [PV] A21:

> *vyāse triguṇe paridhiḥ ṣaṣṭhabhāgavimiśrite/*
> *vyāsārdhaṃ ca paridhyardhaṃ guṇe vṛttaphalaṃ bhavet//*[1] AP 47.22 //
> 'When the diameter is multiplied by three and increased by one-sixth part ⟨of the diameter⟩, ⟨the result⟩ will be the circumference. Half the diameter and half the circumference: in multiplication ⟨the result⟩ will be the area of the circle.'

> *vyāsas triguṇito vṛttir vyāsaṣaḍbhāgasaṃyutaḥ/*
> *vyāsārdhaṃ ca paridhyardham anyonaṃ paritāḍayet//* PV A21 //
> 'The diameter, multiplied by three and increased by one-sixth part of the diameter, is the circumference. One should multiply half the diameter and half the circumference with each other ⟨to obtain the area⟩.'

This value, 19/6, is an approximation to $\sqrt{10}$ of Rule 1 ($\sqrt{10} = \sqrt{3^2+1} \approx 3+\frac{1}{2\cdot 3}$). The same approximation is employed by Śrīdhara in his formula for the volume of a sphere (see under GSK 3.65 below), although the approximation to π that explicitly occurs in Śrīdhara's extant works is $\sqrt{10}$ only. That Śrīdhara used

[1]Read *paridhyardhaguṇaṃ* for *paridhyardhaṃ guṇe*.

22/7 also is logically inferred from Rāghavabhaṭṭa's statement on the several values of π used by Śrīdhara and Bhāskara II. See Hayashi 1995b, 244.

The same approximation, 19/6, to $\sqrt{10}$ is already used by Jaḍivasaha in his calculation of the volume of imaginary (or cosmological) cylindrical granaries (palla, S. palya). See *Tiloyapaṇṇattī* [TP] 1.117–118:

> *samavaṭṭavāsavagge dahaguṇide karaṇiparihio hodi/*
> *vitthāraturiyabhāge parihihade tassa khettaphalaṃ//* TP 1.117 //
> *uṇavīsajoyaṇesuṃ caüvīsehiṃ tahāvaharidesuṃ/*
> *tivihaviyappe palle ghaṇakhettaphalā hu patteyaṃ//* TP 1.118 //
> 'When the square of the diameter of a uniform circle is multiplied by ten, there will be the circumference in *karaṇī* (i.e., in the square power). When one-fourth part of the breadth (i.e., the diameter) is multiplied by the circumference, there will be its area (lit. field-fruit).// When twenty less ⟨by one⟩ *joyaṇas* are divided by twenty-four, there will be the volume (lit. solid field-fruit) of each of the granaries supposed in three ways.'

In verse 117, Jaḍivasaha gives the same two formulas as GSK 3.43 and, in verse 118, 19/24 cubic *joyaṇas* (S. *yojanas*) as the volume of each of the three granaries whose diameter and height are both one *joyaṇa*. This volume must have been obtained by the computation:

$$V = \frac{1}{4} \cdot \sqrt{10 \cdot 1^2} \cdot 1 \approx \frac{1}{4} \cdot \frac{19}{6} \cdot 1 = \frac{19}{24}.$$

GSK 3.46. Area of a segment of a circle.

When a and h denote the chord (*jīvā*) and the height (*sara*, S. *śara*, lit. arrow) respectively of a segment of a circle (*dhaṇukhitta*, S. *dhanuḥkṣetra*, lit. a bow field), its area is obtained by

$$A = \sqrt{\left(\frac{a+h}{2} \cdot h\right)^2 \cdot 10 \cdot \frac{1}{9}}.$$

Cf. Tr 47.

GSK 3.47. Ex.

Calculate the area when $a = 15$, $b = 21$ and $h = 6$.
Answer: $A = \sqrt{4410} = \sqrt{66^2 + 54}$.

Note that the three lengths given here satisfy the two relationships,

$$a^2 + 6h^2 = b^2 \quad \text{and} \quad a + h = b,$$

on which the three formulas given in the next two verses are based. It follows that $b = \frac{7}{5}a$ and $h = \frac{2}{5}a$.

Cf. Tr (86).

GSK 3.48–49. Three lengths of a segment of a circle.

When a, b, and h denote the chord, the arc (*dhaṇu*, S. *dhanus*), and the height respectively of a segment of a circle,

$$b = \sqrt{a^2 + 6h^2}, \quad h = \sqrt{\frac{b^2 - a^2}{6}}, \quad a = \sqrt{\left(b - \frac{b+h}{2}\right)^2 \cdot 4}.$$

The second formula is equivalent to the first but the third is not: it is equivalent to $a = b - h$.

The first two formulas were known to Umāsvāti (5th century or before). Cf. Datta 1929, 124–25. The same two together with another equivalent formula, $a = \sqrt{b^2 - 6h^2}$, are given by Mahāvīra, by Āryabhaṭa II, and by Śrīpati. These formulas are obtained from the approximate relationship, $a^2 + (\pi^2 - 4)h^2 = b^2$ with $\pi = \sqrt{10}$.

It is not known why Pherū rewrote the simple formula, $a = b - h$. Mahāvīra employs the same relationship, $a + h = b$, in his practical (*vyāvahārika*) formula for the perimeter of an elongated circle (*āyata-vṛtta*), and Nārāyaṇa too uses it in his practical formula for the area of a segment of a circle. As Gupta [2002, 13–14] points out, the formula, $b = a + h$, may represent an older tradition, although it has not so far been found in the extant texts prior to Mahāvīra. See Table 3.1. Cf. Gupta 1979, 2001a.

GSK 3.50–51. Irregular figures.

The areas of irregular figures are calculated approximately by reducing them to the nearest regular figures as follow.

(1) A young moon figure (*bāl-iṃda*, S. *bāla-indu*) → 2 trilaterals.
(2) A drum figure (*muruja*, S. *muraja*) → 2 bows + 1 oblong in the middle.
(3) A barleycorn figure (*java*, S. *yava*) → 2 bows.
(4) A thunderbolt figure (*kulisa*, S. *kuliśa*) → 2 quadrilaterals.
(5) An elephant's tusk figure (*gaya-daṃta*, S. *gaja-danta*) → 1 trilateral.
(6) A cart-wheel-circle figure (*sagaḍa-cakka-vaṭṭa*, S. *śakaṭa-cakra-vṛtta*) → 1 quadrilateral.
(7) A moon figure (*caṃda*, S. *candra*) → 1 bow.
(8) A full moon figure (*paripunna-caṃda*, S. *paripūrṇa-*) → 1 circle.

The sixth figure is called 'a rim-like figure' (*nimm-āgāra-khitta*, S. *nemy-ākāra-kṣetra*) in GSK 3.53, which is an annulus.

Cf. PG 116 = Tr 44 (*gaja-danta*, *nemi*, *bāla-indu*, *vajra*), Tr 48 (*muraja*, *yava*, *pañcādi-bhuja*).

GSK 3.52–53. Ex.

Calculate the areas of

(1) a young moon figure whose width and length are 5 and 20 *karas*, respectively;

(2) an elephant's tusk figure whose perpendicular and base are $\frac{1}{2}$ of that of (1) (i.e. 10 *karas*) and 3 *karas*, respectively;

(3) a rim-like figure, both openings of which are 3 *karas* and whose perpendicular is 5 *karas*; and

Table 3.1: Indian Formulas for a Segment of a Circle (in chronological order)

Texts	Chord-Arrow-Arc	Area	π
UTA 3.11	$a^2 + 6h^2 = b^2$		$\sqrt{10}$
BKS 1.38-41; 1.122	$a^2 + 6h^2 = b^2$	$A = \sqrt{10} \cdot a \cdot \frac{h}{4}$	$\sqrt{10}$
TP 4.2401		$A = \sqrt{10} \cdot a \cdot \frac{h}{4}$	$\sqrt{10}$
BAB 2.10 (criticizes)	$\sqrt{10}\left(\frac{a}{4} + \frac{h}{2}\right) = b$	$A = \sqrt{10} \cdot a \cdot \frac{h}{4}$	$\sqrt{10}$
Tr 47		$A = \sqrt{\left(\frac{a+h}{2} \cdot h\right)^2 \cdot 10 \cdot \frac{1}{9}}$	$\sqrt{10}$
GP 25 (allegedly)		$A = \frac{a+h}{2} \cdot h \cdot \left(1 + \frac{1}{20}\right)$	$\frac{63}{20}$?
GSS 7.43b & 45; 7.43a	$a^2 + 5h^2 = b^2$	$A = (a+h) \cdot \frac{h}{2}$	3
GSS 7.21 (elongated circle)	$2a + 2h = 2b$	$2A = 2b \cdot \frac{2h}{4}$	3
GSS 7.73b-75a; 4.70b-71a	$a^2 + 6h^2 = b^2$	$A = \sqrt{10} \cdot a \cdot \frac{h}{4}$	$\sqrt{10}$
GSS 7.63 (elongated circle)	$(2a)^2 + 6(2h)^2 = (2b)^2$	$2A = 2b \cdot \frac{2h}{4}$	$\sqrt{10}$
MS 15.90-91a; 15.89	$a^2 + 6h^2 = b^2$	$A = \sqrt{\left(\frac{a+h}{2} \cdot h\right)^2 \cdot \left(1 + \frac{1}{9}\right)}$	$\sqrt{10}$
MS 15.94b-97a; 93b-94a	$a^2 + \frac{288}{49}h^2 = b^2$	$A = \frac{a+h}{2} \cdot 22 \cdot h \cdot \frac{1}{21}$	$\frac{22}{7}$
SS 13.39-40	$a^2 + 6h^2 = b^2$		$\sqrt{10}$
GSK 3.48-49; 3.46	$\begin{cases} a^2 + 6h^2 = b^2 \\ a + h = b \end{cases}$	$A = \sqrt{\left(\frac{a+h}{2} \cdot h\right)^2 \cdot 10 \cdot \frac{1}{9}}$	$\sqrt{10}$
GK 4.12	$a + h = b$	$A = b \cdot \frac{2h}{4}$	3
PV X20a		$A = \frac{a+h}{2} \cdot h \cdot \left(1 + \frac{1}{18}\right)$	$\frac{19}{6}$
PV A24		$A = \sqrt{\left(\frac{a+h}{2} \cdot h\right)^2 \cdot 10 \cdot \frac{1}{9}}$	$\sqrt{10}$
Keśava*		$A = \frac{(a+h)h}{2} \cdot \left(1 + \frac{1}{20}\right)$	$\frac{63}{20}$?
CCM 70[a]		$A = \frac{a+h}{2} \cdot h \cdot \left(1 + \frac{1}{20}\right)$	$\frac{63}{20}$?
CCM 70		$A = \sqrt{\left(\frac{a+b}{2} + h\right)\left(\frac{a+b}{2} - h\right) \cdot h^2 \cdot 2 \cdot \frac{1}{3}}$	?

*In his half verse cited in Gaṇeśa's commentary on *Līlāvatī* 213.

(4) a thunderbolt figure whose base and top are 5 *hatthas*, and whose width at the middle and perpendicular are 3 and 10 ⟨*hatthas*⟩, respectively.

Answer: (1) $A = 50$ sq. *karas*. (2) $A = 15$ sq. *karas*. (3) $A = 15$ sq. *karas*. (4) $A = 40$ sq. *hatthas*.

Cf. PG (131) = Tr (82), PG (132) = Tr (83), PG (133), Tr (84).

3.4 Procedure for Excavations

GSK 3.54. Dimensions of a pit with irregular sides (the arithmetical mean).

Let a_i, b_i, and h_i be the lengths measured at n_1, n_2, and n_3 places of the width, the length, and the height (*uṃḍḍatta*, S. *aunnatya*), respectively. Then, their respective mean (*sama*) lengths are obtained by

$$\bar{a} = \frac{\sum_{i=1}^{n_1} a_i}{n_1}, \qquad \bar{b} = \frac{\sum_{i=1}^{n_2} b_i}{n_2}, \qquad \bar{h} = \frac{\sum_{i=1}^{n_3} h_i}{n_3}.$$

Cf. Tr 52.

GSK 3.55. Capacities (*khatta-phala*) of a rectangular pit with uniform sides (a, b, h) and of a cubic pit.

$$V = a \cdot b \cdot h, \qquad V = a^3.$$

The capacity of a pit, or the volume of the empty space of a pit, is called *khatta-phala* (S. *khāta-phala*, lit. 'the fruit of an excavation') in contrast to the volume of a solid figure, which is expressed as *pāhāṇa-ghaṇa-hattha* (S. *pāṣāṇa-ghana-hasta*, 'the stones in cubic *hatthas*', GSK 3.61), or *pāhāṇa-saṃkhā* (S. *-saṃkhyā*, 'the number of stones', GSK 3.71–72), or simply *pāhāṇa* (GSK 5.25) or *sela* (S. *śaila*, 'stones', GSK 3.65). On the other hand, the volume of a hole or an empty (hollow) space in a building or in a wall is called 'the fruit of a hole' (*khalla-phala*, 3.76, 3.81, 3.84).

Cf. Tr 53.

GSK 3.56. Ex. 1: Capacity of a pond (*pukkharaṇī*) with irregular sides.

Calculate the capacity of a pond when $a = 5$, $b = 16$ *hatthas*, $h_1 = 2$, $h_2 = 3$, and $h_3 = 4$ *karas*.

Answer: $\bar{h} = 3$ *karas* and $V = 240$ cubic *hatthas*.

Cf. Tr (87) (numerically the same), (88).

GSK 3.57. Ex. 2: Capacities of rectangular and cubic ponds.

(1) Calculate the capacity of a rectangular pond when $a = 10\frac{1}{4}$, $b = 16\frac{1}{2}$, and $h = 8$ *karas*.

Answer: $V = 1353$ cubic *karas*.

We have emended *addhudae* ('one-half for the height') of the Nahatas' text to *aṭṭhudae* ('eight for the height') because the height (depth) $\frac{1}{2}$ *kara* seems to be too small for a pond whose width and length are $10\frac{1}{4}$ *karas* and $16\frac{1}{2}$ *karas*. With this emendation GSK 3.57 comes to numerically agree with Tr (89).

(2) Calculate the capacity of a cubic pond when $a = b = h = 9$ *karas*.

Answer: $V = 729$ cubic *karas*.

Cf. Tr (90).

GSK 3.58. Capacity of a cylindrical well (*kuva*, S. *kūpa*).

When d and h denote the diameter (*vitthāra*, S. *vistāra*) and the depth (*veha*, S. *vedha*) respectively of a cylindrical well, its capacity (*kara-saṃkhā*, S. *kara-saṃkhyā*, lit. 'the number of *karas*') is:

$$V = \left(3d^2 + \frac{d^2}{6}\right) \cdot h \cdot \frac{1}{4}.$$

This formula is obtained from GSK 3.45.

Cf. Tr 54, which gives a correct formula (with $\pi = \sqrt{10}$) for the capacity of a well shaped like a truncated cone (d_i = top and bottom diameters, h = height):

$$V = \frac{\sqrt{\left\{d_1^2 + d_2^2 + (d_1 + d_2)^2\right\}^2 \cdot 10 \cdot h}}{24}.$$

GSK 3.59. Ex.

Calculate the capacity of a cylindrical well when $d = 6$ *hatthas* and $h = 20$ *karas*.
Answer: $V = 570$ cubic *hatthas*.

GSK 3.60. Capacities of various pits with uniform cross-sections.

When A and h denote the area of the bottom (or the horizontal cross-section) and the depth, respectively, of a pit, its capacity is:

$$V = A \cdot h.$$

GSK 3.61. Volume of a rectangular stone.

When a, b, and h respectively denote the length, the width, and the thickness (*piṃḍa*, S. *piṇḍa*), all measured in *aṃgulas*, of a rectangular stone, its volume in *hatthas* is:

$$V = \frac{abh}{13824} \quad \left(= \frac{abh}{24^3}\right) \quad \text{cubic } hatthas.$$

This is simply a conversion of the cubic *aṃgulas* to the cubic *hatthas*.

Another interpretation, though less plausible, is possible. By regarding the phrase, *jiṇa*(24)-*aṭṭha*(8)-*terasehiṃ*(13), as giving not the single divisor, 13824, but the three divisors, 24, 8, and 13,

$$V' = \frac{abh}{24 \cdot 8 \cdot 13} \quad \text{stone cubic } \textit{hatthas}.$$

This is a conversion of the ordinary cubic *aṃgulas* to the 'stone cubic *hatthas*' (*pāhāṇa-ghaṇa-hattha*, S. *pāṣāṇa-ghana-hasta*). Note that $24 \cdot 8 \cdot 13 = 2496$ and

$$V' = \frac{72}{13} V,$$

where V denotes the same volume in the normal cubic *hatthas*.

The use of a special cubic '*hasta*' for stones is mentioned by Śrīdhara (Tr 55: *pāṣāṇa-phala-hastāḥ*, 56b: *karā dṛṣadaḥ*, and 57b: *dṛṣadaḥ phalaṃ*), by Āryabhaṭa II (MS 15.107: *pāṣāṇa-karāḥ*, 108: *pāṣāṇa-hastāḥ*), and by Śrīpati (SS 13.45b: *pāṣāṇa-hastāḥ*), who all give the formula:

$$V' = \frac{9}{4} V \qquad \text{or} \qquad V' = \frac{abh}{6144},$$

where a, b, and h are measured in *aṅgulas*. Note that, since $6144 = 24 \cdot 8 \cdot 32 = 24 \cdot 16 \cdot 16$, the above V' is more than twice as large as this V'.

Cf. *Aparājitapṛcchā* [AP] 47.16, which mentions a standard size of stone slab:

> *navatyaṅguladairghye ca pṛthutve caturviṃśatiḥ/*
> *dvādaśāṅgulapiṇḍaṃ ca śilāmānapramāṇataḥ//* AP 47.16 //
> 'Ninety *aṅgulas* in length, twenty-four ⟨*aṅgulas*⟩ in breadth, and twelve *aṅgulas* in thickness: this is the size of a standard slab.'

That is to say, the size of a standard slab is: $a = 90$, $b = 24$, $h = 12$ *aṅgulas*, and $V = 25920$ cubic *aṅgulas* $= \frac{45}{24}$ cubic *hastas*.

According to Abū al-Faḍl [2004, 194: Book I, Ā'īn 86], 'Pieces of red sandstone (*sang-i gulūla*), broken from the rocks in any shape, are sold by the *phārī*, which means a heap of such stones, without admixture of earth, 3 *gaz* long, $2\frac{1}{2}$ *g.* broad and 1 *g.* heigh.' It is interesting that the standard size was given in rectangular shape even for a heap of small stone pieces.

GSK 3.62. Ex.: A stone slab (*silā*, S. *śilā*).

Calculate the volume (*gaṇiya-phala*) of a rectangular stone when $a = 3\frac{1}{2}$, $b = 5\frac{1}{3}$, $h = \frac{1}{2}$ *hattha*.

Answer: $V = 9\frac{1}{3}$ cubic *hatthas* or $V' = 51\frac{9}{13}$ stone cubic *hatthas*.

Cf. Tr (92) (numerically the same).

GSK 3.63. Volumes of various stones with uniform cross-sections.

When A and h denote the area of the bottom (or the horizontal cross-section) and the thickness respectively of a stone, its volume is:

$$V = A \cdot h.$$

Note that the conversion factor, $24 \cdot 8 \cdot 13$, is not mentioned here.

Cf. Tr 57.

GSK 3.64. Ex.: A stone slab like a grindstone (*gharaṭṭa-paṭṭa*).

Calculate the volume of a cylindrical stone when $d = 10$ and $h = 1\frac{1}{2}$ *karas*.

Answer: According to GSK 3.45, $C = 95/3$ *karas* and $A = 475/6$ sq. *karas*. Therefore, $V = 118\frac{3}{4}$ cubic *karas* or $V' = 657\frac{9}{13}$ stone cubic *karas*.

Cf. Tr (94) (numerically the same).

GSK 3.65. Volume and surface area of a spherical (*gola*) stone.

When d and C denote the diameter (*udaya*, lit. 'height') and the circumference (*parihi*, S. *paridhi*) respectively of a great circle of a sphere, its volume (*sela*, S. *śaila*) and surface area (*khitta*, S. *kṣetra*) are:

$$V = \frac{d^3}{2}\left(1 + \frac{1}{9}\right), \qquad S = \frac{C}{4} \cdot C \cdot \left(1 + \frac{1}{9}\right).$$

The latter formula is repeated verbatim in GSK 5.25 and utilized in GSK 3.74. Compare the first formula with the following:

$$V = \frac{d^3}{2}\left(1 + \frac{1}{18}\right) \qquad \text{(Tr 56a, MS 15.108, SS 13.46)}$$

$$V_v = \left(\frac{d}{2}\right)^3 \cdot \frac{1}{2} \cdot 9 \quad \left\langle = \frac{d^3}{2}\left(1 + \frac{1}{8}\right)\right\rangle \qquad \text{(GSS 8.28b)}$$

$$V_s = V_v \cdot \frac{1}{9} \cdot 10 \quad \left\langle = \frac{d^3}{2}\left(1 + \frac{1}{4}\right)\right\rangle \qquad \text{(GSS 8.29a)}$$

$$V = \frac{d^3}{2}\left(1 + \frac{1}{21}\right) \qquad \text{(L 203)}$$

The interpretation of the second formula (V_s) of the GSS, which gives a volume worse than V_v, is based on the manuscripts used for the edition of Raṅgācārya, who however changes the text so as to obtain a better result:

$$V_s = V_v \cdot \frac{1}{10} \cdot 9 \quad \left\langle = \frac{d^3}{2}\left(1 + \frac{1}{80}\right)\right\rangle.$$

For the calculations of the volume of a sphere in India see Gupta 1991.

The second formula is peculiar to Pherū. Its origin may be explained as follows. Mahāvīra (GSS 7.25) gives a formula for the surface area of a segment of a sphere:

$$S_p = \frac{p}{4} \cdot b,$$

where p and b denote the circumference (*paridhi*) and the curvilinear width (*viṣkambha*) respectively of the convex segment of a sphere. This formula seems to have been obtained by applying to the convex segment his formula for the area of a circle,

$$A = C \cdot \frac{d}{4} \quad \text{or} \quad A = \frac{C}{4} \cdot d,$$

which he employs also for the area of an elongated circle (GSS 7.21 and 63). If one applies this formula to a hemisphere, where $p = C$ and $b = C/2$, one obtains the result:

$$S_p = \frac{C}{4} \cdot \frac{C}{2}.$$

By doubling this, one obtains a hypothetical formula for the entire surface:

$$S^* = \frac{C}{4} \cdot C.$$

Perhaps, Pherū noticed (by experiments?) that S^* gave smaller results and tried to improve it by assuming $n = 9$ in the equation:

$$S = S^* \left(1 + \frac{1}{n}\right).$$

The reason he has chosen 9 for n is not clear but it may be related to the same correction factor in his formula for the volume.

Or, the hypothetical formula S^* may have been obtained by a method similar to the one used for obtaining the formula for the area of a circle:

$$A = \frac{C}{2} \cdot \frac{d}{2}.$$

If one cuts the surface of a sphere into many small pieces like orange peels along longitudinal lines and makes each piece flat with its mid-width and length kept unchanged, one obtains a number of narrow barleycorn figures touching each other at the middle of each convex side. By cutting the whole figure along the equatorial line and making the two halves engaged, one obtains an oblong whose width and length are $C/4$ and C, respectively, but whose area is a little smaller than the sum of the areas of all barleycorn figures because two adjacent pieces slightly overlap each other due to their convex sides. This explains the last equation.

For the calculations of the surface area of a sphere in India, see Hayashi 1997.

GSK 3.66. Ex.

Calculate the volume (*gaṇiya*) and the surface area (*khitta*) of a sphere when $d = 6$ *karas*.

Answer: $V = 120$ cubic *karas*. According to GSK 3.45, $C = 19$ *karas*, and therefore $S = 100\frac{5}{18}$ sq. *karas*.

Cf. Tr (93).

GSK 3.67–68. Weights per cubic *kaṃbiya* of various kinds of stones (specific gravities).

Stones	Weights (*maṇas/kaṃbiya*[3])
Stones from Delhi	50 *maṇas* + 24 *tullas*[*1]
vaṃsī (red sandstone)	48
mammāṇī (marble)	60
kasiṇa (black stone)	62
jajjāvaya	49
kannāṇaya[*2]	49
kuḍukkaḍa	60

[*1] 800 *tullas* (*tolas*) = 1 *maṇa* (see GSK 1.9–10a). GSK 5.26 reads '2400 *tullas*' instead of '24 *tullas*'. Sharma [1959, 318] accepts the reading of GSK 5.26 and thinks that 1 *maṇa* is equated there to 2400 *tullas*.

[*2] Kannāṇā is Pherū's birth-place.

The post-colophonic statement of one of the manuscripts consulted by L. C. Jain for his edition of the GSS (Appendix 5, p. 57) says: 'A cube (*sama-corasa*) of 1 carpenter's *gaja* of the stone *ujala* ⟨weighs⟩ 48 *maṇas*; 1 ⟨cubic⟩ *gaja* of the stone *pāleva* 60 *maṇas*; and 1 ⟨cubic⟩ *gaja* of the stone *ṣāra* 40 *maṇas*.'

These two verses are repeated in GSK 5.26–27.

3.5 Procedure for Piling

The purpose of calculating the volume of wall space is to estimate the number of bricks needed for different types of architectural innovations being introduced by Muslim Sulṭāns in northern India, especially Delhi, such as arches, domes and minarets. In fact the first successful dome with pointed arches was erected by Pherū's employer, ᶜAlā' al-Dīn Khaljī in 1311; the building in known as ᶜAlā'ī Darwāza. It is about the same time that Pherū was composing the GSK.

GSK 3.69. Nine kinds of walls (*bhittī*).

There are nine kinds of walls or constructions: (1) a dome (*gommaṭṭa*, Pe. *gumbad*), (2) a square spiral stairway (*caürasa-pāyascva*), (3) a circular spiral stairway (*vaṭṭa-pāyaseva*), (4) a minaret (*munāraya*, Pe. *mīnār*, Ar. *manār*), (5) an arch (*tāka*, Ar. *ṭāq*), (6) a stair (*sovāṇa*, S. *sopāna*), (7) a bridge (*pula*, Pe. *pul*), (8) a cylindrical well (*kūva*, S. *kūpa*), and (9) a stepwell (*vāvī*, S. *vāpī*).

GSK 3.70–71. A rectangular wall.

Let a, b, and h denote the length, width, and height, measured in *hatthas*, of a rectangular wall, and V' the volume of the space for windows, doors, etc. that are made of, or enclosed by, wood. Then, the net volume of the wall is

$$V = (abh - V')\left(1 - \frac{1}{10}\cdot\frac{3}{2}\right) \text{ cubic } hatthas,$$

where the negative term, $\frac{1}{10}\cdot\frac{3}{2}$ or $\frac{3}{20}$, is for the volume of the clay used as mortar.

GSK 3.72. Ex. 1: Net volume.

Calculate the volume (*pāhāṇa-saṃkhā*) of a wall when $a = 10$, $b = 2$, and $h = 5$ *karas* and it has a window whose width and height are 2 and 3 *karas*, respectively.

Answer: $V = (10 \cdot 2 \cdot 5 - 2 \cdot 2 \cdot 3)\left(1 - \frac{1}{10} \cdot \frac{3}{2}\right) = 74\frac{4}{5}$ cubic *hatthas*.

GSK 3.73. Ex. 2: Number of bricks.

Calculate the number of bricks (*iṭṭā*, S. *iṣṭakā*) when the size of a brick is $\frac{1}{2} \times \frac{1}{3} \times \frac{1}{8}$ and that of a wall is $10 \times 1\frac{1}{2} \times 4$.

Answer: $(10 \cdot \frac{3}{2} \cdot 4) \cdot (1 - \frac{1}{10} \cdot \frac{3}{2})/(\frac{1}{2} \cdot \frac{1}{3} \cdot \frac{1}{8}) = 51/\frac{1}{48} = 2448$ bricks.

The extant GSK does not have a rule for this example, but GSK 3.87 for timber may be applied.

Cf. Tr 58, (95)–(96).

GSK 3.74. Volume and surface area of the wall of a dome.

Let C_1 and C_2 be the inner and the outer circumferences, respectively, of the wall of a dome measured at its foot. Then, the volume (*cayaṇa*, S. *cayana*, lit. piling) of the wall and its surface area (*khitta*, S. *kṣetra*) can be calculated by

$$V_g = \frac{C_1}{2} \cdot \frac{C_1}{4} \cdot \left(1 + \frac{1}{9}\right), \quad S_g = \frac{C_2}{2} \cdot \frac{C_2}{4} \cdot \left(1 + \frac{1}{9}\right).$$

Both formulas are obtained by halving S, for which see GSK 3.65 above.

Apparently, V_g is not a volume but an area unless we regard it as being multiplied by the unit of length. See verse 76 below.

GSK 3.75. Ex.: Dome.

Calculate the volume of the wall and the surface area of a dome when $C_1 = 19$, $d_1 = 6$, and $C_2 = 24$. No unit of length mentioned.

Answer: $V_g = 50\frac{5}{36}$ and $S_g = 80$.

One of the given numbers, 6, is called *vitthara* ('the breadth'). Although the thickness of a wall may be called *vitthara* (see verse 85), here it seems to be the diameter of the inner circumference because (1) the word *piṃḍa* would be more preferable if 6 be the thickness, (2) the word *vitthara* (or *vitthāra*, S. *vistāra*) is sometimes used for the diameter of a circle (see verse 58), (3) 6 and 19 satisfy Pherū's formula for the circumference of a circle (verse 45), (4) Pherū's formula for V_g (verse 74), as it stands now, does not include the thickness as a factor, and (5) if 6 be the thickness of the wall, 19 and 24 cannot be the inner and the outer circumferences of the same dome. If, on the contrary, we assume 19 and 24 are the inner and the outer circumferences of the same dome, as they are likely, the thickness of the wall is: $(d_2 - d_1)/2 = (c_2 - c_1)/(2\pi) = (24-19)/(2 \cdot \frac{19}{6}) = 15/19$. This does not seem to cause any contradiction.

GSK 3.76. Capacity of the empty space of a dome.

Let C_1 and d_1 be the inner circumference and the diameter, respectively, of the wall of a dome measured at its foot. Then, the capacity of the empty space (*khalla*) of the dome can be calculated by

$$V_{gk} = C_1 \cdot \frac{d_1}{2} \cdot \left(1 + \frac{1}{9}\right).$$

Like V_g in verse 74, V_{gk} is not a volume but an area unless we regard it as being multiplied by the unit of length.

GSK 3.77–78. Volumes of a square wall and of a spiral stairway in it.

$$V = abh \;\; kaṃviyas, \qquad V_{cp} = \frac{a_1 + a_2}{2} \cdot b \cdot h,$$

where V_{cp} is the volume of the square spiral stairway (*caürasa-pāyaseva*).

GSK 3.79. Volumes of a circular wall and of a spiral stairway in it.

$$V = \text{ as in a cylindrical well (verse 58 or 85?)},$$

$$V_{vp} = \text{ as in a rectangular solid with sides, 6, 3, 1/2(?)},$$

where V_{vp} is the volume of the circular spiral stairway (*vaṭṭa-pāyaseva*).

GSK 3.80. Volume of the wall of a minaret.

$$V_m = V_{vp} + V',$$

where V' stands for the volume of decorations on the surface of the minaret, which are semi-triangular or semi-circluar cylinders.

GSK 3.81. Volume of the empty space of an arch.

When a and h denote the length and height, respectively, of an arch (*tāka*) accompanied by a pointed apex (*sihā*, S. *śikhā*) in a wall whose thickness is b, the volume of its empty space (*khalla-phala*) can be calculated by

$$V_{tk} = abh\left(1 - \frac{1}{7} \cdot \frac{3}{2}\right).$$

GSK 3.82. Ex.: Arch.

Calculate the volume of the empty space of an arch when $a = 7$ *karas*, $h = 4$ *hatthas*, and $b = 1$ *hattha*.

Answer: $V_t = 22$ cubic *hatthas*.

GSK 3.83a. Volume of a staircase.

Let a_1, a_2, b, and h be the the length at the top, length at the bottom, width, and height, respectively, of a staircase (*sovāṇa*). Then its volume can be calculated by

$$V_s = \frac{a_1 + a_2}{2} \cdot b \cdot h.$$

Note that this formula gives the exact volume. For, when d denotes the length (width) of each step, the volume is

$$V_s = \frac{(a_1 - d/2) + (a_2 + d/2)}{2} \cdot b \cdot h.$$

GSK 3.83b. Ex.: Staircase.

Calculate the volume of a staircase when $a_1 = 1$, $a_2 = 9$, $b = 2$, and $h = 6$.
Answer: $V_s = 60$.

GSK 3.84. Volume of a bridge structure.

When a, b, and h denote the length, the width, and the height, respectively, of a bridge (*pula*), V_{tk} the volume of arches, V_b that of the parapets on both sides (? *bhuva-juva*, lit. a pair of sides; cf. verse 40 above), V_n that of the exits (*niggama*, S. *nirgama*), and V_{pk} that of the empty space. Then, the volume of the bridge structure (*pula-baṃdha*) can be calculated by

$$V_p = abh - V_{tk} + V_b + V_n - V_{pk}.$$

GSK 3.85a. Volume of the wall of a cylindrical well.

When C, b, and h denote the inner circumference, the breadth (i.e., thickness), and the height, respectively, of the wall of a well (*kuva*, S. *kūpa*), the volume of the wall can be calculated by

$$V_k = Cbh.$$

If the phrase, *kuva-bhitti-majjhi parihī*, means 'the circumference at the centre of the wall of a well', then the above formula is exact; otherwise it only gives an approximate volume.

GSK 3.85b. Ex.

Calcualte the volume of the wall of a cylindrical well when $C = 18$, $b = 2$, and $h = 10$ *karas*.
Answer: $V_k = 360$ cubic *karas*.

GSK 3.86. Volumes of the walls and the bases of various stepwells.

The volumes of the walls and the bases of various stepwells (*vāvī*, S. *vāpī*) are calculated according to the method most appropriate for the shapes of the walls and of the bases. The text does not give the details of the calculation, but naturally we have to calculate:

$$V_v = V_1 + V_2,$$

where V_1 is the volume of the wall and V_2 the volume of the base. V_1 can be calculated by the formulas given in verses 77–78, 79, 83, and 85a, and V_2 by the formula of verse 63. For the area of a conch-shell-like circular figure (*saṃkha-vatta* or a spiral figure, S. *śaṅkha-vṛtta*), see GSS 7.23 and 65–66, and GK 4.10–11. Cf. Hayashi 1992.

3.6 Procedure for Sawing

GSK 3.87. Number of planks cut out of a piece of timber.

When a, b, and h denote the length, the width, and the thickness (*dala*) of a piece of timber, and a', b', and h' those of a piece of plank, the number of planks that can be cut out of the timber is calculated by

$$n = \frac{abh}{a'b'h'}.$$

GSK 3.88. Ex.

Calculate the number of planks when $a = 8$, $b = \frac{1}{2}$, $h = \frac{1}{3}$ *kara*, and $a' = 1$, $b' = \frac{1}{4}$, $h' = \frac{1}{9}$ *kara*.
Answer: $n = 48$.

GSK 3.89. Wages for sawing.

When a and b denote the length measured in *gajas* (= *hatthas*) and the width (i.e., the thickness) in *visuvas* (20 *visuvas* = 1 *gaja*), respectively, of a piece of timber, and n the number of cutting lines, the total length of the cut in *hatthas* (or *gajas*) is,

$$L = an \quad \textit{hatthas}.$$

The wages (W) for sawing (*cirāvaṇī*), expressed in *koḍīs*, is determined by means of L and b according to the price rate given below.
Koḍī the cowry shell, is a unit of currency. See under GSK 1.3–4a above.

GSK 3.90–92. Price rate for sawing.

One *koḍī* is paid as the wages for sawing the following quantities:

Width (b) in *visuvas*:	$1 \sim 1\frac{1}{2}$,	$2 \sim 3$,	4,	$5 \sim 9$,	$10 \sim 13$,	$14 \sim 16$,	$17 \sim 20$
Length (L) in *gajas*:	100,	80,	60,	40,	30,	20,	10

Within each range of width, the wages must be proportional to the total length of cutting (L). Pherū remarks that a piece of timber whose width (or thickness) is more than 20 *visuvas* (= 1 *gaja*) cannot be cut by a saw.

GSK 3.93. Ex.

Calculate the wages for sawing a piece of timber when $a = 7$ *gajas*, $b = 7$ or (and?) 8 *visuvas*, and $n = 10$.
Answer: $L = 70$ *gajas* and, as both 7 and 8 *visuvas* fall in between $5 \sim 9$ *visuvas*, $W = 70/40 = 1\frac{3}{4}$ *koḍīs*.

Cf. Tr (100), which uses the rate of wages, 6 *rūpas* for sawing 100 square *hastas* of *khādira* timber. According to Abū al-Faḍl [2004, 195: Book I, Āʾīn 87], the wages for sawing the *sīsaū* wood is $2\frac{1}{2}$ *dammas/gaz*^{2} and that for the *nāzhū* wood is 2 *dammas/gaz*^{2}. See GSK 3.95 below.

GSK 3.94. Three notes on sawing.

(1) 8 *javas* (S. *yavas*) = 1 *kaṃviy-aṃgula* (S. *kambikā-aṅgula*).

(2) The width of one sawing line on a plank = 1 *java*.

(3) In cutting a cylindrical log of wood, regard its thickness (*piṃḍa*) as the length (a).

GSK 3.95. Factor of hardness of timber.

The wages (W) for sawing (see verses 89–92 above) is multiplied by a factor (α), which is either 1 or $1\frac{1}{4}$ or $1\frac{1}{4}$, according to the hardness of the timber.

Timber (Sanskrit, English, Latin)	Hardness	α
mahuva (*madhuka/madhūka*, honey tree, *Bassia latifolia*)	middle	1
vaḍa (*vaṭa*, Indian fig tree, *Ficus benghalensis*)	middle	1
sāla (*śāla*, sal tree, *Vatica robusta*)	middle	1
sīsama (*śiṃśapā*, Indian rosewood, *Dalbergia sissoo*)	middle	1
niṃva (*nimba*, neem tree, *Azadirachta indica*)	middle	1
sirīsa (*śirīṣa*, sirisa tree, *Acacia sirissa*)	middle	1
khayara (*khadira*, cutch tree, *Acacia catechu*)	hard	5/4
aṃjaṇa (*añjana*, Asiatic barberry, *Berberis asiatica?*)	hard	5/4
kīra (*cīḍha*, chir-pine tree, *Pinus longifolia*)	hard	5/4
seṃvalu (*śālmalī*, silk-cotton tree, *Bombax ceiba*)	soft	3/4
suradāru (*suradāru*, Himalayan cedar, *Pinus deodara*)	soft	3/4

Cf. BSS 12.48–49, GSS 7.63–67, MS 15.112–14.

3.7 Procedure for Mounds of Grain

GSK 3.96. Volume of grain heaped up on the ground.

When C and h denote the circumference and the height measured in *hatthas*, respectively, of a mound of grain heaped up on an even ground, its volume prescribed is

$$V = \left(\frac{C}{6}\right)^2 \cdot h \quad \text{cubic } \textit{hatthas}.$$

One cubic *hattha* is also called *patta*.

Cf. Tr 61–62, L 227.

GSK 3.97. Height of a mound.

The height of a mound of grain may be approximately determined by the formula,

$$h = \frac{C}{\beta},$$

where β is either 9 or 10 or 11 according to the fineness of the grain.

Grain (Sanskrit, English, Latin)	Fineness	β
tila (*tila*, sesamum, *Sesamum indicum*)	fine	9
kuddava (*kodrava*, a kind of cereal, *Paspalum scrobiculatum*)	fine	9
mugga (*mudga*, kidney beans, *Vigna radiata*)	middle	10
gohuma (*godhūma*, wheat, *Triticum aestivum*)	middle	10
vora (unidentified)	coarse	11
kulattha (*kulattha*, horse gram, *Dolichos biflorus*)	coarse	11

GSK 3.98–99. Ex.

Calculate the volumes of mounds of grain (1) when $C = 36$ and $h = 4$ ⟨*karas*⟩; (2) when $C' = C/2$ (piled against a flat wall) and $h = 4$; (3) when $C' = C/4$ (piled against the inside of a corner) and $h = 4$; (4) when $C' = 3C/4$ (piled against the outside of a corner) and $h = 4$.

Answer: (1) $V = 144$ cubic *karas*, (2) $V' = 72$ cubic *karas*, (3) $V' = 36$ cubic *karas*, (4) $V' = 108$ cubic *karas*.

A calculation method for (2) to (4) is given in the next verse.

Cf. Tr (102), (104), (105) (numerically the same as the GSK).

GSK 3.100. Volumes of mounds piled up against walls.

When a partial circumference C' is given, (1) calculate the entire circumference C by

$$C = kC',$$

where the multiplier (k) is either 2 or 4 or 4/3 according to whether the mound is piled up against a flat wall, or the inside of a corner, or the outside of a corner; (2) calculate the volume, V, by the formula of verse 96; (3) divide the V by the respective multipliers:

$$V' = \frac{V}{k}.$$

Cf. Tr 64.

3.8 Procedure for Shadows

GSK 3.101. Height of a pillar, etc.

When d and c_d denote the height (length) of a staff (*daṃḍa*) and its shadow, both measured in *aṃgulas*, and h and c the height and the shadow of a pillar (*thaṃbha*, S. *stambha*), etc., both measured in terms of the staff,

$$h = \frac{cd}{c_d} \quad \textit{daṃḍas}.$$

This verse is cited and paraphrased in Sanskrit prose in PM A1.

> *stambhādibhitticchāyā daṇḍena mapitvā daṇḍena guṇyā/ tasyaiva daṇḍasya cchāyayā bhāge yat phalaṃ tad udayamānaṃ/ yathā//*
> prose part of PM A1 //

> The shadow of a pillar, etc. or of a wall, having been measured by a staff, should be multiplied by the ⟨length of the⟩ staff. When ⟨the product is⟩ divided by the shadow of that staff, the quotient is the value of the height. For example (this phrase introduces the verse of PM A2 = GSK 3.102).

GSK 3.102. Ex.

Calculate the height of a pillar when d = 24 *aṃgulas*, $c = 3\frac{1}{4}$ *daṃḍas*, $c_d = 18\frac{1}{4}$ *aṃgulas*.

Answer: $h = 4\frac{20}{73}$ *daṃḍas* $= 102\frac{42}{73}$ *aṃgulas*.

This verse is cited in PM A2 and paraphrased and solved in Sanskrit prose.

> *caturviṃśatyaṅgulena daṇḍamānena stambhasya cchāyā sapādadaṇḍatrayamitā/ daṇḍasya cchāyā 18 aṅgulā/ kiṃ stambhodayamānam// stambhacchāyāṅgulīkṛtā 78/ daṇḍāṅgulaiś caturviṃśatyā guṇitā 1872/ daṇḍacchāyā-18-aṅgulair bhakte labdhaṃ 104 stambhodayaḥ//* prose part of PM A2 //
>
> 'With twenty-four *aṅgulas* for the length of a staff, the shadow of a pillar is three and a quarter staffs long. The shadow of the staff is 18 *aṅgulas* ⟨long⟩. What is the height of the pillar? The shadow of the pillar made into *aṅgulas*, 78, is multiplied by the *aṅgulas* of the staff, twenty-four: 1872. When ⟨this is⟩ divided by the 18 *aṅgulas* of the shadow of the staff, the quotient is 104, which is the height of the pillar.'

Note that c_d = 18 *aṃgulas* and therefore h = 104 *aṃgulas* in this example.

GSK 3.103. Construction of a gnomon.

This verse briefly describes the construction of a gnomon (*saṃka*, S. *śaṅku*) 12 *aṃgulas* long but the details are not clear.

Cf. GSS 9.2–4.

GSK 3.104. Noon shadows and the time of a day.

The first line of the verse gives six 'constant quantities' (*dhuva-rāsī*, S. *dhruva-rāśi*), which seem to mean the noon shadows (s_n) when the sun is at the beginning of each zodiacal sign. Three of the six values are for the signs beginning with Makara and the remaining three for those beginning with Karkaṭa. The given values seem to be corrupted as they have two peaks in one year. It is therefore difficult to restore the whole set of the 'constant quantities'. See the following table (I = Makara, VII = Karkaṭa).

Signs:	I	II	III	IV	V	VI	VII	VIII	IX	X	XI	XII
s_n:	4	4	1				5	3	1			

The second line of verse 104 gives a formula for calculating the time from a shadow. When x denotes the shadow, the portion of time elapsed before noon or remaining after noon is calculated by

$$t = \frac{7}{x+7} \pm 1 \quad \text{day?,} \quad \text{or} \quad t = 1 \pm \frac{7}{x+7} \quad \text{day?,}$$

according to the Nahatas' text but neither of these works well. We naturally expect s_n in this formula as it is given in the first line of the same verse. The formula intended by Pherū may be this:

$$t = \frac{7 \cdot \frac{1}{2}}{x - s_n + 7} \quad \text{day},$$

which is equivalent to the formula of PV X24 (see below). In this formula, the length of the gnomon is assumed to be 7 *aṃgulas*, which contradicts the previous verse, where it is said to be 12 *aṃgulas* long. But that Pherū was acquainted with formulas of this type is known from another work of his.

In JS 3.35, Pherū gives a similar formula corresponding to a gnomon of 12 *aṃgulas*. It reads:

> *jā jahi kāle chāyā diṇ-addha-chāyā-vihīṇa saṃka-juyā/*
> *diṇa-māṇaṃ chac ca guṇaṃ teṇa phalaṃ divasa-gaya-sesaṃ//* JS 3.35 //
> 'Whatever be the shadow at any time is decreased by the noon shadow and increased by the gnomon (12). The length of daylight is multiplied by six and divided by that ⟨sum obtained above⟩. The quotient is the elapsed or the remaining ⟨length⟩ of daylight.'

That is to say,

$$t = \frac{6d}{x - s_n + 12} \quad \langle \textit{ghaḍiyas}, \text{ S. } \textit{ghaṭikās} \rangle,$$

which is equivalent to the formula given in the YJ (see below). Pherū, in the next verse (JS 3.36), also gives an inverse formula, i.e., a formula for obtaining the shadow from a given time:

> *divas-addhaṃ saṃka-guṇaṃ gayaghaḍieṇa phala majjha-chāya-juyaṃ/*
> *saṃk-ūṇa sesa aṃgula jah-iccha-kālassa chāya-varaṃ//* JS 3.36 //
>
> *guṇaṃ*] *juyaṃ* SGS; *gaya-ghaḍieṇa phala*] *gayaghaḍiya-phaleṇa* SGS.
>
> 'Half of the length of daylight is multiplied by the gnomon (12), divided by the elapsed *ghaḍiyas*, increased by the noon shadow, and decreased by the gnomon (12). The remainder in *aṃgulas* is the shadow for any desired time.'

That is to say,

$$x = \frac{12 \cdot (d/2)}{t} + s_n - 12 \quad \textit{aṃgulas}.$$

Compare the following.

$$t = \frac{1/2}{x/g + 1} \quad \text{day} \qquad \text{(BSS 12.52)}$$

$$t = \frac{g}{2(x + g)} \quad \text{day} \qquad \text{(Tr 65, MS 15.118, SS 13.53)}$$

$$t = \frac{1}{2(x/g + 1)} \quad \text{day} \qquad \text{(GSS 9.8b–9a)}$$

$$t = \frac{g}{2(x - s_0 + g)} \text{ day} \qquad \text{(GSS 9.15b–16a)}$$

$$t = \frac{12 \cdot (d/2)}{x - s_n + 12} \textit{ nāḍīs} \qquad \text{(YJ 79.32)}$$

$$t = \frac{7 \cdot (d/2)}{x - s_n + 7} \textit{ ghaṭikās} \qquad \text{(PV X24)}$$

In these formulas, g is the length of the gnomon, the 'day' means the length of daylight, d denotes the length of daylight measured in *nāḍīs* or *ghaṭikās*, and s_0 and s_n denote the equinoctial and the noon shadows, respectively.

Chapter 4

Four Special Topics

4.1 Region

GSK 4.1. Introduction.

Pherū says that his object in this section is to teach *desalehapayaḍī* (S. *deśa-lekhā-paddhati*), which probably means 'regional method (or local manner) of writing and counting.'

GSK 4.2. Proportionate distribution of money.

When a_i denotes the amount of money in *dammas* allotted to each person (for investment?) and M the total gain in *jīvalaïs*, the share of each can be calculated by

$$p_i = \frac{M a_i}{\sum_{j=1}^{n} a_j} \textit{ jīvalaïs}.$$

The *jīvalaï* (also called *jivalā* in the next verse) seems to be a monetary unit.

Cf. GSK 3.7.

GSK 4.3. Ex.

Calculate each share when a_i = 3000, 13000, 5000, 2000, 7000 *jīvalaïs* and $M = 9000$ ⟨*jīvalaïs*⟩.

Answer: p_i = 900, 3900, 1500, 600, 2100 *jivalās*.

GSK 4.4. Percentage of *upakkhaï* in *caṭṭī*.

$$\text{Percentage } (sa\bar{\imath}) = \frac{upakkha\ddot{\imath} \times 100}{ca\d{t}\d{t}\bar{\imath}}.$$

The word *caṭṭī* seems to mean an account or the total amount of money recorded in it, while *upakkhaï* seems to be an Apabhraṃśa form either of *upakṣaya* or *upakṣati*, both meaning 'loss'. For *caṭṭī* see GSK 4.5, 4.8, 5.17, and 5.33.

GSK 4.5. Ex.

Calculate the percentage when *caṭṭī* = 950000 and *upakkhaï* = 70000.
Answer: Percentage = $7\frac{7}{19}$.

GSK 4.6. Easy division of *dammas* by 100.

When a_0 and a_1 denote positive integers below ten or zero and a_2 any positive integer,

$$\frac{(a_2 10^2 + a_1 10 + a_0)\ \textit{dammas}}{100} = a_2\ \textit{dammas} + \frac{a_1 10 + a_0}{5}\ \textit{visuvas}.$$

The number of *visuvas* is obtained by the relationship, 1 *damma* = 20 *visuvas* (see GSK 1.4a).

The rules of GSK 4.6–10 teach mechanical shortcuts in commercial arithmetic. Cf. GSK 5.18–21.

GSK 4.7. Easy division of *dammas* by 1000, by 10000, and by 100000.

Since the conversion ratio between any two consecutive sub-units of the *damma* is 20 (GSK 1.3–4a),

$$\begin{aligned} &\frac{(a_3 10^3 + a_2 10^2 + a_1 10 + a_0)\ \textit{dammas}}{1000} \\ &= a_3\ \textit{damm} + \frac{a_2}{10} \cdot 20\ \textit{visu} + \frac{a_1}{100} \cdot 20^2\ \textit{viss} + \frac{a_0}{1000} \cdot 20^3\ \textit{paḍiv} \\ &= a_3\ \textit{damm} + 2a_2\ \textit{visu} + 4a_1\ \textit{viss} + 8a_0\ \textit{paḍiv} \end{aligned}$$

Likewise,

$$\begin{aligned} &\frac{(a_4 10^4 + a_3 10^3 + a_2 10^2 + a_1 10 + a_0)\ \textit{dammas}}{10000} \\ &= a_4\ \textit{damm} + 2a_3\ \textit{visu} + 4a_2\ \textit{viss} + 8a_1\ \textit{paḍiv} + 16a_0\ \textit{kāiṇ}. \end{aligned}$$

$$\begin{aligned} &\frac{(a_5 10^5 + a_4 10^4 + a_3 10^3 + a_2 10^2 + a_1 10 + a_0)\ \textit{dammas}}{100000} \\ &= a_5\ \textit{damm} + 2a_4\ \textit{visu} + 4a_3\ \textit{viss} + 8a_2\ \textit{paḍiv} + 16a_1\ \textit{kāiṇ} + 32a_0\ \textit{paḍik}. \end{aligned}$$

GSK 4.8. Ex.: Division by 100.

Divide the original money (*mūla-dhaṇa*) of *caṭṭī*, 209534 ⟨*dammas*⟩, by 100.
Answer: 209534/100 = 2095 *dammas* 34/5 *visuvas* = 2095 *dammas* 6 *visuvas* 16 *vissaṃsas*.

GSK 4.9. Easy division of *dammas* by 10 and by 100.

$$\frac{(a_1 10 + a_0)\ \textit{dammas}}{10} = a_1\ \textit{dammas} + 2a_0\ \textit{visuvas}.$$

The second line of the verse gives the same rule as verse 6 (division by 100).

GSK 4.10. Easy division of *dammas*, etc. by 20.

$$\frac{(a_1 10 + a_0)\ \textit{dammas}}{20} = \frac{a_1}{2}\ \textit{dammas} + a_0\ \textit{visuvas}.$$

$$\frac{b_1 \text{ } visuvas}{20} = b_1 \text{ } vissaṃsas, \quad \text{and so on.}$$

GSK 4.11a. Wages for elapsed days.

When M_u denotes the annual wages called *mukkātaï* (or *mukātaï*, Ar-Pe. *muqāṭa*^{c}*a*) and D the number of days in a year, the wages for the elapsed days, d, is calculated by

$$m = \frac{M_u d}{D}.$$

GSK 4.11b. Ex.

Calculate the wages when $M_u = 5000$ and $d = 4$ months 9 days.
Answer: Since $d = 129$ days and $D = 360$ days (GSK 1.11), $m = 1791\frac{2}{3}$.

GSK 4.12. Daily Wages.

When M_a denotes the monthly wages (*maseliya*) in *dammas*, the daily wages in *visuvas* can be calculated by

$$m_0 = M_a \left(1 - \frac{1}{3}\right) \text{ } visuvas/\text{day}.$$

For, since 1 *damma* = 20 *visuvas* (GSK 1.4a) and 1 month = 30 days (GSK 1.11), we have the relationship, $m_0 = (20M_a)/30$ *visuvas*/day.

GSK 4.13. Meeting of two runners 1.

When v_1 and v_2 ($v_1 < v_2$) denote the constant speeds per day of two runners and t_1 the number of the days intervening between their departures, the time (in days) necessary for their meeting is calculated by

$$t_2 = \frac{v_1 t_1}{v_2 - v_1} \text{ days.}$$

For, we have the equation, $v_1(t_1 + t_2) = v_2 t_2$.

This verse is cited in PM A3.

GSK 4.14. Ex.

Calculate the time for the meeting of two runners when $v_1 = 4$ *joyaṇas*/day, $v_2 = 7$ *joyaṇas*/day, and $t_1 = 8$ days.
Answer: $t_2 = 10\frac{2}{3}$ days.
This verse is cited in PM A4 and paraphrased and solved in Sanskrit prose.

> *caturyojanagāt paścān navamadine saptayojanī gantā calitaḥ/ kiyad-dinair milati/ nyāsaḥ/ 4/ 9/ 7/ caturbhir nava guṇyante 36/ catur-ṇāṃ saptānāṃ cāntareṇa bhāge/ 12/ dvādaśabhir dinair militaḥ//*
> prose part of PM A4 //
> 'On the ninth day after ⟨the departure of⟩ a four-*yojana* goer, a seven-*yojana* goer started. In how many days will he meet ⟨the former⟩? Setting-down: 4, 9, 7. Nine is multiplied by four: 36.

> When ⟨this is⟩ divided by the difference between four and seven: 12. ⟨Therefore⟩, he meets ⟨him⟩ in twelve days.'

Note that $t_1 = 9$ days and therefore $t_2 = 12$ days in PM A4. This seems to have been caused by the misunderstanding of the expression, 'on the ninth day'.

GSK 4.15. Ex. for the next rule.

Calculate the time for the meeting of two camels when $a = 5$ *joyaṇas*/day, $d = 2$ *joyaṇas*/day^2, and $v = 14$ *joyaṇas*/day.

Answer: $t = 10$ days.

This verse is cited in PM A5 and is given a *nyāsa* (a tabular statement of the numerical data) without a solution (the answer is given after the quotation of the next verse).

> *nyāsaḥ/ 5/ 2/ 14//* prose part of PM A5 //
> 'Setting-down: 5, 2, 14.'

GSK 4.16. Meeting of two runners 2 (camels).

One runner (camel) goes the distance a for the first day and increases his daily journey by d every day. Another goes the distance v every day constantly. When they start at the same time, the time for their meeting can be calculated by

$$t = \frac{v-a}{d} \cdot 2 + 1 \text{ days.}$$

For, we have the equation, $A(t) = vt$, where $A(t)$ is the sum of the arithmetical progression, for which see GSK 3.26 above.

It is interesting that this formula is expressed in terms of the positions of the three quantities, a, d and v, in the *nyāsa*. See the Translation.

This verse is repeated verbatim in GSK 4.54 below, and is cited in PM A6 with the answer to the previous example.

> *labdhāṅka-10-dinair milati//* prose part of PM A6 //
> 'The digits obtained are 10 days, in which she (the female camel) meets ⟨the male camel⟩.'

GSK 4.17. Conversion of Vikrama and Hijrī dates.

Pherū is the only Indian mathematician to discuss the conversion of eras from one to another.

Let a given day be dated (y_v, m_v, d_v) in the Vikrama calendar and (y_h, m_h, d_h) in the Hijrī calendar.

1. Pherū's procedure for converting a Vikrama date into a Hijrī date.

(1) Convert the Vikrama years and months into solar days: $(y_v, m_v) \rightarrow d_{vym}$.
(2) Calculate the intercalary months elapsed: $m_{va} = d_{vym}/976$.
(3) Express its integer part in years and months: $[m_{va}] = (y', m', 0)$.
(4) Add it to the Vikrama date: $(y'_v, m'_v, d'_v) = (y_v, m_v, d_v) + (y', m', 0)$.
(5) Subtract from it the difference between the two eras in lunar years to obtain the corresponding Hijrī date: $(y_h, m_h, d_h) = (y'_v, m'_v, d'_v) - (699, 2, 2)$.

Here, (y_v, m_v) are expired and (y_h, m_h) are current. $(699, 2, 2)$ should be $(699, 2, 1)$.

2. Pherū's procedure for converting a Hijrī date into a Vikrama date.

(1) Convert the Hijrī years and months into lunar days: $(y_h, m_h) \rightarrow d_{hym}$.
(2) Calculate the intercalary months elapsed: $m_{ha} = d_{hym}/1006$.
(3) Express its integer part in years and months: $[m_{ha}] = (y', m', 0)$.
(4) Subtract it from the Hijrī date: $(y'_h, m'_h, d'_h) = (y_h, m_h, d_h) - (y', m', 0)$.
(5) Add it to the difference between the two eras in the Vikrama calendar to obtain the corresponding Vikrama date: $(y_v, m_v, d_v) = (y'_h, m'_h, d'_h) + (679, 0, 2)$.

For the first three steps, Pherū simply says, 'Do the same'. Here, y_h and y_v are 'expired', and m_h and m_v current. $(679, 0, 2)$ should be $(679, 4, 1)$.

Cf. Subbarayappa and Sarma 1985, 59–61; Sarma 1990.

4.2 Cloth

This section entitled the 'topic of cloth' enumerates first (GSK 4.18) several varieties of silk and cotton cloth. The next verse (4.19) mentions the rate of shrinkage in sewing (1.5 %), in cutting (1 %) and in washing (1 to 3 % depending on the material). For Indian textile terminology, see Habib 2004 and 2008.

The rest of the section (4.20–37) deals with the calculation of the area of the cloth required for constructing various types of tents (*khīma*, Ar. *khaima*), starting with simple ones with one tent pole (*iga-thaṃbha*) or two tent poles (*du-thaṃbha*) (4.21, 30) to larger tents serving as assembly halls (*vārigaha*, from Pe. *bārgāh*) (4.29, 31). Tents were an important element of residential architecture of the Muslim rulers. Abū al-Faḍl [2004, 89–90: Book I, Ā'īn 21] describes the tents erected for the use of Emperor Akbar and his entourage, while Pope [1977, vol. 3, chap. 43] discusses the history of tents based mainly on Persian painting. Pherū's account of the tent construction in the early fourteenth century Delhi is of great value, but unfortunately his peculiar diction is not conducive to clear understanding.

The following are the formulas that we can pick up from the verses.

$$\text{GSK 4.22:}\quad A_1 = \frac{a+b}{2} \cdot h, \quad A_2 = A_1 + cd\left(1 + \frac{1}{8}\right).$$

$$\text{GSK 4.26:}\quad A = \frac{a+b}{2} \cdot h.$$

$$\text{GSK 4.27:}\quad A = \frac{p}{2} \cdot h.$$

$$\text{GSK 4.28:}\quad A = ah.$$

$$\text{GSK 4.29:}\quad A = \frac{a+b}{2} \cdot h.$$

4.3 Magic Squares

This is the first discussion of magic squares in a mathematical text in India. Pherū calls a magic square *jaṃta* (S. *yantra*, lit. a device, an apparatus, a figure, a diagram). He differentiates three types of magic squares, namely, odd (*visama*, S. *viṣama*), even (*sama*), and evenly-odd (*samavisama*), according to the number of cells (*kuṭṭha*, S. *koṣṭha*; or *giha/geha*, S. *gṛha*, lit. a house) on each side of the square. In the eight verses that follow, Pherū gives one construction method each, with examples, for the first two types and one example only for the last. For the sake of easy expression, we call 'a magic square with n cells ⟨on each side⟩' (*n-giha-jaṃta*) simply 'a magic square of order n'. Pherū never refers to the constant sum, $n(n^2+1)/2$, that makes a square magic.

For a survey of Indian magic squares with a full bibliography see Hayashi 2008.

GSK 4.38. Classification of magic squares and examples of order 4.

Lines 1–3: An example of magic square of order 4.

Pherū simply enumerates the sixteen numbers (12, 3, 6, 13; 14, 5, 4, 11; 2, 9, 16, 7; 8, 15, 10, 1), the first two quarters in the first line and the rest in the third line, without specifying how to arrange them. After a few minutes of trial and error, however, we find that the first two quarter sequences should be arranged in the same direction and the remaining two in the reversed direction.

12	3	6	13
14	5	4	11
7	16	9	2
1	10	15	8

Note that this square not only satisfies the conditions of a magic square but also has a number of quadruplets whose sum equals the constant sum, 34, and that the diagonals of every quarter square of four numbers each has the sum, 17, which is just half the constant, although it is not pan-diagonal.

Line 4: Rearrangement of the given magic square.

The details are not clear but eventually the following seems to be what Pherū intended.

(1) Divide the square into four quarters of four numbers each.

I	II
III	IV

(2) Rearrange the four quarters optionally, rotating or reversing each of them.

For example, when we rotate the four quarters clockwise by 90°, rotating II and III severally and reversing I and IV severally, we obtain the new magic square:

1	7	12	14
10	16	3	5
8	2	13	11
15	9	6	4

Cf. GSK 4.42ab below for another example.

Line 5: Not understood.

Line 6: Classification.

Here, Pherū refers to the classification of magic squares into the three types, odd, even, and evenly-odd, according to the number (n) of cells on each side. He does not give definitions but simply mentions that 4 is an even number. But we have no reason to deny that he knew the definitions themselves:

$$n \text{ is} \begin{cases} \text{odd} & \text{if it cannot be halved} : n = 2k+1, \\ \text{even} & \text{if it can be halved twice or more} : n = 4k, \\ \text{evenly-odd} & \text{if it can be halved only once} : n = 4k+2. \end{cases}$$

In the last line of verse 41, Pherū describes the 'even' n as 'beginning with four cells and increasing by four'.

GSK 4.39–40. Magic squares of the evenly-odd order ($n = 4k+2$): an example.

Pherū only shows an example ($n = 6$) for magic squares of the evenly-odd order probably because he did not know general construction method.

He first gives two arithmetical progressions ($(a, d, n) = (1, 7, 6)$ for the first and $(6, 5, 6)$ for the second), which are to be arranged in the two diagonals, and then enumerates the remaining numbers (32, 34, 33, 5; 7, 19, 18, 25; 35, 3, 4, 2; 12, 13, 24,30; 28, 27; 14, 17; 10, 9; 20, 23). Here, with a single phrase, 'beginning in the east (top)', Pherū hints that they should be arranged clockwise in the square: east (top) → south (right) → west (bottom) → north (left) → east → ..., that is, in the *pradakṣiṇa* manner.

1	32	34	33	5	6
30	8	28	27	11	7
24	23	15	16	14	19
13	20	21	22	17	18
12	26	9	10	29	25
31	2	4	3	35	36

GSK 4.41. Magic squares of the even order ($n = 4k$): a construction method.

The details of Pherū's rule are not clear enough but it may be summarized as follows. The example we give for illustration is the case of $n = 8$ ($k = 2$).

(1) Divide the square into k^2 component squares of sixteen cells each ($n^2 = 16k^2$).

(2) Fill the first 'quarter square' (*pā-jaṃta*, S. *pāda-yantra*) with the numbers of the first 'quarter sequence' (*pā-oli* or *pāya-oli*, S. *pāda-āvali*), $1 \sim 4k^2$, by putting 4 consecutive numbers in each component square according to the pattern of the first quarter sequence (of 4 numbers) of any model square of order 4, for which see the next verse. Note that the order of the component squares may be determined arbitrarily. Note also that the 'quarter square' means not any real square but one-quarter part of the square consisting of $4k^2$ cells scattered in the square (the cells occupied by numbers in the following diagram).

			4				8
	2				6		
1				5			
		3				7	
			12				16
	10				14		
9				13			
		11				15	

(3) Fill the second 'quarter square' by the second 'quarter sequence', $(4k^2 + 1) \sim 8k^2$, by putting 4 consecutive numbers in each component square according to the pattern of the second quarter sequence (of 4 numbers) of the model square. This time, the order of the component squares must be reversed.

		30	4			26	8
32	2			28	6		
1	31			5	27		
		3	29			7	25
		22	12			18	16
24	10			20	14		
9	23			13	19		
		11	21			15	17

(4) Do the same for the remaining half and the following magic square is obtained.

63	33	30	4	59	37	26	8
32	2	61	35	28	6	57	39
1	31	36	62	5	27	40	58
34	64	3	29	38	60	7	25
55	41	22	12	51	45	18	16
24	10	53	43	20	14	49	47
9	23	44	54	13	19	48	50
42	56	11	21	46	52	15	17

GSK 4.42ab. Model square of order 4.

Pherū simply enumerates the sixteen numbers: 15, 9, 6, 4; 8, 2, 13, 11; 1, 7, 12, 14; 10, 16, 3, 5. There is no doubt that these were arranged in the most natural way, that is, all quarter sequences from left to right. Note that this square can be constructed by the method told in verse 38, line 4, above.

15	9	6	4
8	2	13	11
1	7	12	14
10	16	3	5

GSK 4.42cd. Ex.: a magic square of order 8.

Construct a magic square of 64 cells or of order 8 'in the order of one-three-two-four'.

Answer: The 'order of one-three-two-four' presumably indicates the order of filling component squares with the first quarter sequence of numbers. The most natural way of giving serial numbers to the component squares seems to be from left to right on each row as in the case of writing Indian scripts. Then, the square obtained is as follows.

63	33	30	4	55	41	22	12
32	2	61	35	24	10	53	43
1	31	36	62	9	23	44	54
34	64	3	29	42	56	11	21
59	37	26	8	51	45	18	16
28	6	57	39	20	14	49	47
5	27	40	58	13	19	48	50
38	60	7	25	46	52	15	17

GSK 4.43–44. Magic squares of the odd order ($n = 2k + 1$): a construction method.

Here, Pherū's rule is very clear. It is summarized as follows.

(1) Construct the central column with the 'first sequence' (*paḍham-oli*, S. *prathama-āvali*) that consists of the arithmetical progression whose first term is one, common difference $(n + 1)$, and number of terms n.

n^2
$\vdots$
$2n+3$
$n+2$
1

(2) Fill the next left (or right) column by putting the number (q) that equals the sum, $(p+n)$, in the cell reached by the horse move (in chess, 'knight's move' in modern parlance) (*asu-kama*, S. *aśva-krama*) from the cell of p in the centre column. Note that the horse move to the second next column would not make a magic square.

q	
	p

or

	q
p	

where $q = p + n$.

(3) If the sum, $(p+n)$, is greater than n^2, then subtract n^2 from it.

(4) Repeat the same for the next columns. If the left (or the right) side is reached, go to the last (or the first) column, although this is not mentioned by Pherū.

GSK 4.45. Ex.: Magic squares of orders 3 and 9.

First, Pherū enumerates the nine numbers: 4, 9, 2; 3, 5, 7; 8, 1, 6. These numbers constitute the following square, which can, of course, be constructed by the method told above. Here, Pherū proceeds toward left from the central column.

4	9	2
3	5	7
8	1	6

Second, he asks for a magic square of 81 cells or of order 9.

Answer: Proceeding toward left from the centre column as in the above square of order 3, we obtain the following square.

37	48	59	70	81	2	13	24	35
36	38	49	60	71	73	3	14	25
26	28	39	50	61	72	74	4	15
16	27	29	40	51	62	64	75	5
6	17	19	30	41	52	63	65	76
77	7	18	20	31	42	53	55	66
67	78	8	10	21	32	43	54	56
57	68	79	9	11	22	33	44	46
47	58	69	80	1	12	23	34	45

4.4 Miscellaneous

GSK 4.46. Offering of flowers at a temple.

Type of problem: At each entrance of a temple (*sura-giha*, S. *sura-gṛha*), a pious man (*dhammiya*, S. *dhārmika*) offers one flower (*kusuma*) to Yakṣa, half of the remaining flowers to the main image, and again one flower to Yakṣa. If the temple has n entrances and the number of the flowers that remain in his hand at the end of his tour is c, how many flowers (x_0) did he originally have?

$$\frac{x_0 - 1}{2} - 1 = x_1, \ \ldots, \ \frac{x_{i-1} - 1}{2} - 1 = x_i, \ \ldots, \ \frac{x_{n-1} - 1}{2} - 1 = c.$$

Rule: (1) Starting from c, repeat the operations, 'double and add three', n times. (2) If $c = 0$, then assume the initial number for the operations to be 3. That is to say,

$$x_{n-1} = 2c + 3, \ \ldots, \ x_{i-1} = 2x_i + 3, \ \ldots, \ x_0 = 2x_1 + 3.$$

This rule can be obtained by applying the rule of inversion (see verse 56 below) after rewriting each equation as $(x_{i-1} - 3)/2 = x_i$.

This verse is cited and paraphrased in Sanskrit prose in PM A7.

> *śeṣapuṣpāṇi dvārasaṃkhyaṃ dviguṇīkṛtya tribhir yuktāni sarvapuṣpa-saṃkhyā/ puṣpaśeṣo yatra na syāt tatrādau puṣpatrayaṃ prakalpya prāgvat prakalpyā/ yathā//* prose part of PM A7 //
> 'The ⟨number of the⟩ remaining flowers, doubled and increased by three as many times as the number of the entrances, is the number of all the flowers. Where no flower remains, having assumed three flowers ⟨as the initial number⟩, one should settle ⟨the number of all the flowers⟩ as before. For example (this phrase introduces the next example, PM A8).'

GSK 4.47. Ex.

Calculate the original number of flowers when $n = 4$ and $c = 20$.

Answer: $x_0 = 365$.

This verse is paraphrased and the problem is solved in Sanskrit prose in PM A8.

> *kaścid dhārmikaś caturdvāracaitye pratidvāraṃ praveśanirgamayor yakṣasyaikaikaṃ kusumaṃ dattvā śeṣakusumārdhena devārcaṃ vidhāya śeṣāṇi viṃśatiṃ darśayati/ tatra kā sarvapuṣpasaṃkhyā// nyāsaḥ/ śeṣapuṣpa/ 20/ dviguṇa/ 40/ tribhir yukte/ 43/ dviguṇe/ 86/ triyukte/ 89/ dviguṇe/ 178/ triyukte/ 181/ dviguṇe 362 triyukte/ 365/ evaṃ sarvakusumasaṃkhyā//*

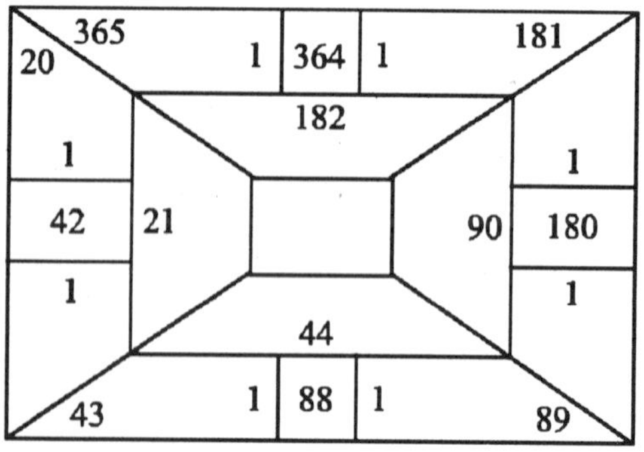

// prose part of PM A8 //

'A certain pious man, at a temple with four entrances, having offered one flower each to Yakṣa at each entrance while going in and out through the entrance and having made worship of the ⟨main⟩ god with half of the remaining flowers, shows the remaining twenty ⟨flowers⟩. What is the number of all the flowers in that case? Setting-down: the remaining flowers 20. Doubled, 40; increased by three, 43; doubled, 86; increased by three, 89; doubled, 178; increased by three, 181; doubled, 362; and increased by three, 365. Thus the number of all the flowers. (Figure)'

GSK 4.48. Equal distribution of the plucked mangos.

Type of problem: One m-th part of the soldiers that constitute an army went to a forest and the i-th soldier among them plucked $(a+(i-1)d)$ mangos. On coming back to the camp, they divided the mangos they obtained equally among all the soldiers of the army and every soldier received the same number (p) of mangos. How many soldiers (z) went to the forest? How many soldiers (y) were there in the army? How many mangos (x) did they pluck in total?

$$z = \frac{y}{m}, \quad x = A(z), \quad p = \frac{x}{y}.$$

Rule:

$$z = \frac{mp-a}{d} \cdot 2 + 1, \qquad x = A(z), \qquad y = \frac{x}{p}.$$

For $A(z)$, see GSK 3.26 above.

This verse is cited in PM A9. The first formula of this rule is given also in PM A31 and an example for it is given in PM A32 (a troop of monkeys). A Sanskrit mathematical anthology deals with a simpler case of the same type of problem, in which the i-th person plucks i pieces of fruit; see Hayashi 2006, 207–08.

GSK 4.49. Ex.

Calculate the number of mangos (x), the number of all the soldiers (y), and the number of the soldiers (z) who went to the forest when $m = 8$, $a = 4$, $d = 6$, and $p = 20$.

Answer: $z = 53$, $x = 8480$, and $y = 424$.

This verse is cited in PM A10 and the problem is solved in Sanskrit prose.

nyāsaḥ/ aṣṭamāṃśaḥ 8/ ādiḥ 4/ vṛddhiḥ 6/ prāptaphalam 20/ phalam aṃśaguṇaṃ $\begin{array}{|c|}8\\20\\\hline\end{array}$ *jātaṃ/ 160/ ādi-4-hīne 156/ vṛddhyā ṣaṭkena bhāge/ 26/ dviguṇe 52/ rūpayute 53/ iyaṃ sainyāṣṭamāṃśajanasaṃkhyā/ atha śreḍhīsaṅkalite nyāsaḥ/ 4/ 6/ 53/ gaṇite labdham āmramānaṃ 8480/ asya viṃśāṃśaḥ/ 424/ kaṭakajanasaṃkhyā//* prose part of PM A10 //

'Setting-down: one-eighth part 8, the first term 4, the increase 6, the fruit obtained 20. The fruit is multiplied by the part: $\begin{array}{|c|}8\\20\\\hline\end{array}$ What is produced is 160. Decreased by the first term, 156; divided by the increase, i.e., by six, 26; doubled, 52; and increased by unity, 53. This is the number of persons in the one-eighth part of the army. Now, setting-down for the sum of the series: 4, 6, 53. When calculated, the number of the mangos obtained is 8480. One-twentieth of this is 424; ⟨this is⟩ the number of persons in the army.'

GSK 4.50. Equal number of *varisolas* eaten.

Type of problem: A mother-in-law offers a confectionary called *varisola* to her n sons-in-law, and each one eats the same number of pieces. Every time before offering, she increases the number of pieces remaining on the plate by a_i ($i = 1, 2, ..., n$). The last man eats all the pieces on the plate. What is the original number of pieces (x_0) and what is the number of pieces (y) eaten by each person?

$$a_1 x_0 - y = x_1, \ ..., \ a_i x_{i-1} - y = x_i, \ ..., \ a_n x_{n-1} - y = 0.$$

Rule: (1) Write down the given multipliers (a_i) in order from the bottom upwards. (2) Starting from the lowest term, which is replaced with unity, perform the following operation successively: Multiply the upper term by the lower and add one to the product. The uppermost term obtained (p_n) is the original quantity (x_0).

$$\begin{array}{ll} a_n & p_n = a_n p_{n-1} + 1 \\ a_{n-1} & p_{n-1} = a_{n-1} p_{n-2} + 1 \\ \vdots & \\ a_3 & p_3 = a_3 p_2 + 1 \\ a_2 & p_2 = a_2 p_1 + 1 \\ a_1 & p_1 = 1 \end{array}$$

(3) Multiply the given multipliers (a_i) with each other, ignoring the unity, with which the first multiplier is replaced in the previous computation. The result is the quantity eaten by each man (y).

$$y = \prod_{i=2}^{n} a_i .$$

Note that the given relationships can be reduced to an indeterminate equation with the two unknown quantities, x_0 and y:

$$\left(\prod_{i=1}^{n} a_i\right) x_0 = \left\{\sum_{k=2}^{n}\left(\prod_{i=k}^{n} a_i\right) + 1\right\} y.$$

The above rule gives a solution to this equation by assuming

$$x_0 = \sum_{k=2}^{n}\left(\prod_{i=k}^{n} a_i\right) + 1, \qquad y = \prod_{i=1}^{n} a_i \quad \left(\prod_{i=2}^{n} a_i \ \text{ if } a_1 = 1\right).$$

This verse is cited and paraphrased in Sanskrit prose in PM A11.

> *krameṇaikādayo guṇakāḥ sthāpyante/ pratyekaṃ pūrvapūrveṇa saṃguṇya rūpayutaṃ kāryam/ eṣā janopabhoktavyasolasaṃkhyā/ ete 'pi guṇakasthāpitāṅkāḥ pratyekaṃ pūrvapūrveṇa guṇyā bhakṣitasaṃkhyā bhavati// yathā//* prose part of PM A11 //
>
> 'In the ⟨given⟩ order, the multipliers beginning with unity are put down. Each term should be multiplied by the previous one and increased by unity. This is the number of ⟨all the⟩ *solas* to be eaten by ⟨all⟩ the persons. Also, these numbers (lit. digits) which were put down as the multipliers should each be multiplied by the previous one; ⟨the result⟩ will be the number ⟨of *solas*⟩ eaten ⟨by each person⟩. For example (this phrase introduces PM A12).'

In order to obtain the same solution (meant, however, for the problem of the number of lotus flowers used for a daily rite), a different procedure has been given by an anonymous author cited in the two manuscripts that contain Nārāyaṇa's part of the *Kriyākramakarī* (ed. by K. V. Sarma, pp. 264–66). He arranges the given multipliers vertically from top to bottom and multiplies them successively in the reverse order. Then, the uppermost term obtained is y and the sum of the other terms obtained plus one is x_0.

$$\begin{array}{ll} a_1 & a_1 a_2 \cdots a_{n-1} a_n \\ a_2 & a_2 \cdots a_{n-1} \\ \vdots & \vdots \\ a_{n-1} & a_{n-1} a_n \\ a_n & a_n \end{array}$$

The anonymous author gives two examples.

(1) $n = 9$ and $\{a_i\} = \{1, 2, 3, 4, 5, 6, 7, 8, 9\}$. Answer: $x_0 = 623530$ and $y = 362880$, or cancelled by 10, $x_0 = 62353$ and $y = 36288$.

(2) $n = 9$ and $\{a_i\} = \{3, 2, 1, 6, 5, 4, 9, 8, 7\}$. Answer: $x_0 = 254584$ and $y = 362880$, or cancelled by 8, $x_0 = 31823$ and $y = 45360$.

GSK 4.51. Ex.

Calculate the original number of *varisolas* and the number eaten by each son-in-law when $n = 5$ and $\{a_i\} = \{1, 2, 3, 4, 5\}$.

Answer: $x_0 = 206$ and $y = 120$.

This verse is paraphrased and the problem is solved in Sanskrit prose in PM A12. The author of the PM arranges $\{a_i\}$ from the top downward. He also gives a verification of the answer by calculating x_1, x_2, x_3, and x_4 one after another.

ke 'pi pañca jāmātaraḥ śvaśrūgṛhe gataḥ/ teṣv ādyasya śvaśrvā varasolakāni dattāni tadbhakṣitaśeṣāṇi dviguṇīkṛtya dvitīyasya tadbhakṣitaśeṣāṇi triguṇīkṛtya tṛtīyasya tadbhakṣitaśeṣāṇi catrurguṇīkṛtya caturthasya tadbhakṣitaśeṣāṇi pañcaguṇīkṛtya pañcamasya/ tena sarvāṇy api bhakṣitāni/ tāni samāni jātāni// nyāsaḥ | *1* | *2* | *3* | *4* | *5* | *pūrvapūrveṇa guṇanam/ ekena dvike guṇite ekayute/ 3/ tena trikena trike guṇite ekayute/ 10/ caturguṇe saike ca/ 41/ pañcaguṇe saike ca/ 206/ iyanti prathamajāmātur dattāni// punaḥ pūrvaguṇakānāṃ pūrvapūrvaguṇanam/ ekena dvike guṇite 2 dvikena trike gu° / 6/ ṣaṭkena catuṣke gu° / 24/ caturviṃśatau pañcagu° 120/ etāni tanmadhyād ādyena bhuktāni śeṣa/ 86/ dviguṇa/ 172/ dvitīyasya dattāni/ 120/ bhuktāni śeṣa/ 52/ triguṇa/ 156/ tṛtīyena 120 bhuktāni śeṣa 36/ caturguṇa/ 144/ bhukta 120 śeṣa pañcaguṇa/ 120 bhuktāni//* PM A12 //

'Certain five sons-in-law went to their mother-in-law's house. To the first of them were given ⟨a certain number of⟩ *varasolakas* by the mother-in-law; the ⟨pieces⟩ left over by him were doubled ⟨in number⟩ and ⟨given⟩ to the second; the ⟨pieces⟩ left over by him were made three times more than before and ⟨given⟩ to the third; the ⟨pieces⟩ left over by him were made four times more than before and ⟨given⟩ to the fourth; and the ⟨pieces⟩ left over by him were made five times more than before and ⟨given⟩ to the fifth. All ⟨the pieces given to the fifth⟩ were eaten by him. They (i.e. the number of pieces eaten by each son-in-law) happened to be the same. Setting-down: | 1 | 2 | 3 | 4 | 5 | Successive multiplication by the preceding term: when two is multiplied by one and increased by one, 3; when three is multiplied by this three and increased by one, 10; when ⟨this is⟩ multiplied by four and increased by one, 41; and when ⟨this is⟩ multiplied by five and increased by one, 206. These ⟨206 pieces⟩ were given to the first son-in-law. Of the previous multipliers, another successive multiplication by the preceding term: when two is multiplied by one, 2; when three is multiplied by two, 6; when four is multiplied by six, 24; and when twenty-four is multiplied by five, 120. These ⟨120 pieces⟩, out of those ⟨206 pieces⟩, are eaten by the first; the leftovers, 86, are doubled, 172, and given to the second, ⟨by whom⟩ 120 ⟨pieces⟩ are eaten; the leftovers, 52, are multiplied by three, 156, ⟨and given to the third⟩; by the third 120 ⟨pieces⟩ are eaten; the leftovers, 36, are multiplied by four, 144, ⟨and given to the fourth⟩, ⟨by whom⟩ 120 ⟨pieces⟩ are eaten; and the leftovers, ⟨24⟩, are multiplied by five, 120, which are all eaten ⟨by the fifth⟩.'

GSK 4.52. Equal division of a square piece of cloth.

Type of problem: When a square piece of cloth whose side is x *karas* is equally divided among y persons, the share of each is p sq. *karas*. When these y persons are made to stand at equal distances along the four sides of the cloth, n persons hold one *kara* length.

$$x^2 = py, \qquad y = 4nx.$$

Rule:

$$x = 4np, \quad y = 4nx.$$

This verse is cited in PM A13 and paraphrased in Sanskrit prose.

> *ye janā hastaṃ gṛhṇanti te caturguṇāḥ kāryās tato labdhavastra-karair guṇyās tad vastrasya dairghyavistārarūpaṃ pramāṇaṃ jñeyam/ idam eva karajanair guṇyaṃ caturguṇaṃ ca sarvasaṃkhyā bhavanti//* prose part of PM A13 //
> '⟨The number of⟩ those persons who hold one *hasta* ⟨of the cloth⟩ should be multiplied by four and then also by the ⟨square⟩ *karas* (*hastas*) of a piece of cloth obtained ⟨by each person⟩. This should be known as the size of the cloth which assumes the forms of the length and the width. The same, multiplied by ⟨the number of⟩ the persons per *kara* and by four, is the number of all ⟨persons⟩.'

GSK 4.53. Ex.

Calculate the side-length of a square piece of cloth and the number of people when $n = 3$ per *kara* and $p = 9$ sq. *karas*.
Answer: $x = 108$ *karas* (*hatthas*) and $y = 1296$.

This problem is solved (without the verse) in Sanskrit prose in PM A14.

> *nyāsaḥ/ hasta/ 1/ jana/ 3/ labdhakara/ 9/ janāḥ/ 3/ caturguṇāḥ/ 12/ labdhakara-9-guṇite/ 108/ idaṃ janais tribhir guṇitaṃ/ 324/ punaś caturguṇāḥ/ 1296/ eṣā sarvajanasaṃkhyā// dairghye/ 108/ vistareṇa/ 108/ guṇite/ 11664/ eṣā vastrakarasaṃkhyā//* A14 //
> 'Setting-down: ⟨the number of⟩ persons per 1 *hasta*, 3; ⟨square⟩ *karas* obtained, 9. ⟨The number of⟩ persons, 3, is multiplied by four, 12. When ⟨this is further⟩ multiplied by the ⟨square⟩ *karas* 9 obtained, 108. This, ⟨which is the side of the square piece of cloth⟩, is multiplied by the three persons, 324, and further multiplied by four, 1296. This is the number of all persons. When the length, 108, is multiplied by the width, 108, ⟨we have⟩ 11664. This is the number of ⟨square⟩ *karas* of the cloth.'

GSK 4.54. = GSK 4.16.

GSK 4.55. Ex.

Calculate the time for the meeting of two runners when $a = 4$ *joyaṇas*/day, $d = 3$ *joyaṇas*/day^2, and $v = 16$ *joyaṇas*/day.

Answer: $t = 9$ days.

GSK 4.56. Solution of an equation by the rules of inversion.

Problem in modern notation:

$$\left[\{(x \times a) \div b\}^2 + c\right]^{1/2} - d = e.$$

Rule in modern notation:

$$x = \left\{(e+d)^2 - c\right\}^{1/2} \times b \div a.$$

For the actual procedure, see the next verse.

Note that all the six basic rules of inversion are used in this solution.

$$\text{If } x - a = b, \text{ then } x = b + a; \quad \text{if } \sqrt{x} = b, \text{ then } x = b^2;$$

$$\text{if } x + a = b, \text{ then } x = b - a; \quad \text{if } x^2 = b, \text{ then } x = \sqrt{b};$$

$$\text{if } x \div a = b, \text{ then } x = b \times a; \quad \text{if } x \times a = b, \text{ then } x = b \div a.$$

GSK 4.57. Ex.

$$\left[\{(x \times 5) \div 9\}^2 + 9\right]^{1/2} - 2 = 3.$$

Answer: $x = \frac{36}{5} = 7\frac{1}{5}$.

The actual procedure must have been like this. (1) Write down the names of the operations with the numbers given in the problem exactly in the same order as stated. (2) Write down the names of the inverse operations below them. (3) Starting from the last term called 'remainder' (which is usually called 'visible', *dṛśya*), perform the inverse operations, as indicated in the above rule, successively in the inverse order. The number obtained at the end is the solution.

multiplier	divisor	square	additive	root	subtractive	remainder
5	9		9		2	3
divisor	multiplier	root	subtractive	square	additive	↓
36/5 ←	36 ←	4 ←	16 ←	25 ←	5 ←	3

GSK 4.58. Mind reading 1.

Ask the audience to think of a positive integer (x) smaller than 105 (although this restriction is not stated in the text) and to perform the following operations upon it. (1) Diminish it severally by three, by five, and by seven, repeatedly until no more subtraction is possible; (2) multiply the three remainders (r_1, r_2, r_3) by 70, 21 and 15, respectively; (3) add them up and increase the sum by 105. Ask him/her to tell the result (p). Divide it by 105. Then, the remainder (r) is the number thought of by the audience. That is to say,

$$x = 3q_1 + r_1,\ x = 5q_2 + r_2,\ x = 7q_3 + r_3;\ p = 70r_1 + 21r_2 + 15r_3 + 105.$$

$$\text{If } p = 105q + r \ \ (0 \le r < 105), \text{ then } x = r.$$

For,

$$p = 105\{x - (2q_1 + q_2 + q_3) + 1\} + x.$$

Note that the addition of 105 is not necessary from the mathematical point of view. Perhaps it is for the sake of avoiding the case $p < 105$, when the answer is trivial ($x = p$).

This verse is cited and paraphrased in Sanskrit prose in PM A18, but the author (compiler) of the PM mistakenly renders *una* of *ti^una* into S. *guṇa* ('multiplied'); the correct rendering is *ūna* ('diminished'). In the following quotation, we leave the erroneous number 51 as it is in the manuscript; it should be 15.

> *triguṇe kṛte yac cheṣaṃ tat 70-guṇaṃ* ⟨*pañcaguṇe kṛte yac cheṣaṃ tat*⟩ *21* ⟨*-guṇam*⟩*/ saptaguṇe kṛte yac cheṣaṃ tat 51-guṇam// eṣām aikye śatamadhye pañcādhikaśataṃ yojyam/ śatopari pātyam// iti paracintitāṅkajñānam//* prose part of PM A18 //
>
> 'When ⟨the number in question is⟩ multiplied by three, the remainder is multiplied by 70; when ⟨it is⟩ multiplied by five, the remainder is multiplied by 21; and when ⟨it is⟩ multiplied by seven, the remainder is multiplied by 51. When the sum of these ⟨three products⟩ is within (i.e., less than) one hundred, one hundred and five should be added; when ⟨it is⟩ more than one hundred, it should be subtracted. Thus how to know the digits thought of by the other person.'

GSK 4.59. Mind reading 2.

Ask the audience to think of any two positive integers (x, y), called 'son' and 'daughter' (number? or age?), and to perform the following operations upon them. (1) Add two to the 'son' (x), (2) double it, (3) add one to it, (4) multiply it by five, (5) add the 'daughter' (y) to it, and (6) multiply it by ten. Ask him/her to tell the result (p). Subtract 250 from it and remove '0' in the units' place. Then, the (new) tens' and units' places are the 'son' (x) and the 'daughter' (y), respectively. For,

$$p - 250 = [[\{(x+2) \times 2 + 1\} \times 5] + y] \times 10 - 250 = (10x + y) \cdot 10.$$

Note that y must be smaller than ten. There is no restriction for x. Note also that the text is not clear about the time when the audience tells the result of his/her computation to the mind reader.

This verse is cited and paraphrased in Sanskrit prose in PM A17.

> *cintitaputrāṅkā dvayayutā dviguṇā ekayutāḥ pañcaguṇitāḥ putryaṅkasahitā daśaguṇāḥ 250-hīnāḥ kriyante/ śūnyaṃ vinā śeṣāṅkābhyāṃ kramāt putraputrījñānam//* prose part of PM A17 //
>
> 'The *son* digits ⟨which were⟩ thought of are increased by two, multiplied by two, increased by one, multiplied by five, increased by the *daughter* digits, multiplied by ten, and decreased by 250. One knows the *son* and the *daughter* respectively from the remaining digits except zero.'

GSK 4.60. Restoration of erased digits (*maliy-aṃka*, S. *mardita-aṅka*).

Not understood.

GSK 4.61. Number consisting of the same digits.

When 12345679 is multiplied by any optional (single) digit (a) and by 9, the result is a number consisting of 9 a's.

$$12345679 \times a \times 9 = aaaaaaaaa.$$

GSS 2.10 sets the problem, 12345679 × 9, as an example of multiplication, calling the product, 111111111, 'a royal necklace' (*nara-pāla-kaṇṭhikā-ābharaṇa*).

GSK 4.62. Equal divisions of cows (or a common multiple).

When the number (x) of cows can be divided by each of the integers, a_i ($i =, 1, 2, ..., n$), the number (actually, a candidate for it) can be obtained by multiplying all terms of the column, $\{a_1, a_2, \cdots, a_n\}$, with each other (*uvarup-pari*, S. *upary-upari*, lit. upwards). That is to say,

$$x = \prod_{i=1}^{n} a_i.$$

This is a solution of the system of indeterminate equations,

$$\frac{x}{a_i} = y_i \quad \text{(an integer for every } i\text{)}.$$

This verse is cited and paraphrased in Sanskrit prose in PM A15. According to the author (compiler) of the PM, the multiplications are made from the top downwards.

> *uparito 'dhastāṅkaṃ yāvad anyonyam aṅkā guṇyāḥ sarvagosaṃkhyāḥ//*
> *yathā//* prose part of PM A15 //
> 'The digits ⟨arranged vertically and⟩ mutually multiplied from the top to the bottom are the number of all cows. For example (this phrase introduces PM A16).'

GSK 4.63. Ex.

Calculate the number of cows when $\{a_i\} = \{4, 5, 7, 9, 6, 8\}$.
Answer: $x = 60480$.

This verse is paraphrased and the problem is solved in Sanskrit prose in PM A16. Cf. PM B14.

> *4/ dvārair niḥsaranti gāvaḥ/ pañcasarassu pibanti/ 7/ vṛkṣeṣu sthitāḥ/ 9/ pratolīṣu praviśanti/ 6/ vaneṣu caranti/ 8/ gopai rakṣyante/ sarvatra samasaṃkhyāḥ// nyāsaḥ/ 4/ 5/ 7/ 9/ 6/ 8/ anyonyaṃ guṇyante/ catuṣkena pañcake gu° / 20/ 20/ saptake gu° / 140/ anena navake gu° / 1260/ anena ṣaṭke/ 7560/ anenāṣṭake/ 60480/ iyaṃ*

sarvagosaṃkhyā// PM A16 //
'⟨A certain number of⟩ cows go out ⟨of a town⟩ through 4 gates, drink water at five lakes, rest under 7 trees, enter into 9 roads, roam about in 6 forests, and are protected by 8 cowherds. Everywhere ⟨they are⟩ the same in number. Setting-down: 4, 5, 7, 9, 6, 8. ⟨These are⟩ multiplied mutually: when five is multiplied by four, 20; when seven is multiplied by 20, 140; when nine is multiplied by this, 1260; when six ⟨is multiplied⟩ by this, 7560; and when eight ⟨is multiplied⟩ by this, 60480. This is the number of all cows.'

GSK 4.64. Distribution of cow's milk (*duddha*, S. *dugdha*).

Not understood.

Chapter 5

Quintet of Topics

5.1 Yields of Grains

GSK 5.1. Introduction.

After saluting the Creator, the author (whose name is not mentioned here) enumerates 'the five topics (*uddesa*)' to be treated in this part. These are concerned with the 'yield' (*nippatti*, S. *niṣpatti*) or 'fruit' (*phala*) of five things, whose names are given in two compounds and one word, all in the genitive plural case.

(1) 'of grains, sugarcane juice, and oil',
(2) 'of regional tax and price',
(3) 'of measures (or measurement)'.

The 'five' refers to (1) grains, (2) sugarcane juice and oil (both are products of extraction), (3) regional tax, (4) price, and (5) measurement. An extra section on masonry is appended after the 'measurement.'

Although the last verse (GSK 5.33) contains the name of Pherū, this chapter presumably did not originally belong to the GSK, because it is not included in the '45 gateways' to mathematics that constitute the GSK (see GSK 1.15) and also because it repeats several verses of the foregoing chapters. But it had been added by the time when the manuscript used by the Nahatas for the edition of the GSK was copied by Purisaḍa in VS 1404 (AD 1347).

GSK 5.2. Irrigated regions.

The author remarks that there is much difference in the harvest according to the quality of the soil and that Delhi (*ḍhilliya*), Hansi (*āsiya*), and Narhar (*narahaḍa*) are irrigated regions (*varuṇa paesa*, S. *varuṇa-pradeśa*). Hansi: 29°6′ N 75°58′ E, Hissar District, Haryana State. Narhar (or Nahar): 28°25′ N 76°22′ E, Bhiwani District, Haryana State.

GSK 5.3. Land measure.

Line 1 (GSK 5.3a):

$$\text{Area of land} = (\text{length}) \times (\text{width}) \ \textit{viggahayas}.$$

The area in *viggahayas* (S. *vigrahakas*) is here called 'the land-number' or 'the land-measure' (*bhū-saṃkhā*, S. *bhū-saṃkhyā*). The linear measures to be used for the length and the width are mentioned in the second line.

Line 2 (GSK 5.3b): = GSK 1.4b.

GSK 5.4–8. Farm produce.

The author here lists the yields in *maṇas* of 24 kinds of farm produce per *vīgahaya* (S. *vigrahaka*), which consists of 20 *visuvas*. The reason he mentions the *visuva* here is not clear, as there is no use of it in this context. Abū al-Faḍl [2004, 585: Book III, Ā'īn 10] records a vigesimal subdivision of the *bigha* (i.e., *vīgahaya*) including *biswah* (i.e., *visuva*):

20 ***answānsah*** = 1 ***tapwansah***,
20 ***tapwansah*** = 1 ***taswānsah***,
20 ***taswānsah*** = 1 ***biswānsah***,
20 ***biswānsah*** = 1 ***biswah***,
20 ***biswah*** = 1 ***bigha***.

For a similar vigesimal subdivision of the monetary unit *damma*, see GSK 1.3–4a above.

The farm produce mentioned and their yields per *vīgahaya* are shown in Table 5.1.

The data given by Pherū on the annual produce per *bīghā* of several crops in Delhi-Haryana region is very valuable for the economic history of the period. Recognizing the significance of this data, Dashrath Sharma published it under the title 'Some Light on the Economic Conditions from Ṭhakkura Pherū's Works' as Appendix N (p. 318) of his book *Early Chauhan Dynasties*, which came out in 1959. Later, in 1982, Irfan Habib made a brief analysis of these figures in the chapter on 'Agrarian Economy' of *The Cambridge Economic History of India* (vol. 1, chap. iii, p. 50) which he jointly edited with Tan Raychaudhury.

Prof. Habib has now drawn our attention to similar data recorded in a mid-fourteenth-century Persian work on accountancy (*ḥisāb*), entitled *Dastūr al-Albāb fī ᶜIlm al-Ḥisāb* by Hājjī ᶜAbd al-Ḥamīd Muḥarrir Ghaznavī (MS: Raza Library, Rampur: Riyazī Fārsī No. 32), written in AH 760 (AD 1358–59) for his son Rukn al-Dīn Abū al-Faṭ'h Aḥmad. In this work, Muḥarrir enumerates the rates of annual yield of different crops and fruits per *bīghā*. While the rest of the text is in prose, this part is in verse, presumably for easy memorization. Prof. Habib has very kindly translated for us the passage reproduced below.

* * * * * * * * * * * * * * *

[f. 102b]

Chapter 7. Knowledge of Yield of Food Grains, etc., that is grown on a *bīghā*. Poem.

[f. 102a]

I offer to your wisdom like a slave
In a versified form, easy to remember
Whatever I say is an addition to your art
It is necessary for every clerk to know so much
So that you know ... every foot (?)
From ... this yield is obtained ...
The crop of food grain will appear from it
Of fruit I will speak later.
Sixteen [*mans* is the yield of] the *sawan*, and *kangumi* millets (*shāmākh o gāl*)
Horse gram, til, safflower, mustard (*kulth, kanjad, ma^{c}ṣafar, sarsaf*), the same.
Korī (inferior rice) ... *man.*
Forty [*man*] is wheat, and ten *man* is cotton and sunn-hemp (*puba o san*)
As to *mung* (*māsh*), *moth* and gram, see to it yourself,
Along with green pea and *mung.*
With *masur* you should know *lobiya* to be at thirty *man,*
As for barley, it is fifty-five *mans,* O youth!
In gram, when greener, see two hundred *mans.*
In the sown field, it is four hundred *man,* brave one!
Candied sugar (*qand*) [or gur ?] is eighty *mans*, or somewhat less.
That [comes out of] four hundred and thirty-two *mans* of syrup.
... In a salt pan (*khārī*), thirty-two of what?
Two hundred and eighty-eight of jars,
One jar carries one and a half *man,*
So said Hājjī, of the idiom of this country.
He said, and in reply as to the yield (*wafā*) of a *bīghā*
[It is] two hundred and fifty *mans,* O excellent one.
Water-melon (*kharbuza*) is eighty *mans,* without fear,
Cucumbers (*khayyār o badrang*) amount to eight thousand.
In the house (?), know this too, my friend,
A *bīghā* of opium-poppy (*koknar*) yields twenty thousand [*jīluls*?]
One whole *man* and two and a half *ser* come from safflower.
Depend on estimation (*qayās*) for mango, lemon and *ber* (jujube).
In onion and in select *ber* (?), remember:
[The yield] is different in the village and in the orchard: do go around
Whatever is obtained as its weight,
It is necessary to test it by calculation.
For 'Muḥarrir' (pen-name of the author) is God-fearing, free of sorrow,
With foresight, he is fearful of calamity.
The peasant should be allowed to gain (*naf^{c}a*) from this.
So remains in possession of his cattle and seed.
If you wish to get more out of this calculation (*ḥisāb*)
You will yourself ruin the orchard as well as the village.

* * * * * * * * * * * * * * *

Table 5.1: Agricultural Produce, Yield per *Bīghā* (GSK 5.4–8)

Farm produce Apabhraṃśa (Sanskrit, English, Latin)	Yields (*maṇas/vīgahaya*)
Autumn harvest (*savāṇiya*)	
kuddava (*kodrava*, a kind of cereal, *Paspalum scrobiculatum*)	60
maüṭṭha (–, moth beans, *Phaseolus aconitifolius*)	24
caüla (–, chaula beans, *Vigna catiang* var. *sinensis*)	22
tila (*tila*, sesamum, *Sesamum indicum*)	16
muggamāsa (*mudga-māṣa*, kidney beans, *Vigna radiata*)	18
kaṃguṇiya (*kaṅgunika*, Italian millet, *Panicum italicum*)	20
cīṇaya (*cīnaka*, common millet, *Panicum miliaceum*)	15
kūrīsa (*kūra-īśa*, rice king)	7*
kappāsa (*karpāsa*, cotton, *Gossypium herbaceum*)	16
juvāra (–, Indian millet, *Sorghum vulgare*)	40
saṇa (*śaṇa*, flax, *Crotalaria juncia*)	10
ikkhu (*ikṣu*, sugarcane, *Saccharum officinarum*)	10
Spring harvest (*āsāḍhiya*)	
gohuva (*godhūma*, wheat, *Triticum aestivum*)	45
kalāva (*kalama*, rice, –)	32
massūra (*masūra*, lentil, *Ervum lens*)	32
caṇaya (*caṇaka*, chickpea, *Cicer arietinum*)	32
java (*yava*, barley, *Hordeum*)	56
sarisama (*sarṣapa*, mustard, *Brassica nigra*)	10
alasī (*atasī*, linseed, *Linum usitatissimum*)	10
karaḍa (*karaṭa*, safflower, *Carthamus tinctorius*)	10
vaṭulā (*vartula*, val pulse, *Dolichos lablab*?)	14
tori (–, Indian rape, *Brassica çumpestris* var. *toria*)	14
kulattha (*kulattha*, horse gram, *Dolichos biflorus*)	14
jīra (*jīra*, cumin, *Cuminum cyminum*)	10
dhaṇiya (*dhānya*, coriander, *Coriandrum sativum*)	10

*Accoring to Sharma [1959, 318], 20 mds for 'kāngarī (?)' and $15\frac{3}{4}$ mds for 'rice (kārī).' He seems to have read '*panaraha ... savāīyā*' as $15\frac{3}{4}$. He does not mention *cīṇaya* between *kaṃguṇiya* (his 'kāngarī') and *kūrīsa* (his 'kārī').

Table 5.2: Comparison of the Figures of Pherū and Muḥarrir

Grain	Pherū	Muḥarrir
Sesamum	16	16
Italian millet	20	16
Rice (*kūrīsa/korī*)	7	?
Cotton	16	10
Flax	10	10
Wheat	45	40
Lentil	32	30
Barley	56	55
Mustard	10	16
Safflower	10	16
Horse gram	14	16

It is of interest that such figures for the annual crop yield and similar matters were current among professionals like accountants, clerks and traders and that these were passed on from father to son, as Pherū and Muḥarrir state expressly. A comparison of the figures given by Pherū and Muḥarrir shows substantial agreement. See Table 5.2.

GSK 5.9. Tax on farm products(?)

Not understood.

5.2 Yields of Sugarcane Juice and Oil

GSK 5.10–12. Quantities of the substances produced from sugarcane juice.

The author here enumerates the quantities of the products obtained from sugarcane juice, after mentioning '9 *khāris*' and '50 *maṇas* of sugarcane juice', which seems to mean that 50 *maṇas* (in weight) of sugarcane juice is extracted from 9 *khāris* (in volume) of sugarcane. The '*khāri*' is presumably equivalent to one cubic *hasta*, usually called 'the Māgadha Khārikā' (cf. L 7), although it is not defined anywhere in the present work. If this identification is accepted, '9 *khāris*' cannot be the volume of the '50 *maṇas* of sugarcane juice' itself, because it would make the specific gravity (γ) of sugarcane juice less than 0.8. For, if $\gamma = 50/9$ *maṇas/khāris*, since 24000 grams $<$ 1 *maṇa* $<$ 34000 grams and 63 cm $<$ 1 *hasta* (or *kaṃviya*) $<$ 72 cm (see GSK 5.28 below), we have $0.357 < \gamma < 0.756$. Sugarcane juice must be heavier than water ($\gamma > 1$).

In the following list, we have supplied the weight in *maṇas* of each product obtained from 50 *maṇas* of sugarcane juice and the percentage.

Products	Quantities	Weights (*maṇas*)	%
Jaggery (*gula*)	$\frac{1}{5}$	10	20
Brown sugar (*sakkara*)	$\frac{1}{6}$	$8\frac{1}{3}$	$16\frac{2}{3}$
Sugar candy (*khaṃḍa*)	$\frac{1}{16}$	$3\frac{1}{8}$	$6\frac{1}{4}$
Thickened juice (*ravva*)	$\frac{3}{32}$ *1	$4\frac{11}{16}$	$9\frac{3}{8}$
Watery marrow (*nīravasa*)	$\frac{3}{32}\pm$ *2	$4\frac{11}{16}\pm$	$9\frac{3}{8}\pm$
? (*navi*)	$\frac{3}{32}\pm$	$4\frac{11}{16}\pm$	$9\frac{3}{8}\pm$
? (*nivāta*)	$\frac{1}{24}$ *3	$2\frac{1}{12}$	$4\frac{1}{6}$
? (*varisolaga*)	$\frac{3}{64}$ *4	$2\frac{11}{32}$	$4\frac{11}{16}$
Scum (*sīra*)	$\frac{1}{16}$	$3\frac{1}{8}$	$6\frac{1}{4}$
Total		$43\frac{7}{96}\pm$	$86\frac{7}{48}\pm$

*1 $\frac{1}{16}\cdot(1+\frac{1}{2})=\frac{3}{32}$.

*2 '±' means 'a little more or less than' the given number.

*3 $\frac{1}{16}\cdot(1-\frac{1}{3})=\frac{1}{24}$.

*4 $\frac{1}{16}\cdot(1-\frac{1}{4})=\frac{3}{64}$.

GSK 5.13. Quantities of oil obtained from sesamum, etc.

Raw materials	Products	Quantities
Sesamum (*tila*)	Oil (*tilla*)	$\frac{9}{20}$
Mustard (*sarisama*)	Oil (*tilla*)	$\frac{7}{20}$
Safflower (*karaḍa*)	Oil (*tilla*)	$\frac{5}{20}$
Milk (*duddha*)	Butter (*lūṇia*)	$\frac{1}{9}$ or $\frac{1}{8}$
Milk (*duddha*)	Ghee (*ghia*)	$\frac{1}{12}$ or $\frac{3}{32}$ *

* $\frac{1}{9}(1-\frac{1}{4})=\frac{1}{12}$ or 8.3%, $\frac{1}{8}(1-\frac{1}{4})=\frac{3}{32}$ or 9.4%. Pherū does not specify whether it is cow's milk or buffalo's milk. Perhaps the former is from the cow's milk and the latter from the buffalo's milk. In the above-mentioned *Dastūr al-Albāb fī ᶜIlm al-Ḥisāb*, Hājjī ᶜAbd al-Ḥamīd Muḥarrir Ghaznavī gives very detailed information about the yield of milk and ghee of cows and buffaloes. According to him, 1 *man* of buffalo's milk yields 3 *sers* of ghee (i.e. 7.5%) whereas 1 *man* of cow's milk produces $2\frac{1}{2}$ *sers* of ghee (i.e. 6.3%). Pherū's figures are somewhat higher.

5.3 Yields of Regional Tax

This section probably has some bearing on the price regulations and tax reforms introduced by ᶜAlā' al-Dīn Khaljī but is hardly intelligible.

GSK 5.14. Tax on cattle and household goods.

Substances	Tax[*1]
A cow	10
A she-goat	10
A she-buffalo	20
A plough	4[*2]
Fire of hearth	hearth tax[*3]
Barber goods? (*nāviya*)	0
Necklaces? (*valahāra*)	0
Dowry? (*mahara*)	0

[*1] None of the monetary units defined in GSK 1.3–4a is mentioned here but the unit of the tax may be *vayalla*. See the next note.

[*2] *cahu vayalli halo*, where *vayalla* may be the unit of the tax.

[*3] *kuḍhiyā*. Cf. OHED *kuḍī*, a hearth; historically hearth tax.

GSK 5.15. Exemption from tax.

The author lists several things on which movable (? *cara*) tax is not levied, but the details are not clear.

GSK 5.16. Exceptions.

Substances	Tax (in *ṭaṃkas*)
A plough of 32 *ṭaṃkas*	1 (hearth tax 1)[*]
	$1\frac{1}{2}$ (hearth tax 2)
	2 (hearth tax 3)
An old she-buffalo	1
An old cow	$\frac{1}{2}$
An old bull	1

[*] See the 'hearth tax' in the table under verse 14 above.

GSK 5.17. Variable nature of this item.

In line 1, the author refers to the possibility that this item may change according to the particular account (*caṭṭiya*). Line 2 is not understood.

5.4 Yield of Price

This section deals with shortcuts in commercial arithmetic. Cf. GSK 4.6–10.

GSK 5.18. Commodities per *damma* and per *ṭaṃka* 1.

When *a pāīs* of a certain thing can be bought for one *damma*, 3*a seīs* and 2*a pāīs* of the same thing can be bought for one *ṭaṃka*. That is,

$$\frac{a\ p\bar{a}\bar{\imath}s}{1\ damma} = \frac{3a\ se\bar{\imath}s + 2a\ p\bar{a}\bar{\imath}s}{1\ \d{t}a\d{m}ka},$$

since 16 *pāīs* = 1 *seī* (GSK 1.8) and 50 *dammas* = 1 *ṭaṃka* (GSK 1.4a).

GSK 5.19. Commodities per *damma* and per *ṭaṃka* 2.

$$\frac{a \textit{ seras}}{1 \textit{ damma}} = \frac{a(1+\frac{1}{4}) \textit{ maṇas}}{1 \textit{ ṭaṃka}},$$

and conversely,

$$\frac{a \textit{ maṇas}}{1 \textit{ ṭaṃka}} = \frac{a(1-\frac{1}{5}) \textit{ seras}}{1 \textit{ damma}},$$

since 40 *seras* = 1 *maṇa* (GSK 1.9) and 50 *dammas* = 1 *ṭaṃka* (GSK 1.4a).

GSK 5.20. Prices per *maṇa* and per *sera*.

$$\frac{a \textit{ dammas}}{1 \textit{ maṇa}} = \frac{\frac{a}{2} \textit{ visovas}}{1 \textit{ sera}},$$

since 20 *visovās* = 1 *damma* (GSK 1.4a) and 40 *seras* = 1 *maṇa* (GSK 1.9).

GSK 5.21. Prices per score (*koḍiyā*) and per article

$$\frac{a \textit{ dammas}}{1 \textit{ koḍiyā}} = \frac{a \textit{ visovas}}{1 \text{ article}},$$

since 20 *visovas* = 1 *damma* (GSK 1.4a) and 1 *koḍiyā* (score) = 20 articles.

5.5 Measurement

GSK 5.22. Circumference and area of a circle.

When d denotes the diameter of a circle,

$$C = 3d + \frac{d}{6}, \qquad A = C \cdot \frac{d}{4}.$$

GSK 5.22a (for C) = GSK 3.45a and GSK 5.22b (for A) gives the same formula as GSK 3.43b.

GSK 5.23. Calculations of the areas of the square/circle inscribed in a circle/square, respectively.

When A_c denotes the area of a circle and A_{is} that of the square inscribed in it,

$$A_{is} = \left(\frac{12}{20} + \epsilon\right) \cdot A_c,$$

where '$+\epsilon$' is expressed in the verse as 'with a (small) difference' (*sa-visesa*, S. *sa-viśeṣa*) or 'and slightly more', and, in the figure attached to the text, by an *avagraha*-like symbol, 'S'.

Here and in the next verse, $\frac{a}{20}$ is expressed as '*a visova(ga)s*' (S. *viṃśopaka*, 'one twentieth'). In the figure attached to the text, '12 *visovas*' is expressed as '0||2', which means '$(0+\frac{2}{4})$ unit and 2 *visovas*'.

When A_s denotes the area of a square and A_{ic} that of the inscribed circle,

$$A_{ic} = A_s\left(1 - \frac{1}{5}\right).$$

In the figure attached to the text, the coefficient is expressed as '0|||1', which means '$(0 + \frac{3}{4})$ unit and 1 *visova*' (= 16 *visovas*). Note that $(1 - \frac{1}{5}) = \frac{4}{5} = \frac{16}{20}$.

These approximate relationships were presumably obtained as follows. According to GSK 5.22,

$$A_c = \frac{19d^2}{24} = \frac{19}{12}\cdot\frac{d^2}{2} = \frac{19}{12}A_{is}.$$

Hence follows the above formula. Likewise,

$$A_{ic} = \frac{19d^2}{24} \approx \frac{20d^2}{25} = \frac{4A_s}{5}.$$

GSK 5.24. Calculations of the areas of the circle/triangle inscribed in a triangle/circle, respectively.

When A_t denotes the area of an equilateral triangle and A_{ic} that of the inscribed circle,

$$A_{ic} = \frac{12\frac{1}{2}}{20}A_t.$$

When A_c denotes the area of a circle and A_{it} that of the inscribed equilateral triangle,

$$A_{it} = \frac{7\frac{1}{2}}{20}A_c.$$

In the geometric figures attached to the text, these coefficients are expressed as '0||2||' or '$(0 + \frac{2}{4})$ unit and $(2 + \frac{2}{4})$ *visovas*' and '0|2||' or '$(0 + \frac{1}{4})$ unit and $(2 + \frac{2}{4})$ *visovas*', respectively.

Note that these formulas seem to involve more than one approximation to $\sqrt{3}$; it seems impossible to explain the derivations of both formulas by assuming one and the same approximation.

GSK 5.25. Volume and surface area of a sphere.

When d denotes the diameter (*udaya*, lit. height) of a sphere, its volume (*pāhāṇa*, lit. stone) and the surface area (*khitta*) are calculated by

$$V = d^3\left(1 - \frac{1}{4}\right)\left(1 - \frac{1}{4}\right), \quad S = \frac{C}{4}\cdot C\cdot\left(1 + \frac{1}{9}\right).$$

The first formula (for V) is very close to, but different from, the one given in GSK 3.65a, and is equivalent to one of Mahāvīra's formulas (GSS 8.28b).

$$V = \frac{d^3}{2}\left(1 + \frac{1}{9}\right) = \frac{80}{144}d^3 \qquad \text{(GSK 3.65a)}$$

$$V = d^3\left(1-\frac{1}{4}\right)\left(1-\frac{1}{4}\right) = \frac{81}{144}d^3 \qquad \text{(GSK 5.25a)}$$

$$V = \left(\frac{d}{2}\right)^3 \cdot \frac{1}{2} \cdot 9 = \frac{81}{144}d^3 \qquad \text{(GSS 8.28b)}$$

See also under GSK 3.65.

GSK 5.25b (for S) is identical with GSK 3.65b except for the substitution of a synonym (*juya* for *ahiya*).

In the prose passage that follows verse 25, a solution of the same example as GSK 3.66 ($d = 6$) is worked out. The answer should be $V = 121\frac{1}{2}$ and $S = 100\frac{5}{18}$, but the Nahatas' text reads $V = 120$ and $S = 100SS6$. The former value ($V = 120$) may be simply a corruption but, interestingly, it is correct according to the formula of GSK 3.65a. The latter ($S = 100SS6$) includes the fractional portion expressed in *visovas* ('twentieth') and rounded up ($S = 100$ units $+5\frac{5}{9}$ *visovas*), although a better expression of the same quantity seems to be '100|1'. Cf. Strauch 2002, 136–137, where *viśopaka* (sic) occurs as a land measure.

5.6 Masonry

GSK 5.26–27. Weights per cubic *kaṃviya* of stones (specific gravities).

≈ GSK 3.67–68.

GSK 5.28. Weights per cubic *kaṃviya* of clay, etc. (specific gravities).

Substances	Weights (*maṇas*/*kaṃviya*^{3})
Clay (*maṭṭī*)	25
Kodrava grains (*tus-aṃnna*)	8
Wild grains (*vaṇ-aṃnna*)	12
Oil (*tilla*)	10
Ghee (*ghaya*)	10
Salt (*lavaṇa*)	16

From these data, we can estimate the size of a *kaṃviya*. Let it be x cm. The weight of a *maṇa* varied from time to time and from region to region but we are informed that 600 grams $<$ 1 *sera* $<$ 850 grams (Srinivasan 1979, 95). As 40 *seras* make one *maṇa* (GSK 1.10a), we have the relations, 24000 grams $<$ 1 *maṇa* $<$ 34000 grams. The specific gravity of sesamum oil, for example, is nearly 0.92. Therefore, we can safely assume the relations,

$$\frac{10 \times 24000}{x^3} < 0.92 < \frac{10 \times 34000}{x^3}.$$

Hence follow the limits of x:

$$63 \text{ cm} < x < 72 \text{ cm}.$$

GSK 5.29. Quantities of the things necessary for constructing a brick wall 1.

One mason (*rāju*), who constructs a brick wall with 3 workers (*jaṇas*, S. *janas*), gathers the following things for every 12 cubic *gajas*.

Things	Quantities for 12 *gaja*^{3}
Bricks (*iṭṭā*)	1400
Water pots (*jala-gaggarī*)	30

GSK 5.30. Quantities of the things necessary for constructing a brick wall 2.

Here, the quantities are given per cubic *gaja*.

Things	Quantities for 1 *gaja*^{3} (*maṇas*)
? (*hakka*)	27
Lime (*cunna*)	9
Mortar (*khora*)	18
? (*cijja*)	9

GSK 5.31–32. Quantities of the things necessary for plastering.

Things	Quantities
Pure lime (*kevaṇa cunna*)	$\frac{3}{4}$ *maṇa*
Flax (*saṇa*)	$\frac{1}{4}$ *sera*
Mortar (*khora*) for dry and watery places	$\frac{1}{3}$, $\frac{1}{2}$ ⟨*maṇa?*⟩
Cow-dung (*chāṇaya*)	40 *maṇas*
Limestone (*kakkara*) roasted and powdered	60 ⟨*maṇas*⟩
Carried-with-protection (*rakkha-pavāhiya*) sand?	60 ⟨*maṇas*⟩
Carried-without-protection (*arakkha-*) sand?	40 ⟨*maṇas*⟩

GSK 5.33. Concluding remark.

The author (here named Pherū) mentions the 'yield of regional tax' and the 'occasion of ⟨making⟩ an account (*caṭṭiya*)'.

Appendix A

Concordance of the *Gaṇitasārakaumudī* and Other Works

Here we give a concordance of parallel topics in the first three chapters of the *Gaṇitasārakaumudī* and other *pāṭī* works. As the remaining two chapters are unique to Pherū, we excluded them from this concordance.

The numbering of the verses in the BM is due to Hayashi [1995a], of the GT to Hayashi [2005], of the GK to Singh [1998–2002], and of the PV to Hayashi [1991]. We use a pair of parentheses in order to distinguish the numbers for examples from those for rules if they are numbered separately. For example, PG 1 = Rule 1 of the PG, PG (1) = Example 1 of the PG, and BM C10.(10) = Example 10 for rule C10 of the BM.

GSK	AB	BSS	BM	PG	Tr	GSS	MS	SS	GT	L	GK	PV
1.3–4a				9	iv				4	2	1.4	
1.4b										6b	1.7b–8a	
1.5–6				12	7	1.25–31			8–9	5–6	1.6b–7a	
1.7						1.29						
1.8				11	6	1.36–38				7–8	1.10b–11a	
1.9–10				10	5	1.39–41			5	7	5–6a	
1.11	3.1			13	8	1.32–35					1.11b–12a	
1.12–14	2.2			7–8	2–3	1.63–68			2–3	10–11	1.2–3	1'–2'
1.15		12.1		2–6								
1.16–17		12.19		14a	1a			13.21a		117	3.11	X2
1.18												
1.19				(1)	(1)					118		
1.20a				15	2							
1.20b				14	1							A3b
1.21–22				16	3							X3
1.23				(2)	(2)							
1.24				17	4							
1.25												
1.26												
1.27				18–19a	5–6a	2.1	15.3	13.2	17	14	1.13	X5
1.28				(3)	(3)	2.2–17			19,20	17	1.(2)	
1.29		12.55		20a	7a	2.1			18	14–15	1.14–15	X8
1.30							18.67–70				11.11	

GSK	AB	BSS	BM	PG	Tr	GSS	MS	SS	GT	L	GK	PV
1.31		18.32–35		21	8	1.49	15.10b–11a	14.6	52	45–46	1.30	X9
1.32										47	(1.12)	
1.33				22	9	2.18	15.4	13.3	21	18	1.16b	X 10
1.34		12.63		23	10	2.30–31			23	19	1.17	
1.35	2.3a	12.63b		24	11	2.29	15.6a	13.4a	24	20	1.18	X11a
1.36				(4)	(5)	2.32-35			25	21	1.(3)	
1.37–38	2.4			25–26	12–13	2.36	15.6b–7	13.5	26	23	1.19–20	X12–13
1.39–40		12.6		27–28	14–15	2.47			28	24–25	1.21–22	
1.41	2.3b	12.62		28		2.43	15.6	13.4	29	24	1.23	X11b
1.42				(5)	(6)				31	27	1.(4)	
1.43–45	2.5	12.7		29–31	16–18	2.53	15.8–10a	13.6–7	32–33	28–29	1. 24–25	
1.46		12.61		32	19				35			
1.47a		12.3a		39a, 40a	24a							
1.47b		12.2, 8		36	23a		15.13	13.8	53	30	1.26	
1.48–49		12.2, 61		32a	19a		15.14a	13.8	35, 38	37	1.28a	
1.50a		12.2a		32a	19a		15.14a	13.8	35	37	1.28a	
1.50b				(6)	(7)				36, 37	38	1.(9)	
1.51		12.2b		32b	19b		15.14b	13.8	38	37	1.28a	
1.52				(8)	(8)				39	38	1.(9)	
1.53		12.3		33a	20a	3.2	15.15	13.9	40	39	1.28b	X14
1.54				(10)	(9)	3.3–7			41	40	1.(10)	
1.55		12.4, 60	Q4	33b	20b	3.8	15.15	13.10	42	41	1.29a	X14

GSK	AB	BSS	BM	PG	Tr	GSS	MS	SS	GT	L	GK	PV
1.56				(11)	(10)	3.9–12			43	42		
1.57		12.5a		34a	21b	3.13	15.16a	13.9b	44	43	1.29b	
1.58				(12)	(11)				45	44	1.(11)	
1.59		12.5b		34b	21b	3.13	15.16b		46	43	1.29b	
1.60				35a	22a	3.13	15.17a		48	43	1.29b	
1.61				(13)	(12)				49		1.(11)	
1.62				35b	22a	3.13	15.17b		50	43	1.29b	
1.63	2.26	12.10	C10	43	29	5.2a	15.24–25a	13.14	95	73	1.60	X15
1.64												
1.65											1.(68)	
1.66			C10.(4)	(25)	(30)				96	75	1.(67)	
1.67			C10.(34)	(26)	(31)	5.9			97	74	1.(66)	
1.68				(27), (28)	(32), (33)					76		
1.69				(29)	(34)							
1.70				(30)	(35)							
1.71			C10.(10)	(31)	(36)	5.5			103		1.(69)	
1.72		12.11b–12		45	31	5.32	15.26–27	13.15	107	82	1.62	
1.73				(39)	(44)				108	83		
1.74				(40)	(45)				109	84		
1.75				(43)	(49)							
1.76				(44)	(50)						1.(77)	

GSK	AB	BSS	BM	PG	Tr	GSS	MS	SS	GT	L	GK	PV
1.77				(41), (42)	(46), (47), (48)							
1.78				(45)	(51)					85	1.(78)	
1.79				(46), (47)	(52), (53)					86		
1.80										87		
1.81		12.11a		44b	30a		15.25b		105	77	1.61	X15
1.82				(34)	(38)							
1.83				(35)	(41)					81		
1.84				(38)	(42)					80		
1.85				(37)	(43)							
1.86			54, 55			6.167					2.1b–3a	
1.87			54.(1)			6.168						
1.88												
1.89							15.28b–29					
1.90		12.13a		46a	32a	6.18	15.28a	13.16a	112	88	1.63a	
1.91				(48)	(54)	5.37, 38 6.19–20				89	1.(81)	
1.92				46b	32b			13.16b	115	78	1.63b	
1.93				(50), (51)	(56)	5.39, 40			116, 117	79	1.(82)	

GSK	AB	BSS	BM	PG	Tr	GSS	MS	SS	GT	L	GK	PV
2.1a		12.8a		36	23a	3.55	15.13b–14	13.11–12	53	30	1.26a	
2.2a		12.8b		38a	23b	3.99	15.13a	13.11–12	55	32	1.26b	
2.2b				(15)	(14), (15)				56	33	1.(6)	
2.3		12.9a		38b	25a	3.99	15.19a					
2.4				(17)	(19)							
2.5				(16)	(18)	8.34				97	2.(39)	
2.6		12.9b		39	24	3.113	15.11b–12	13.10	57	34	1.27	
2.7				(19)	(17)	3.114–121			58, 59	35, 36	1.(7)	
2.8		12.9b		40		3.126	15.11b–12	13.10	60	34	1.27	
2.9				(21)		3.127–31			61	35, 36	1.(7)	
2.10				42	25b–26a	3.138						
2.11				(23)	(20)	3.139–40						
2.12			Q6	41	26b–27a		15.18		62			
2.13				(22)	(22)				63			
2.14				74a	(27b)	4.4a	15.20		64		1.37b–38	
2.15				(96)	(23), (24), (25)				65, 66	53		
2.16					(26), (28), (29)						1.(19)–	
2.17					(27)				66		–1.(24)	

GSK	AB	BSS	BM	PG	Tr	GSS	MS	SS	GT	L	GK	PV
3.1		12.14b		47	33	6.21, 23	15.31	13.17	118	90		
3.2				(52), (53)	(57), (58)	6.22, 24			119	91	2.(12)	
3.3				48	34			13.18	120			
3.4				(54)	(59)				121			
3.5				51	35	6.77			127			
3.6				(57–58)	(60–61)	6.78			128*			
3.7		12.16a	N1	59a	38a	6.79	15.36	13.19a		94	2.1a	X16
3.8				(71)	(67)					95	2.(1)	
3.9					(68–69)							
3.10												
3.11				59b	38b	6.87–89	15.37–38a	13.19b		98	2.3b	
3.12												
3.13				(75)	(70)					100	2.(7)	
3.14				(73–74)	(71–72)					99	2.(8)	
3.15			27	52b	36	6.169	15.38b-39			103	2.18	X17
3.16			27.(1)	(61)	(63)	6.170–71				104	2.(25)	
3.17			27.(2)	(62)								
3.18				53	37	6.175	15.39,40			103	2.18	X17
3.19				(63)	(65)							
3.20				(64)	(66)							

* The GT is availabe up to the middle of the ‘procedure for mixture’ (*miśraka-vyavahāra*).

GSK	AB	BSS	BM	PG	Tr	GSS	MS	SS	GT	L	GK	PV
3.21			28	54		6.175–76	15.39a			106	2.19	
3.22			28.(1)	(65)		6.177–79				107	2.(26)	
3.23				55		6.176	15.41			108	2.20	
3.24							6.182–83			109	2.(27)	
3.25						6.184						
3.26	2.19	12.17	Q12	85a	39	2.64	15.47	13.20		121	3.1–2a	A4
3.27				(103)	(73)					122, 123	3.(1), (2)	
3.28				86a	40a	2.74a	15.48	13.23a		124	3.2b	
3.29				86b	40b	2.74b	15.49	13.23b		126	*	
3.30	2.20	12.18	N18	87	41	2.69, 70	15.50	13.24		128	3.3	
3.31	2.21	12.19b		103b				13.21		117b	3.14	
3.32				(116), (117)						118		
3.33	2.22	12.20		102b–03a		6.296, 301		13.22		119	3.15	
3.34				102a		6.296					3.13b	
3.35a	2.9a				42a							
3.35b				(122)	(75)							
				(123)	(76)							
3.36		12.21b		117	43a	7.50a	15.69	13.28		169	4.33b–34a	
3.37				(128)	(80)					170	4.(33)	
3.38				(124)	(81)							

* The rule for the common difference is missing in the printed edition of the GK.

GSK	AB	BSS	BM	PG	Tr	GSS	MS	SS	GT	L	GK	PV
3.39	2.8b			115	42b		15.78	13.30		173	4.34b	X19a
3.40				(127–30)	(77–79)					174	4.(34), (35)	
3.41					49–51a		15.76a				4.38	
3.42	2.17a	12.24			51b		15.54b	13.33		136		
3.43	2.7	12.40			45	7.60	15.88	13.35		201	4.9	X19b
3.44					(85)					202	4.(5)	
3.45												
3.46					47		15.89					A24
3.47					(86)							
3.48–49*						7.73b–75a	15.90-91a				4.12	
3.50–51				116	44,48	7.73b–75a	15.101				4.13	
3.52–53				(131), (132)	(82), (83)						4.(9–10)	
3.54		12.44b	N14		52	8.4b	15.105a	13.43		214	5.1	B21U
3.55		12.44a			53	8.4a	15.104a	13.43		214	5.2a	B21U
3.56					(87), (88)	8.5-8				215, 216	5.(1), (2)	
3.57					(89), (90)							
3.58												
3.59												
3.60												

* For GSK 3.48–49 see Table 3.1 in Part IV.

GSK	AB	BSS	BM	PG	Tr	GSS	MS	SS	GT	L	GK	PV
3.61												
3.62					(92)							
3.63					57							
3.64					(94)							
3.65					56a	8.28b, 29a	15.108	13.46		203		X23
3.66						(93)						
3.67–68												
3.69												
3.70–71												
3.72												
3.73					58, (95–96)						6.(9)	
3.74												
3.75												
3.76												
3.77–78												
3.79												
3.80												
3.81												
3.82												
3.83a												
3.83b												
3.84												

GSK	AB	BSS	BM	PG	Tr	GSS	MS	SS	GT	L	GK	PV
3.85a												
(3.85b)												
3.86						7.23					4.10–11	
3.87												
3.88												
3.89												
3.90–92												
3.93					(100)							
3.94												
3.95		12.48–49				7.63–67	15.112–14			225	6.9	
3.96–97		12.50						13.51		227		A29–31
3.98–99												
3.100		12.51			64		15.117	13.52		229	7.11	A29–31
3.101												
3.102												
3.103												
3.104		12.52										

Appendix B

The Type Problems

The authors of *pāṭī* works like Śrīdhara and Śrīpati treat special 'types' (or 'classes', *jātis*) of linear and quadratic equations, giving algorithms designed specifically for each type. Bhāskara (in the L) and Nārāyaṇa (in the GK) tried to treat them systematically according to some principles. Pherū only treats the most basic type (which we call the 'original-part type') under the name 'pillar-part class' (GSK 2.14). We give below a classification of these 'type problems' in modern notation. The English names of those 'types' used here are based on their Sanskrit names (which are given in parentheses after each reference under each item) but not always literal translations. Note that the verse numbers of the GT employed here for reference are not those of Kāpadīā's edition but those proposed in Hayashi 2005.

1. Original-part type

$$\begin{cases} x - \frac{b_1}{a_1}x = y_1 \\ \vdots \\ y_{i-1} - \frac{b_i}{a_i}x = y_i \\ \vdots \\ y_{n-1} - \frac{b_n}{a_n}x = d \end{cases}$$

Tr 27b (*stambha-uddeśa*), PG 74a (*stambha-uddeśaka*), GSS 4.4a (*bhāga-jāti*), GT 64 (*dṛśya-jāti*), GSK 2.14 (*thaṃbh-aṃsaka-jāti*):

$$x = \frac{d}{1 - \sum_{i=1}^{n} \frac{b_i}{a_i}}.$$

L 53 gives an example of this type and solves it by means of the optional quantity operation (*iṣṭa-karman*).

GK 1. exs. 18, 19, and 22 give examples for

$$y_{i-1} \pm \frac{b_i}{a_i}x = y_i,$$

calling this type *dhana-ṛṇa-aṃśa-jāti* (positive-negative-part class), and solve them by means of the optional quantity operation (GK 1.37b–38).

2. Part-difference type

$$\begin{cases} x - \frac{b_1}{a_1}x = y_1 \\ y_1 - \frac{b_2}{a_2}x = y_2 \\ y_2 - \frac{b_3}{a_3}\left|\frac{b_1}{a_1}x - \frac{b_2}{a_2}x\right| = y_3 \\ \vdots \\ y_{n-1} - \frac{b_n}{a_n}|\cdots| = d \end{cases}$$

PG 74b (*viśeṣa-uddeśaka*), GT 70 (*viśleṣa-jāti*): Calculate b'_i/a'_i such that

$$\frac{b_i}{a_i}|\cdots| = \frac{b'_i}{a'_i}x,$$

and apply the rule for *stambha-uddeśaka* (see the original-part type above).

L 55 gives an example of this type and solves it by means of the optional quantity operation (*iṣṭa-karman*).

GK 1. exs. 18, 21, and 24 give examples for

$$\begin{cases} x \pm \frac{b_1}{a_1}x = y_1 \\ y_1 \pm \frac{b_2}{a_2}x = y_2 \\ y_2 \pm \frac{b_3}{a_3}\left|\frac{b_1}{a_1}x - \frac{b_2}{a_2}x\right| = d \end{cases}$$

calling this type *dhana-ṛṇa-viśleṣa-jāti*, and solve them by means of the optional quantity operation (GK 1.37b–38).

3. Remainder-part type

$$\begin{cases} x - \frac{b_1}{a_1}x = y_1 \\ \vdots \\ y_{i-1} - \frac{b_i}{a_i}y_{i-1} = y_i \\ \vdots \\ y_{n-1} - \frac{b_n}{a_n}y_{n-1} = d \end{cases}$$

PG 74a (*śeṣa-uddeśaka*): Calculate the right-hand side of the equation,

$$\frac{b_i}{a_i}y_{i-1} = \frac{b_i}{a_i}\left\{\prod_{j=1}^{i-1}\left(1 - \frac{b_j}{a_j}\right)\right\}x,$$

for $i = 2, 3, ..., n$ and apply the rule for *stambha-uddeśaka* (see the original-part type above).

GSS 4.4b (*śeṣa-jāti*):

$$x = \frac{d}{\prod_{i=1}^{n}\left(1 - \frac{b_i}{a_i}\right)}.$$

MS 15.20 (*śeṣa-jāti*), GT 67 (*śeṣa-jāti*):

$$x = \frac{d}{\frac{\prod(a_i - b_i)}{\prod a_i}}.$$

L 54 gives an example of this type and solves it by means of the optional quantity operation (*iṣṭa-karman*).

GK 1. exs. 18, 20, and 23 give examples for

$$y_{i-1} \pm \frac{b_i}{a_i}y_{i-1} = y_i,$$

calling this type with the plus and the minus signs *dhana-sva-aṃśa-jāti* and *bhāga-apavāha-jāti*, respectively, and solve them by means of the optional quantity operation (GK 1.37b–38).

4. Sum-part type

$$\begin{cases} x + \frac{b_1}{a_1}x = y_1 \\ \vdots \\ y_{i-1} + \frac{b_i}{a_i}y_{i-1} = y_i \\ \vdots \\ y_{n-1} + \frac{b_n}{a_n}y_{n-1} = d \end{cases}$$

MS 15.21a (*yoga-jāti*):

$$x = \frac{d}{\frac{\prod(a_i+b_i)}{\prod a_i}}.$$

5. Root-remainder type

$$\begin{cases} x - c_0\sqrt{x} = y_1 & y_1 - \frac{b_1}{a_1}y_1 = x_1 \\ x_1 - c_1\sqrt{x_1} = y_2 & y_2 - \frac{b_2}{a_2}y_2 = x_2 \\ \vdots & \vdots \\ x_{n-1} - c_{n-1}\sqrt{x_{n-1}} = y_n & y_n - \frac{b_n}{a_n}y_n = x_n \\ x_n - c_n\sqrt{x_n} = d & \end{cases}$$

PG 75 (*mūlādi-śeṣa-uddeśaka*), GT 73 (*śeṣa-mūla-jāti*): Apply the following algorithms successively:

$$x_i = \left(\frac{\sqrt{c_i^2 + 4y_{i+1}} + c_i}{2}\right)^2, \qquad y_i = \frac{x_i}{1 - \frac{b_i}{a_i}},$$

for $i = n, (n-1), ..., 1, 0$ where $y_{n+1} = d$ and $x_0 = x$.

GK 1.40 (*śeṣa-mūla-jāti*): Starting from the 'visible' quantity, d, apply GK 1.39–40a (part-root($\mp$) type) for *niraṃśa-jāti* and 1.37b–38 (remainder-part type) alternately.

6. Remainder-root type

$$\begin{cases} x - \frac{b}{a}x = y \\ y - c\sqrt{y} = d \end{cases}$$

GSS 4.40 (*śeṣa-mūla-jāti*):

$$x = \frac{\left\{\frac{c}{2} + \sqrt{\left(\frac{c}{2}\right)^2 + d}\right\}^2}{1 - \frac{b}{a}}.$$

7. Root(∓) type

$$x \mp c\sqrt{x} = d$$

L 65 (*guṇa-karman* 1):

$$x = \left\{ \pm\frac{c}{2} + \sqrt{\left(\frac{c}{2}\right)^2 + d} \right\}^2 .$$

8. Part-root type

$$x - \frac{b}{a}x - c\sqrt{x} = d$$

PG 76 (*bhāga-mūla-agra-uddeśa*), GT 76 (*mūla-agra-bhāga-jāti*): Calculate

$$c' = \frac{c}{1 - \frac{b}{a}}, \qquad d' = \frac{d}{1 - \frac{b}{a}}.$$

Then

$$x = \left\{ \frac{c'}{2} + \sqrt{\left(\frac{c'}{2}\right)^2 + d'} \right\}^2 .$$

GSS 4.33 (*mūla-jāti*):

$$x = \left\{ \frac{\frac{c}{2}}{1 - \frac{b}{a}} + \sqrt{\left(\frac{\frac{c}{2}}{1 - \frac{b}{a}}\right)^2 + \frac{d}{1 - \frac{b}{a}}} \right\}^2 .$$

9. Part-root(∓) type

$$x \mp \frac{b}{a}x \mp c\sqrt{x} = d$$

L 66 (*guṇa-karman* 2): Calculate

$$c' = \frac{c}{1 \mp \frac{b}{a}}, \qquad d' = \frac{d}{1 \mp \frac{b}{a}}.$$

Then

$$x = \left\{ \pm\frac{c'}{2} + \sqrt{\left(\frac{c'}{2}\right)^2 + d'} \right\}^2 .$$

GK 1.39–40a (*mūla-sva-ṛṇa-jāti*): Calculate

$$s = \sqrt{(c')^2 + 4d'}.$$

Then

$$x = \frac{\pm c' + s}{2}.$$

The GK calls the case with $d = 0$ *adṛśya-jāti* (no-visible class) and the case with $b = 0$ *niraṃśa-jāti* (no-fraction class).

10. Part-partial-root(∓) type

$$x \mp \frac{b_1}{a_1}x \mp c\sqrt{\frac{b_2}{a_2}x} = d$$

L 71 (*guṇa-karman* 3): Calculate

$$c' = \frac{\frac{b_2}{a_2}\cdot c}{1 \mp \frac{b_1}{a_1}}, \qquad d' = \frac{\frac{b_2}{a_2}\cdot d}{1 \mp \frac{b_1}{a_1}}.$$

Then

$$x = \frac{\left\{\pm\frac{c'}{2} + \sqrt{\left(\frac{c'}{2}\right)^2 + d'}\right\}^2}{\frac{b_2}{a_2}}.$$

L 71 is not a formula but a problem, which is solved by means of the above algorithm.

GK 1.41b–42a (*guṇa-mūla-jāti*): Calculate

$$c' = \frac{b_2}{a_2}\cdot c, \qquad d' = \frac{b_2}{a_2}\cdot d.$$

To these apply GK 1.39–40a (part-root(∓) type) and divide the result by $\frac{b_2}{a_2}$.

11. Two-visibles type

$$\begin{cases} x - d_1 = y_1 \\ y_1 - \frac{b_1}{a_1}y_1 = y_2 \\ \vdots \\ y_{n-1} - \frac{b_{n-1}}{a_{n-1}}y_{n-1} = y_n \\ y_n - c\sqrt{x} = d_2 \end{cases}$$

PG 77 (*ubhaya-agra-mūla-śeṣa-uddeśa*), GSS 4.47 (*dvir-agra-śeṣa-mūla-jāti*): Calculate

$$c' = \frac{c}{\prod\left(1 - \frac{b_i}{a_i}\right)}, \qquad d' = \frac{d_2}{\prod\left(1 - \frac{b_i}{a_i}\right)} + d_1.$$

Then

$$x = \left\{\frac{c'}{2} + \sqrt{\left(\frac{c'}{2}\right)^2 + d'}\right\}^2.$$

Cf. *bhāga-mūla-agra-uddeśa* of the PG (see type 8 above).

GT 80 (*ubhaya-agra-dṛśya-jāti*): Calculate

$$c' = \frac{c}{\prod\left(1 - \frac{b_i}{a_i}\right)}, \qquad d' = \frac{d_2}{\prod\left(1 - \frac{b_i}{a_i}\right)}.$$

Then

$$x = \left\{ \frac{c'}{2} + \sqrt{\left(\frac{c'}{2}\right)^2 + (d' + d_1)} \right\}^2$$

GK 1.41a (without a name): Calculate y_n when $y_1 = 1$; let it be p. Put

$$d' = pd_1 + d_2,$$

and apply GK 1.39–40a (part-root($\mp$) type) for *niraṃśa-jāti*.

12. Partial-root type

$$x - c\sqrt{\frac{b}{a}x} = d$$

GSS 4.51 (*aṃśa-mūla-jāti*): Calculate

$$c' = \frac{b}{a} \cdot c, \qquad d' = \frac{b}{a} \cdot d.$$

Then

$$x = \frac{\left\{ \frac{c'}{2} + \sqrt{\left(\frac{c'}{2}\right)^2 + d'} \right\}^2}{\frac{b}{a}}.$$

GT 87 (*bhāga-mūla-jāti*):

$$x = \left\{ \frac{c + \sqrt{c^2 + 4d/\frac{b}{a}}}{2} \right\}^2 \times \frac{b}{a}.$$

13. Partial-product type

$$x - \frac{b_1}{a_1}x \times \frac{b_2}{a_2}x = d$$

GSS 4.57 (*bhāga-saṃvarga-jāti*):

$$x = \frac{\frac{a_1a_2}{b_1b_2} \pm \sqrt{\left(\frac{a_1a_2}{b_1b_2} - 4d\right) \times \frac{a_1a_2}{b_1b_2}}}{2}.$$

GK 1.44a (*bhāga-saṃguṇya-jāti*): Calculate

$$s = \frac{1}{\frac{b_1}{a_1} \cdot \frac{b_2}{a_2}}, \qquad p = \frac{d}{\frac{b_1}{a_1} \cdot \frac{b_2}{a_2}}.$$

Then

$$x = \frac{s \pm \sqrt{s^2 - 4p}}{2}.$$

14. Partial-square type

$$x - \left(\frac{b}{a}x \mp d_1\right)^2 = d_2$$

GSS 4.61 (*ūna-adhika-aṃśa-varga-jāti*):

$$x = \frac{\left\{\left(\frac{a/b}{2} \pm d_1\right) \pm \sqrt{\left(\frac{a/b}{2} \pm d_1\right)^2 - (d_1^2 + d_2)}\right\}}{b/a}.$$

GT 90 (*hīna-varga-jāti*):

$$x = \frac{\frac{a/b}{2} + d_1 + \sqrt{\left(\frac{a/b}{2}\right)^2 + (a/b) \cdot d_1 - d_2}}{b/a}.$$

This is a solution only for the negative sign of the above equation.

GK 1.42b–43 (*hīna-varga-jāti*): Calculate

$$s = 2d_1 \cdot \frac{a}{b} + \left(\frac{a}{b}\right)^2, \qquad p = \left(d_1^2 + d_2\right) \cdot \left(\frac{a}{b}\right)^2.$$

Then

$$x = \frac{s \pm \sqrt{s^2 - 4p}}{2}.$$

15. Root-sum type

$$\sqrt{x} + \sqrt{x \mp d_1} = d_2$$

GSS 4.65 (*mūla-miśra-jāti*):

$$x = \left(\frac{d_2^2 + d_1}{2d_2}\right)^2.$$

16. Visible-part type

$$x - \frac{b_1}{a_1}x \times \frac{b_2}{a_2}x = \frac{b_3}{a_3}x$$

GSS 4.69 (*bhinna-dṛśya-jāti*), GT 84 (*bhinna-bhāga-dṛśya-jāti*), GK 1.44b (*bhinna-saṃdṛśya-jāti*):

$$x = \frac{1 - \frac{b_3}{a_3}}{\frac{b_1}{a_1} \cdot \frac{b_2}{a_2}}.$$

Appendix C

Index to the Numbers in the Text

The references are to the verse numbers. The verse number put in a pair of parentheses indicates that the number is given indirectly. The letter 'A' attached to a verse number indicates that the number is not given in the verse but is to be obtained as the answer to the problem of that verse. The three abbreviations, p = *praveśa* (introduction), v = *vāsanā* (explanation), and s = *samāpti* (conclusion), attached to a verse number indicate the introductory prose sentence to, the explanatory prose sentence for, and the concluding prose sentence occuring after, that verse, respectively. A digit followed by a period, except for the last digits of items, indicates the number of the chapter in which the verse occurs. We excluded ordinal numbers from this index. The value to be obained as an answer to a problem (indicated by 'verse number + A') is, in this index, expressed in the highest denomination, with fractional part if any.

−79/9: 1.52A?.
0: 1.31, 32, 32A, 4.59.
1/120: 2.2bA?.
1/100: 4.9.
1/64: 1.61A.
1/36: 1.58A?.
1/27: 1.61A.
64/1437: 1.68A.
1/60: 2.2bA?.
3/62: 3.13A.
1/20: 4.10.
1/18: 1.52A?.
1/16: 2.16, 5.10.
3/38: 3.25A.
1/12: 1.52A, 2.2bA?, 15, 4.23.
1/11: 3.97.
1/10: 2.2bA?, 5A, 3.71, 97, 4.9.
1/9: 1.50b, 52, 58A, 2.17, 3.65, 74, 76, 88, 97, 5.13, 25.
1/8: 2.16, 17, 3.73, 4.22, 25, 30, 31, 49, 5.13.
5/38: 3.25A.
1/7: 1.71, 2.13, 17, 3.81.
3/20: 3.14A.
1/6: 1.50b, 52, 54, 70, 71, 2.2b, 7, 9, 15–17, 3.45, 48, 58, 96, 5.10, 22.
1/5: 1.50b, 2.2b, 16, 4.6, 9, 5.10, 19, 23.
2/9: 1.56A.
1/4: 1.58A?, 61, 2.2bA?, 5, 7, 9, 9A, 11, 16, 3.4, 12A, 43, 65, 74, 88, 98, 100, 4.41, 5.12, 13, 17, 22, 25, 25v, 31.
8/31: 3.13A.
11/40: 3.14A.
9/31: 3.13A.
1/3: 1.50b, 52, 58, 61, 2.5, 7, 9, 11, 17, 3.73, 88, 4.12, 32, 5.12, 17, 31.
(7 + 1/2)/20 (= 3/8): 5.24 (expressed as 'seven and a half twentieths'), 24v (expressed as '0|2||').
15/38: 3.25A.
14/31: 3.13A.
1/2 (half, halve): 1.16, 17, 20, 22, 25, 38, 52, 54, 58, 2.2b, 5, 7, 9, 11, 15, 17, 3.4, 12A, 13, 28, 29, 34, 36, 39, 45, 49, 52, 62, 65, 73, 74, 76, 78, 79, 83a, 88, 98, 100, 103, 4.10, 22, 25-29, 37, 38, 41, 47, 5.16, 17, 20.
25/49: 3.10A.
1/3 + 1/4 (= 7/12): 1.61.
3/5: 2.5A.
A little more than 12/20 (= 3/5): 5.23 (expressed as 'twelve twentieths with difference'), 24v (expressed as '0||2S').
(12 + 1/2)/20 (= 5/8): 5.24 (expressed as 'twelve and a half twentieths'), 24v (expressed as '0||2||').
1 − 1/3 (= 2/3): 1.74.
1 − 1/4 (= 3/4): 3.13, 95, 99, 100, 5.31.
1 − 1/5 (= 4/5): 5.24v (expressed as '0|||1').
73/90: 1.50bA.
40/49: 3.10A.
33/40: 3.14A.
5/6: 2.7A.
1/2 + 1/3 (= 5/6): 1.58.
1 − 1/10 (= 9/10): 1.69.
15/16: 2.9A.
1 (one, unity, single, once): 1.4–6, 8–12, 16, 19–22, 25, 34, 36 (one to nine), 36A, 39, 42 (one to nine), 42A, 43, 46, 62, 68, 69, 73, 73A, 77, 80, 87, 2.3–5, 11, 14, 3.2, 4, 5p, 5, 6, 10, 12–14, 25, 26, 28, 29, 31, 33, 34, 82, 83b, 90–92, 94, 96, 104, 4.12, 16, 19, 21, 26, 28, 30, 31, 35, 36, 38, 39, 41–43, 45, 47, 48, 50, 52, 53, 59, 61, 5.3, 4, 13, 16, 18–21, 29, 30, 6.6, 16, 18, 21–23, 30, 31, 42, 50–52, 54, 55, 59, 75, 81–84.
1 + 1/9: 1.52A.
55/49 (= 1 + 6/49): 3.10A.
1 + 1/5: 2.5A.
1 + 1/4: 1.66, 3.14, 95, 4.20, 5.19.
1 + 1/3: 1.75, 3.100.
70/49 (= 1 + 21/49): 3.10A.
1 + 1/2: 1.50b, 56, 58, 74, 3.12A, 64, 71, 73, 81, 90, 4.19, 23, 5.11, 16.
1 + 13/20: 3.14A.
85/49 (= 1 + 36/49): 3.10A.
1 + 3/4: 3.93A.
2 − 1/4 (= 1 + 3/4): 1.75.
1 + 4/5: 2.5A.
1 + 5/6: 1.52A?.
1 + 12/13: 1.56A.
1 + 987/1008: 2.13A.
2 (two, double, pair, twice): 1.18, 20,

22, 24, 25, 34, 35, 36 (one to nine), 38, 42 (one to nine), 43, 62, (71), 77, 78, 80, 85, 89–91, 2.4, 13, 3.6, 8, 14, 16, 26, 30, 31, 33, 34, 40, 41, 50, 56, 72, 83b, 84, 85b, 90, 100, 103, 4.6, 7, 9, 15–17, 19, 21, 26, 28, 30, 31, 35, 38, 40, 42, 45, 46, 48, 51, 57, 59, 5.16, 18, 6.8, 13–15, 17, 26, 27, 33-36, 40–42, 45, 49, 53, 73, 78–80.

100/49 (= 2 + 2/49): 3.10A.

2 + 1/6: 1.67.

2 + 1/5: 3.17.

2 + 1/4: 1.58A.

115/49 (= 2 + 17/49): 3.10A.

2 + 2/5: 2.5A.

2 + 19/40: 3.14A.

2 + 154/261: 1.67A.

3 – 1/4 (= 2 + 3/4): 1.50b.

3 (three, thrice): 1.1, 10, 26, 36 (one to nine), 39–41, 42 (one to nine), 44, 45, 63p, 63, 78–80, 81p, 81, 85b, 88, 93, 2.4, 9, 13, 15, 3.6, 8, 11–14, 16, 20, 22, 31, 33, 35b, 45, 52, 53, 56, 56A, 58, 72, 79, 87, 90, 104, 4.7, 19, 25, 38, 40, 42, 45, 46, 51, 53, 55, 57, 58, 5.16, 18, 22, 29, 6.7, 10, 12, 19, 24, 32, 71, 77.

3 + 1/4: 3.102.

3 + 1/3: 3.17.

3 + 15/32: 1.74A.

3 + 1/2: 1.52, 2.7, 3.62.

4: 1.6, 8–10, 15, 36 (one to nine), 36A, 42 (one to nine), 76, 77, 79, 2.5, 13, 17, 3.2, 6, 8, 13, 20, 36, 49, 56, 58, 73, 82, 90, 98, 100, 104, 104s, 4.7, 11b, 14, 23–25, 31, 36, 38, 40–42, 45, 47, 49, 51–53, 55, 63, 64s, 5.14, 6.9, 11, 25, 37, 43, 44, 46, 70, 76, 95.

4 + 1/6: 3.6A.

4 + 20/73: 3.102A.

4 + 1/3: 1.56.

4 + 23/48: 2.11A.

4 + 7/12: 1.50bA.

4 + 3/4: 5.25v (expressed as '4|||').

5 – 1/4 (= 4 + 3/4): 1.54.

5: 1.36 (one to nine), 42 (one to nine), 72, 73p, 73, 78, 79, 85, 88, 91, 93, 2.10, 13, 3.2, 4, 6, 8, 16, 19, 20, 22, 25, 27, 32, 35b, 40, 52, 53, 56, 72, 90, 100s, 104, 4.3, 15, 26, 27, 30, 38, 39, 42, 45, 51, 57–59, 63, 5.1p, 1, 13, 33, 33s, 6.3, 20, 24, 47, 90.

5 + 1/6: 3.17.

5 + 1/3: 1.52, 58, 68, 3.62.

5 + 3/4: 3.17.

6 – 1/6 (= 5 + 5/6): 1.66.

6: 1.9, 36 (one to nine), 42 (one to nine), 78, 80, 2.4, 5, 6, 3.12, 16, 19, 20, 35b, 47, 48, 58, 66, 75, 79, 83b, 4.30, 38–40, 42, 45, 49, 63, 5.22v, 25v, 6.4, 5, 20, 72.

6 + 1/4: 1.56.

7 – 1/4 (= 6 + 3/4): 1.56, 58.

7: 1.36 (one to nine), 42 (one to nine), 64, 70, 70A (7 years = 2520 days), 72, 78p, 78, 79, 88, 91, 2.4, 3.10, 19, 20, 22, 25, 27, 82, 93, 104, 4.14, 38, 39, 42, 45, 58, 63, 5.5, 13, 6.21.

7 + 1/5: 4.57A.

7 + 1/4: 1.67.

7 + 7/19: 4.5A.

7 + 1/2: 1.52A?, 61.

7 + 21/40: 1.69A.

7 + 21/32: 2.7A.

7 + 2/3: 1.52A?.

7 + 5/6: 1.52A?.

7 + 8/9: 1.52A?.

8: 1.5, 15, 20, 36 (one to nine), 42 (one to nine), 42A, 52, 71, 76, 79, 2.17s, 3.6, 16, 17, 19, 20, 22, 30, 57, 61 (in '13-8-24'), 88, 93, 94, 104s, 4.7, 21, 38, 42, 45s, 61, 63, 5.28, 6.29, 38, 98.

8 + 2/25: 3.6A.

8 + 181/1284: 3.17A.

8 + 1/3: 1.54.

8 + 1/2: 1.75.

9 – 1/3 (= 8 + 2/3): 1.67.

9: 1.30, 36 (one to nine), 36A, 42 (one to nine), 65, 72, 78, 79p, 79, 80, 85, 88, 93, 2.4, 3.14, 16, 19, 19A, 20, 22, 25, 46, 57, 69, 83b, 90, 95s, 4.11b, 38, 40, 42, 45, 53, 55A, 57, 61, 63, 5.10, 13, 30, 6.19, 22, 48, 85, 89.

9 + 1/4: 1.75, 3.17.

$9+1/3$: 1.66, 3.13, 62A.
$9+5/16$: 3.16A.
$10-1/4$ ($=9+3/4$): 1.75.
10: 1.9, 12, 19, 19A ($10n$), 23 ($10n$), 82, 87, 93, 3.6, 10, 16, 19, 22, 25, 43, 44, 46, 53, 64, 72, 73, 85b, 91–93, 4.15A, 38, 40, 42, 59, 5.6–8, 14, 28.
$10+1/36$: 5.25v (suggested for 10).
$10+1/4$: 1.84, 3.57.
$10+1/2$: 1.77, 88A.
$10+2/3$: 4.14A.
11: 1.64, 80p, 3.14, 16, 20, 25, 93, 4.38, 42, 6.23.
$11+1/2$: 1.84.
$11+5/8$: 3.22A.
12: 1.10, 11, 69, 76, 79, 2.15A, 3.40, 103, 4.21, 38, 40, 42, 5.28, 29.
13: 3.38, 40, 61 (in '13-8-24'), 91, 4.38, 40, 42.
$13+351/1632$: 1.77A.
14: 3.38, 4.15, 38, 40, 42, 5.8.
$14+1/11$: 3.20A.
15: 1.87, 3.4A, 35bA, 38, 40, 47, 52–53A, 68s, 4.38, 42, 58, 5.5, 6.1, 69.
$15+1/4$: 1.61.
$15+3/11$: 1.64A.
$16-1/4$ ($=15+3/4$): 4.37.
16: 1.8, 10, 36, 36A, 42, 65, 82, 3.56, 91, 4.24, 35, 38, 42, 55, 5.5, 6, 28.
$16+1/2$: 3.57.
$16+7/8$: 1.82A.
17: 1.87, 4.40.
$17+1969/2176$: 1.75A.
18: 3.85b, 86s, 4.39, 5.5, 30.
$18+1/4$: 3.102.
$18+1/3$: 1.83A.
19: 3.75, 4.39, 5.22v, 25v, 6.56, 74.
$20-1/4$ ($=19+3/4$): 1.77.
20: 1.3, 4, (19), 23 ($10n$), 83, 3.19, 27, 52, 58, 91, 92, 4.36, 37s, 40, 47, 49, 5.3, 4, 5, 14.
21: 3.40, 47, 4.58.
22: 3.82A, 4.24, 5.5.
23: 4.40.
24: 1.5, 6, 36, 64, 3.61 (in '13-8-24'), 67, 75, 102, 4.36, 40, 5.4.
25: 1.15, 36A, 65, 93s, 3.37, 4.39, 5.28, 6.2, 28, 39.
$25+5/7$: 2.4A.
27: 1.28, 42A, 82, 4.40, 5.30.
28: 1.36, 4.40.
$28+4/9$: 1.58A.
$28+1/2$: 5.22v (expressed as '28||', suggested for 28).
30: 1.11, (19), 23 ($10n$), 76, 3.4A, 8A, 91, 4.40, 5.29.
$\sqrt{1000}$ ($\approx 31+31/50$): 3.44A.
32: 1.5, 28, 4.39, 5.7, 16.
33: 4.39.
$34+2/7$: 2.4A.
35: 3.27A, 32A, 4.40.
36: 1.36A, 73A, 3.35bA, 98, 98–99A.
$38+298/911$: 3.9A.
39: 3.37.
$39+7/12$: 1.54A.
40: 1.9, (19), 23 ($10n$), 2.4A, 3.52–53A, 90, 5.6, 32.
$40+1/2$: 1.84.
42: 1.88.
$43+5/9$: 1.66A.
$44+4/9$: 1.65A.
45: 1.15, 76, 3.8A, 5.7.
$45+18/41$: 1.84A.
$45+9/16$: 1.58A.
$47+613/911$: 3.9A.
48: 3.68, 88A, 5.27.
49: 1.36A, 3.68, 5.27.
50: 1.(19), 23 ($10n$), 3.27A, 52–53A, 67, 5.10, 26.
$50+5/36$: 3.75A.
$51+9/13$: 3.62A.
52: 3.37.
53: 4.49A.
55: 1.19A, 3.32A.
56: 5.7.
60: 1.4, 11, (19), 23 ($10n$), 28, 70, 73, 73A, 80, 3.4A, 8A, 37, 68, 83bA, 90, 5.3, 4, 27, 32.
62: 3.68, 5.27.
64: 1.36A, 42A, 4.42, 4.64s.
66: 1.83.
$\sqrt{4410}$ ($\approx 66+2/5$): 3.47A.
70: 1.(19), 23 ($10n$), 4.58.
72: 1.83, 3.98–99A.
$74+4/5$: 3.72A.
75: 1.87A, 3.8A.
78: 6.2, 28.

$\sqrt{6250}$ ($\approx 79 + 1/20$): 3.44A.
80: 1.(19), 23 (10n), 3.75A, 90.
81: 1.36A, 4.45.
84: 3.38A.
86 + 2/3: 3.2A.
90: 1.(19), 23 (10n), 78.
90+1/4: 5.25v (expressed as '90|', suggested for 90).
98: 1.(26).
99: 1.(26).
100: 1.12, (19), 26, 91, 2.4, 3.4–6, 90, 4.4, 6, 8, 19, 35, 37, 5.9.
100 + 5/18: 3.66A.
100+6/20: 5.25v (expressed as '100SS6').
100 + 1/3: 1.69.
100 + 1/2: 1.74.
104: 3.104s, 6.38, 98.
105: 4.58.
108: 1.28, 93, 3.98-99A, 4.53A.
118 + 3/4: 3.64A.
120 − 1/4 (= 119 + 3/4): 1.68.
120: 3.66A, 4.51A.
121 + 1/2: 5.25v (expressed as '121||', suggested for 120).
125: 1.42A.
135: 1.80A.
141: 2.16 (suggested for 145).
144: 98–99A.
145: 2.16 (probably an error for 141).
156: 3.40A.
162: 5.25v.
166 + 2/3: 1.91A.
200: 1.85, 93A, 3.6.
206: 4.51A.
210: 1.19A, 3.8.
216: 1.42A, 5.25v.
225: 1.42, 76A, 3.32A.
240: 3.56A.
245: 3.27A.
250: 4.59.
256: 1.36A.
300: 1.91, 3.4A, 6.
309: 1.42.
311: 5.33s, 6.0.
343: 1.42A.
360: 3.85bA.
365: 4.47A.
421 + 7/8: 1.61A.
424: 4.49A.
433 + 1/3: 3.2A.
465: 1.19A.
500: 3.4A.
512: 1.42A.
520: 3.2.
570: 3.59A.
576: 1.36A.
594: 1.85A.
600: 1.79, 3.6, 4.3A.
657 + 9/13: 3.64A.
679: 4.17.
685 + 5/7: 1.79A.
699: 4.17.
720: 2.16A.
729: 1.42A, 3.57A.
784: 1.36A.
787 + 1/2: 1.78A.
820: 1.19A.
900: 4.3A.
905: 3.4.
964: 1.28.
976: 4.17.
1000: 1.12, 23, 4.7.
1008: 2.17.
1275: 1.19A.
1296: 4.53A.
1354: 3.57A.
1400: 5.29.
1404: 5.33s.
1500: 4.3A.
1764: 3.37A.
1791 + 2/3: 4.11bA.
1830: 1.19A.
2000: 1.6, 28, 4.3.
2095 + 17/50: 4.8A.
2100: 4.3A.
2400: 5.26.
2448: 3.73A.
2485: 1.19A.
2520: 1.70A (2520 days = 7 years).
3000: 4.3.
3240: 1.19A.
3546 + 37/64: 1.61A.
3900: 4.3A.
4095: 1.19A.
4096: 1.42A.
4753: 1.26A.
5000: 4.3, 11b.
5050: 1.19A.

6480: 1.28A.
7000: 4.3.
8480: 4.49A.
9000: 4.3.
10000: 1.12, 4.7.
13000: 4.3.
13824: 3.61.
26028: 1.28A.
30000: 4.3.
$39822 + \frac{2}{9}$: 1.71A ($39822 + \frac{2}{9}$ years = 14336000 days).
50445, etc. (differences between the first 1000 terms of the natural series and the first $10m$ terms of the same): 1.23A.
60480: 4.63A.
64000: 1.28A.
70000: 4.5.
100000: 1.12, 4.7.
209534: 4.8.
950000: 4.5.
1000000: 1.12.
11390625: 1.42A.
14336000: 1.71A (14336000 days = $39822 + \frac{2}{9}$ years).
29503629: 1.42A.
10^0 to 10^{24} (powers of ten): 1.12–14.

Appendix D

Glossary-Index to the Text

We have excluded from this glossary-index the pronouns (*jo*, *so*, etc.), the indeclinables having no mathematical connection (*taha*, *puna*, *ya*, *vi*, etc.), and the verb $\sqrt{}$*hava* (S. $\sqrt{}$*bhū*), which occur very often in the text. The words included in this index may be classified as follows: words for mathematical topics such as *saṃkaliya*, *jāi*, *jaṃta*, etc.; words for mathematical operations such as *paḍhama*, *aṃka*, $\sqrt{}$*pāḍa*, etc.; words for weights and measures such as *aṃgula*, *kaṃbiya*, etc.; words for numerical expressions such as *ikki/iga*, *du/dui*, *vāīya*, etc.; words used in connection with the topics of examples such as *aṃba*, *karaha*, etc. We have also included Sanskrit words used in prose parts (see below) except the indeclinables (*atha*, *iti*, etc.) and the verb *āha*, which occur quite often. We have supplied the original or the commonest or the most typical meanings, too, if they are different from the ones intended in the text.

Abbreviations: Ar. = Arabic, D. = *Deśī* (vernacular), H. = Hindī, P. = Prakrit, Pe. = Persian, S. = Sanskrit, N = the Nahatas, the previous editors of the GSK. The symbol, Ts., means either that the word has the same form in Sanskrit (a *tatsama* word) or that it is a Sanskrit word. The references are to the verse numbers. A digit followed by a period, except for the last digits of items, indicates the chapter number in which the verse occurs. The three abbreviations, p = *praveśa* (introduction), v = *vāsanā* (explanation), and s = *samāpti* (conclusion), attached to a verse number indicate respectively the introductory prose sentence to, the explanatory prose sentence for, and the concluding prose sentence occuring after, that verse.

The order of the Indian letters employed in this index: a, ā, i, ī, u, ū, ṛ, ṝ, ḷ, e, ai, o, au, ṃ, ḥ, k, kh, g, gh, ṅ, c, ch, j, jh, ñ, ṭ, ṭh, ḍ, ḍh, ṇ, t, th, d, dh, n, p, ph, b, bh, m, y, r, l, v, ś, ṣ, s, h.

a

aṃka S. aṅka, a mark, sign: a numerical figure, digit 1.14, 14s, 18, 23, 29, 34, 39–41, 43–46, 49, 72, 92, 3.18, 24, 31, 4.6, 7, 9p, 9, 10, 60p, 60, 61p, 61; a number 1.35, 37, 53, 2.14, 4.38, 44, 50, 64.

aṃgula S. aṅgula, a digit: a linear measure 1.6, 71, 3.10, 61, 102, 104. Cf. kaṃviyaṃgula, karaṃgula, pavvaṃgula.

aṃjaṇa S. añjana, Asiatic barberry, *Berberis asiatica* Roxb.? 3.95.

aṃta S. anta, last: the last term in any series 1.63, 81, 86, 90, 103, 4.16, 54, 60, 61; the last term of a mathematical series 1.36, 3.26; the last place (i.e., the unit place) in the decimal place-value notation 1.37, 45, 4.6.

aṃtara S. antara, being in the middle: the difference 1.22, 51, 86, 89, 3.41, 4.13, 5.2; intermediate space 3.78, 79; inside 3.98.

aṃtima S. antima, the last 4.7, 9, 10, 51, 64.

aṃba S. āmra, mango, *Mangifera indica* L.: mango fruit 4.49.

aṃsa[1] S. aṃśa, part 1.50, 52, 66, 67, 69, 70, 71, 2.1, 2, 5, 7, 11, 13, 14p (-ka), 15–17, 3.3, 7 (investment), 11, 14, 17, 45, 58, 65, 71, 73, 74, 76, 81, 88, 96, 97, 4.9, 22, 23, 25?, 30, 31, 49, 5.10, 13, 22, 23, 25, 25v, 31; a numerator 1.46–48, 50, 51, 53, 55, 57, 59, 60, 62, 2.1–3, 6, 8, 10, 12, 14; a fraction 4.48.

aṃsa[2] S. asra or aśra, a side of a geometric figure. See aṭṭhaṃsa, khaḍaṃsa, caüraṃsa, taṃsa.

akka[1] S. aikya, oneness, unity: the sum 3.5. See ikka[2], aikya.

akka[2] = ikka[1] 5.18 (in dammakkihi). Cf. acchiya for the euphonic change, i > a.

akkhara S. akṣara, a letter 1.18.

agga S. agra, the foremost point: the next term 1.17.

aggi S. agni, fire: a bhūtasaṃkhyā for three 4.38.

aggima S. agrima, the foremost point: the next term 1.17, 41.

aggha S. argha, the price 5.1.

agghapamāṇa S. argha-pramāṇa, the price standard 1.65.

agghamāṇa S. argha-māna, the price rate 3.89.

aghaṇa S. aghana, non-cubic (place) 1.43 (-paya), 62 (-paya).

aṃgaja Ts., a son 1.93s, 2.17s, 3.104s, 4.64s, 5.33s.

$\sqrt{}$accha S. $\sqrt{}$ās, to enclose: acchaï 1.79.

acchiya = icchiya 3.87 (in jahacchiya). Cf. akka[2] for the euphonic change, i > a.

accheya S. accheda, having no denominator: an integer 1.46.

aṭṭha S. aṣṭan, eight 1.5, 15, 20, 52, 54, 71, 75, 76, 79, 93, 2.17, 3.6, 9, 16, 17, 19, 20, 22, 57, 61, 88, 93, 94, 4.7, 21, 22, 25, 30, 31, 45, 61, 63, 5.28. See aḍa.

aṭṭhaṃsa S. aṣṭa-asra, octagonal 3.86, 4.25?, 29.

aṭṭhama S. aṣṭama, the eighth 2.16, 3.73, 4.49.

aṭṭhavīsa S. aṣṭāviṃśati, twenty-eight 1.36.

aṭṭhāra/aṭṭhārasa S. aṣṭādaśan, eighteen 3.85, 102, 5.5. See aṭhara.

aṭhara S. aṣṭādaśan, eighteen 4.39. See aṭṭhāra/aṭṭhārasa.

aḍa S. aṣṭan, eight 1.28 (-ahiya), 3.30, 5.13 (-aṃsa). See aṭṭha.

aḍayālīsa S. aṣṭa-catvāriṃśat, forty-eight 3.68, 5.27.

aḍḍha (in saḍḍha 'with a half') S. ardha, half 1.52, 61, 74, 75, 77, 84, 2.7, 3.57, 62, 5.24. See addha.

aḍḍhāiya S. ardha-adhika, (one) increased by a half (i.e., one and a half) 1.56, 58. Cf. divaḍha(ya), saḍḍha.

aṇukama/aṇukkama S. anukrama, order, sequence 4.38, 42, 51, 62.

aṇubhava S. anubhava, experience 3.76.

aṇubhūya S. anubhūya, having expe-

rienced 1.2.
aṇusāra S. anusāra, following 5.17.
atalasa, Ar. aṭlas, satin: a kind of silk 4.18. Cf. Habib 2004, item 7.
attha[1] S. artha, object: for the sake of 3.28, 4.49.
attha[2] S. asta, setting 3.103.
attha[3] (error for hattha?) S. hasta, a hand: a cubic measure 1.7. See hattha.
adaṃtīya S. adantīya, one who does not have teeth 5.15.
addha S. ardha, half 1.16, 20, 52, 54, 58, 2.1, 2, 7, 9, 11, 15, 17, 3.4, 13, 34, 36, 39, 45, 46, 49, 62, 65, 73, 74, 76, 78, 83, 88, 98, 103, 4.10, 22, 25-29, 37, 41, 47, 5.16, 17, 20, 24, 31. See aḍḍha. Cf. aḍḍhāiya, divaḍha(ya).
addhiya/addhīya S. ardhita, halved 1. 38, 49.
adha S. adhas, below, down 4.42.
adhikāra Ts., a topic: the four topics of mathematics or any one of them 4.1p, 17s, 18p, 37s, 38p, 45s, 46p, 64s. See ahigāra.
adhyāya Ts., a chapter 1.93s, 2.17s, 3.104s.
anu-$\sqrt{}$bhū. See aṇubhūya.
anulomavilomagaī S. anuloma-viloma-gati, going along or against hairs: the direct or inverse order 1.27.
anna[1] Ts., grain, food 1.68, 3.96, 5.2, 4, 17?, 6.91.
anna[2] S. anya, other 1.63, 81, 3.22, 5.17?, 18.
annonna S. anyonya, mutual(ly) 1.47.
appagaï S. alpa-gati, one who has the lower speed: the slow runner or traveller 4.13. Cf. bahugaï.
abhinna Ts., not broken: an integer 6.11 (corruption of bhinna).
amuṇiyarāsi S. ajñāta-rāśi, unknown quantity 4.56. Cf. $\sqrt{}$muṇa, muṇe-yavva.
arakkha S. arakṣa, without a protector or protection: a kind of sand? 5.32. Cf. rakkha.
aruṇa Ts., red 4.36, 37.
argha Ts., price 5.21s. See aggha.
alasī S. atasī, linseed, *Linum usitatissimum* L. 5.7.
alīka Ts., false 1.30p.
avara S. apara, other, remainder 1.34, 4.51.
avigaya S. avikṛta, unchanged 1.31.
avva S. arbuda (through avvuya and avvua), a numeral for 10^9 or the name of the tenth decimal place 1.12.
aṣṭan Ts., eight 2.17s, 3.104s, 6.29, 98. See aṭṭha, aḍa.
aṣṭama Ts., the eighth 6.47.
asa S. asra or aśra, a side of a plane figure. See caürasa.
asī S. aśīti, eighty 3.90.
asu S. aśva, a horse 4.43 (in asu-kama, horse's (knight's) move in chess). Cf. H. (Avadhī) asu in OHED.
aha[1] S. adhas, lower, below 2.6.
aha[2], tell (impv.)? 1.73.
ahavā S. avadhā, depression: a projection or a segment of the base of a triangle 3.41.
ahigāra S. adhikāra, a topic: the four topics of mathematics 1.15.
ahiya S. adhika, more, greater 1.34, 39, 3.24, 4.20, 32, 39, 44, 5.11, 17: increased by 1.23, 28, 35, 42, 67, 77, 82, 85, 93, 2.11, 17, 3.4, 48, 58, 65, 76, 84, 4.22, 26, 30, 31, 57, 5.24; additional, additive, increase 3.24, 4.17 (-māsa), 56. Cf. aḍḍhāiya.

ā

āi/āī S. ādi, the beginning 4.17, 40: the first term in any sequence 1.14, 21, 31, 32, 34, 63, 68, 72, 80, 81, 86, 2.10, 3.3, 63, 90, 91, 95, 101, 104, 4.7, 10, 16, 18, 33, 41, 54, 55, 61, 64, 5.9; the first term of a mathematical series 1.19, 21, 23, 36, 42, 3.26–30, 4.15, 39, 42, 43, 45, 48, 50. See ādi.
āima S. ādima, the first 1.63, 89, 4.10.
āesi ? 4.49.
ākāra Ts., form 3.50, 6.64-67. See

āgāra.
āgāra S. ākāra, form 3.53, 4.29.
āṇayaṇa S. ānayana, fetching (an answer to a question), calculation 3. 29.
āṇiya S. ānīta, brought 4.49.
ādāya Ts., income: a positive mixed quantity 1.51. Cf. vaya.
ādi Ts., the beginning 6.59: the first term of a series 3.28p, 4.49. See āi/āī.
ādya Ts., the first, beginning 4.64s, 6.50.
ānayana Ts., fetching (an answer to a question), calculation 3.28p, 29p, 30p, 31p, 33p, 34p, 58p, 61p, 103p, 4.17p, 43p, 46p, 48p, 50p, 52p, 61p, 62p, 6.50-54, 71, 72. See āṇayaṇa.
āmalaya S. āmalaka, myrobalan, *Emblica officinalis* Gaertn. 3.12.
āmra Ts., mango, *Mangifera indica* L.: mango fruit 4.48p. See aṃba.
āyaria S. ācārya, a teacher 1.1.
āyāma Ts., length 1.7, 85, 3.35, 56.
ālaṃva S. ālamba, hanging down, support: the stick of an umbrella? 4. 34 (chatta-).
ālaya Ts., a house 3.70.
√āva S. ā-√yā, to come: āvaṃta 4.47; āvaṃti 4.3, 63.
āvaṭṭa S. āvṛtta, turned round: melted together (gold) 3.17.
āsāḍhiya S. āṣāḍhika, spring (harvest) 5.6. Cf. H. asāḍhī and sāḍhī.
āsiya, the proper name of a region around Hansi, in the modern state of Haryana 5.2.

i

iṃda S. indu, the moon. See bāliṃda.
iṃdiya S. indriya, the sense organs: a bhūtasaṃkhyā for five 4.38, 45.
ikka[1] S. eka, one 1.4, 5, 6, 9, 68, 69, 3.12, 14, 25, 67, 96, 4.12, 35, 36, 47, 50, 5.16, 18–21, 26, 30. See akka[2], ika, ikkikka, iga, ega/egga.
ikka[2] S. aikya, oneness: the sum 3.18, 21, 32.
ikkaṭṭhā S. ekatra, in one place 2.16.
ikkārasa/ikkārasī S. ekādaśan, eleven 1.64, 84, 3.14, 16, 20. See igārasa.
ikkāsī S. ekāśīti, eighty-one 4.45.
ikkikka S. ekaika, one by one, each 1.21, 3.10, 4.50.
ikkhu S. ikṣu, sugarcane 5.1, 6, 10 (-rasa).
ikṣu Ts., sugarcane, *Saccharum officinarum* L., 5.12s (-rasa). See ikkhu.
iga S. eka, one 1.6, 11, 20, 80, 2.5, 3.13, 14, 31, 90, 94, 104, 4.19, 21, 26, 30, 36, 38, 41, 43, 45, 47, 53, 59, 64, 5.12, 29. See ikka, ega/egga.
igayāla S. ekacatvāriṃśat, forty-one 2.16.
igavīsa S. ekaviṃśati, twenty-one 3.40, 47, 4.58.
igārasa S. ekādaśan, eleven 3.25. See ikkārasa, gāra/gārasa.
igārasama S. ekādaśa, eleventh 3.97.
igigi S. ekaikaṃ, one by one 4.38.
igega S. ekaika, one by one, each 1.29, 4.47, 53.
iccāi S. ityādi, beginning with this, and so on 1.14.
iccha/icchā S. icchā, requisite: optional, desired 3.18, 24, 33, 43; any optional natural number 1.16, 18, 35; optionally 4.38. Cf. jahiccha.
icchiya S. iṣṭa, arbitrary, optional 1.41, 4.61. See acchiya, jahacchiya, jahicchiya. Cf. iccha/icchā.
iṭṭa/iṭṭā S. iṣṭakā, a brick 3.73p, 73, 5.29.
iṭṭha S. iṣṭa, optional, arbitrary 1.35.

ī

īṭa = iṭṭa/iṭṭā 6.76.
īsa S. īśa, a leader, lord 1.1.
īsara S. īśvara, a lord, master: a bhūtasaṃkhyā for eleven 4.38.

u

uṃḍḍatta S. aunnatya, height, elevation: the height (or depth) of a pit 3.54, 55, 56, 59.
ukkamaso S. utkramaśas, in the inverse order 4.64.

ucca Ts., high, elevated 3.78 (kamucca).

uccatta S. uccatva, the state of being high: height 3.102.

uṇa[1] (mostly with the one-syllabic numerals, bi-/vi-, ti-, ca- in the sense of two-fold, three-fold, or four-fold) S. guṇa, merit: a multiplier 1.3, 10, 18, 20, 24, 25, 34, 38, 40, 41, 44, 45, 49, 3.30, 33, 34, 41, 45, 49, 58, 4.9, 16, 23, 25, 31, 48, 51, 52, 54, 59, 5.14, 18, 22, 30. See guṇa.

uṇa[2] (in paüṇa unless otherwise indicated) S. ūna, less 1.50, 56, 58, 68 (pā-), 75, 77, 2.9, 3.13, 17, 95, 99, 100, 4.37, 5.12, 13, 25, 25v, 31. See una, ūṇa.

uṇayālīsa S. ūna-catvāriṃśat, forty less (by one), i.e., thirty-nine 3.37.

uṇavanna S. ūna-pañcāśat, fifty less (by one), i.e., forty-nine 3.68, 5.27.

uṇavīsa S. ūnaviṃśati, twenty less (by one), i.e., nineteen 3.75, 4.39.

utta S. ukta, told 2.10, 3.60.

uttara Ts., upper, latter, more: increase 3.8, 4.58; the common difference of an arithmetical progression 1.19, 3.26, 27, 28, 29p, 29, 30, 6.51; north 2.16, 3.103.

utpatti Ts., producing 6.3, 6-10; agricultural produce 5.9s. See uppattī.

udaya Ts., rise 3.103: height 1.7, 3.10, 57, 66, 70, 72, 73, 77, 78, 81–85, 96-99, 101, 4.20–22, 26–31, 6.92; diameter of a sphere (or of a spherical stone) 3.65, 5.25.

udayattha S. udaya-asta, rising and setting 3.103.

udāharaṇa Ts., an example 3.17p.

udesa = uddesa(ga), an example 5.29.

uddeśaka Ts., a mathematical problem, example 4.56p. See uddesa-(ga).

uddeśapaṃcaga/uddesapaṃcaga S. uddeśa-pañcaka, the quintet of topics 5.1p, 33, 33s.

uddesa(ga) S. uddeśa(ka), an illustration: a mathematical problem, example 4.57, 6.37; an example 4.35, 5.17, 28; a topic 5.1.

una S. ūna, less, diminished 4.58 (ti-). See uṇa[2].

upaa (or upaya) S. upari(ma), upper 4.42. See upara, uvara.

upakkhaï/uppakkhaï S. upakṣati or upakṣaya, loss? 4.4, 5. See saï and caṭṭiya/caṭṭī, with which upakkhaï occurs.

upaya. See upaa.

upara S. upari(ma), upper 4.62.

uppattī S. utpatti, the product, profit 1.86, 5.33.

uppanna S. utpanna, produced, product 3.21, 22, 23.

uvakama S. upakrama, beginning 4.41.

uvama S. upama, similar to 3.51, 52, 53, 55.

uvara/uvari/uvarima S. upari(ma), upper: above, upper (top) part 1. 27, 33, 46, 49, 89, 2.6, 8, 12, 3.10, 78, 81, 83, 91, 92, 4.21, 22, 26, 27, 29, 41, 62; the uppermost term of a series 1.25, 26. See upaa and upara.

uvaruppari S. upary-upari, upwards 1.41, 4.50, 62.

uvahi S. udadhi, the sea: a bhūtasaṃkhyā for seven 4.38.

√uvvara S. ud-√vṛt, to remain: uvvaraï 4.46; uvvarahi 4.46, 47.

ū

ūṇa S. ūna, less: decreased 1.35, 54, 66, 67, 74, 77, 2.9, 3.23, 26, 28, 30, 48, 81, 84, 4.12, 23, 59, 5.12, 23; subtrahend 4.56. See uṇa[2].

ūṇaya S. ūnaka, a subtrahend 4.17.

ṛ

ṛṇa. See riṇa.

e

eka Ts., one 6.59: one and the same 1.63; the sum 3.33p, 6. 54, 55. See ega/egga.

ekatra Ts., in one place: in total 3.104s. See egattha.

ekapatrīkaraṇa Ts., reduction of several bonds into one 3.5p.

ekādaśarāśika Ts., of eleven quantities: rule of eleven, eleven-quantity operation 6.23.

ekādaśarāśaya/ekādasarāsika S. ekā-daśa-rāśika, of eleven quantities: rule of eleven, eleven-quantity operation 1.72p, 80p.

ega/egga S. eka, one, a numeral for 1, or the name of the first decimal place 1.8, 11, 12, 16, 19–22, 25, 36, 41, 42, 46, 2.4, 3.4, 6, 10, 26, 28, 29, 33, 34, 82, 83, 88, 94, 4.16, 19, 31, 39, 42, 43, 50, 54, 61, 5.16, 20, 21, 6.42: one and the same 1.81, 4.51; one of the two 3.38. Cf. ikka, iga.

egattha S. ekatra, in one place 2.11: into one (alloy) 3.16 (-gāliya); the sum, total 3.54, 4.3.

egapatrīkaraṇa S. eka-patrī-karaṇa, reduction of several bonds into one 6.42.

egaya S. ekaka, one 1.21.

ai

aikya Ts., oneness, unity: the sum 3.31p, 33p, 34p, 6.53. See akka^1, ikka^2.

o

oli S. āvali, a sequence 4.41, 43.

k

kaṃguṇiya S. kaṅgunika, Italian millet, *Panicum italicum* L. 5.5.

kaṃcolu ? 3.10.

kaṃbala/kaṃvala S. kambala, a blanket 1.78, 85.

kaṃbiya/kaṃviya/kaṃvī S. kambī/-kambikā, a branch or joint of a bamboo: a linear measure 1.4, 2. 15, 3.67, 77, 5.3, 26.

kaṃviyaṃgula S. kambikā-aṅgula, a bamboo-joint digit: a linear measure 3.94.

kakkaḍa S. karkaṭa, a crab: the zodiacal sign Cancer 3.104.

kakkara S. karkara, limestone 5.32.

kajja S. kārya, to be done 4.1.

kaṭaka/kaṭakka Ts., an army 4.49.

kaṭṭhā S. kāṣṭa, wood 3.70.

kaḍa S. kaṭa, grass 4.23.

√kaḍḍha OIA √*kardh, to obtain: kaḍḍha 4.3, 6. Cf. Pali kaḍḍhati. Cf. also HGA p. 366.

kaṇa Ts., a grain or a seed 5.8.

kaṇaya S. kanaka, gold 1.69, 84, 3.16-19, 23.

kattaraṇa S. kartana, cutting 4.19.

kanna S. karṇa, an ear: the hypotenuse of a right-angled triangle 3.42; the diagonal of a square or of a rectangle 4.39.

kannacalā S. karṇa-cala, one who moves one's ears (or one who goes diagonally?) 5.15.

kannāṇaya, of Kannāṇā (birth-place of Pherū): stone from Kannāṇā 3.68, 5.27.

√kappa S. √kḷp, to assume: kappijjā 3.50.

kappaḍa S. karpaṭa, cloth 1.64, 4.19-21, 32.

kappāsa S. karpāsa, cotton, *Gossypium herbaceum* L. 5.6. Cf. Habib 2004, item 45.

kama S. krama, a step: order, sequence 1.3, 19, 34, 47, 72, 80, 90, 2.5, 15, 3.3, 14, 36, 73, 78, 100, 4.7, 10, 19, 38, 41, 42, 50, 59, 64, 5.13; a linear measure 1.4, 5.3; the way of motion 4.43 (asu- S. aśva-). Cf. sukama and parakama.

kamara Pe. kamar, waist? 4.21, 22, 24, 25, 31.

kamaso S. kramaśas, in due order, successively 3.20.

kammayara S. karma-kara, a worker, an artisan 1.76.

kaya^1 S. kṛta, made 1.22, 24, 49, 3.96. See kiya.

kaya^2 S. kraya, purchase 3.11: the buying rate 1.86. Cf. vikaya.

√kara S. √kṛ, to make: karavi 1.29, 51, 2.1; karahi 2.16; kari 1.29, 38, 40, 4.48; kariūṇa 1.55; karijja 4.4; karivi 1.40 (ghaṇaṃ-), 4.17; karevi 2.10; kareviṇu 1.48, 4.51; kāuṃ 3. 29, 46, 54; kāriṇi 4.3; kujjā 1.60. See √kīra.

kara Ts., making, a hand: making 5.1 (siṭṭhi-), 17; a linear measure 1.64, 78, 79, 3.35, 37, 52, 53, 56–59, 62, 64, 66, 72, 79, 82, 85, 88, 99, 103, 4.19, 24, 27, 52, 53; a square measure 4.52, 53; a cubic measure 2.5; a bhūtasaṃkhyā for two 4.38, 42, 59; a kind of fine cotton? 4.18; a tax 5.1 (desa-), 9, 15, 17s (deśa-), 33 (desa-). See kora[2].

karaṃgula S. kara-aṅgula, a hand-digit: a linear measure 1.5.

karaḍa S. karaṭa, safflower, *Carthamus tinctorius* L. 5.7, 13.

karaṇa Ts., making 3.5p, 24, 94, 4.1 (bhūya-), 6.42: mathematical operation, procedure 1.59, 62, 90, 92p, 2.10, 3.28p, 29p (-sūtra), 61p (-sūtra), 87p (-sūtra), 101p (-sūtra), 103p, 4.17s, 38p.

karapuḍa S. kara-puṭa, hands joined and hollowed: a volume measure 1.8, 80.

karabha Ts., a camel 4.54p. See karaha/karahī.

karavatta S. kara-patra, a saw 3.89p, 89, 92, 94, 6.87 (-ī).

karaha/karahī S. karabha/karabhī, a male/female camel 1.93, 4.15, 16, 54, 55.

kala S. kalā, a digit of the moon: a bhūtasaṃkhyā for sixteen 4.42.

kalāva S. kalama, rice (sown in May–June) 5.7.

kalāsavarṇana Ts., making a fraction into the same colour or homogenization of a fraction: reduction of a composite fraction to a simple one 2.1p (-savarṇṇana), 6.30. See savaṃnaṇa/savannaṇa/savarṇana.

kaliya S. kalita, mixed 4.32, 5.32.

kavāḍaśaṃdhī S. kapāṭa-sandhi, door-junction: a multiplication method 1.27.

kavilīya S. kapilīya, a tawny one 5.15.

kasiṇa S. kṛṣṇa, black 4.37: black stone 3.68, 5.27.

√kaha S. √kath, to tell: kahasu 1.64, 82, 85; kahi 1.79; kahijjaï 1.18; kahu 1.87.

kāiṇi S. kākiṇī, a monetary unit 1.3.

kāyavva S. kartavya, to be made 1.44.

kāla Ts., time 1.70, 71, 73, 2.5, 3.1, 3, 5.

kāṣṭha Ts., wood 6.86, 88.

kimisa ?, 4.36.

kiya S. kṛta, made 1.79, 3.48. See kaya[1].

kiri S. kila sambhāvanāyām (in the sense of assumption) 1.77.

kīḍaya S. kīṭaka, a worm 1.71.

√kīra S. √kṛ, to make: kīraï 1.48; kīrae 1.22, 3.15; kīraṃti 3.87. See √kara.

kīra, S. cīḍha, chir-pine tree, *Pinus longifolia* Roxb. 3.95.

kuṭṭha S. koṣṭha, an apartment: a cell of a magic square 4.38, 45. Cf. giha/geha.

kuḍukkaḍa, a kind of stone 3.68, 5.27.

kuḍhiyā/kuḍhī, hearth tax? 5.14, 16. Cf. H. kuḍī, 'a hearth, historically hearth tax' (OHED).

kuddava S. kodrava, a kind of cereal, *Paspalum scrobiculatum* L. 3.97, 5.4. Cf. tusaṃnna.

kulattha Ts., horse gram, *Dolichos biflorus* L. 3.97, 5.8.

kulisa S. kuliśa, thunderbolt: a figure like a thunderbolt (which consists of two trapeziums) 3.50, 53.

kuva. See kūva.

kuvvāya ?, 4.34.

kusuma Ts., a flower 4.46p, 46, 47.

kūṇa S. koṇa, a corner 3.98, 99.

kūpa Ts., a cylindrical well 3.58p, 85p, 6.71, 83. See kūva/kuva.

kūrīsa S. kūra-īśa, rice king (the best kind of rice?) 5.5.

kūva/kuva S. kūpa, a cylindrical well 3.58, 59, 69, 79, 85.

kevaṇa S. kevala, pure? 5.31 (-cunna).

koḍi S. koṭi, a numeral for 10^7 or the name of the eighth decimal place 1.12.

koḍinīla S. koṭi-nīla, one crore of nīlas: a numeral for 10^{24} or the name of the twenty-fifth decimal place

1.14.

koḍiya/koḍī, a unit of currency 3.89, 90, 91, 92, 93. Cf. H. kauḍī, a cowry shell, S. varāṭa(ka).

koḍiyā, a score, twenty 5.21. Cf. H. koḍī.

koṇa Ts., an angle, corner. See kūṇa, tikoṇa, trikoṇa-, paṃcakoṇa-.

kora[1], a kind of grain 5.9. Cf. vora.

kora[2], probably the linear measure kara lengthened metri causa 4.19.

kosa S. krośa, cry (of a cow): a linear measure 1.6.

kaumudī Ts., moon-light: the title of the work 1.93s, 2.17s, 3.104s, 5.33s.

krakaca Ts., a saw 3.87p, 95s, 6.85.

kraya Ts., buying 3.11p. See kaya.

krayavikraya Ts., buying and selling 1.86p, 6.25.

kṣetra Ts., field: a plane figure 3.35p, 53s, 5.22v, 24s, 25v, 6.56, 60, 61. See khitta, khetta.

kh

kha Ts., sky: a bhūtasaṃkhyā for zero 1.31, 4.59.

√khaṃḍa S. √khaṇḍ, to break: khaṃḍivi 1.29.

khaṃḍa S. khaṇḍa, broken, a break: cutting 3.94; candied sugar 5.10, 12. Cf. sakkara.

khaḍ S. ṣaṣ, six 4.39. See cha/chac-/chat. Cf. khaḍaṃsa[1].

khaḍaṃsa[1] S. ṣaḍ-aṃśa, one-sixth 1. 52, 54, 66, 2.9, 15, 17, 3.58, 86, 96.

khaḍaṃsa[2] S. ṣaḍ-asra, hexagonal 3. 86, 4.29 (in khaḍaṭṭhaṃsā).

khatta S. khāta, dug: a pit, excavation 3.55, 56, 59, 60.

khattaphala S. khāta-phala, the capacity (volume) of a pit 3.55, 59, 60.

khaddha S. khādita, eaten 4.51.

khayara S. khadira, cutch tree, *Acacia catechu* Willd. 3.95.

kharigaha, Pe. khargāh, a folding tent with one or two doors. 4.27. Cf. Abū al-Faḍl [2004, 90: Book I, Ā'īn 21; Plate X].

khalla D., empty space 3.76, 81, 82, 84. Cf. '*bāḍa kā chidra, vilāsa, khāsī, rikta*' (PSM). Apte gives 'a pit' as the second meaning of *khalla*.

khavva S. kharva, crippled: a numeral for 10^{11} or the name of the twelfth decimal place 1.13.

khāta Ts., dug: excavation 3.54p, 68s, 6.69, 70. See khatta.

khāri Ts., a measure of grain 5.10.

khitta S. kṣetra, field 5.3: a plane figure 3.37–40, 43–45, 47, 52, 53, 60, 63, 5.22; the area of a plane figure 3.36, 39, 60, 65, 66, 74, 5.24, 25. See khetta.

khittagaṇana S. kṣetra-gaṇanā, geometric computation 3.42.

khittaphala S. kṣetra-phala, the area of a plane figure 3.37–39, 43–45, 60, 63, 5.22.

khima/khīma Ar. khaima, Pe. khaim, a tent 4.20, 23, 24.

khetta S. kṣetra, field: the area 3.35, 75. See khitta.

kheva S. kṣepa, throwing: an additive 1.31.

khora, mortar 5.30, 31. Cf. H. khoā in OHED and KHOA in Hobson-Jobson.

g

ga Ts., being, remaining 1.5.

ga = gaṇanā 6.77, 82.

gaï/gaī S. gati, going, motion 1.27: motion per day 4.13.

gaggarī S. gargarī, a pot 5.29.

√gaccha S. √gam, to go: gacchei 1.70; gacchihaï 1.71.

gaccha Ts., the number of terms of a series 3.26–29, 30p, 30, 6.52.

gaja Pe. gaz, a linear measure 3.90–93, 4.35–37: a cubic measure (one cubic gaja) 5.29, 30.

gajadaṃta S. gaja-danta, elephant's tusk 6.65. See gayadaṃta.

√gaṇa S. √gaṇ, to count, calculate: gaṇahu 3.79; gaṇiūṇa, 3.32, 44; gaṇijja 1.37; gaṇiyaṃti 3.86, 5.8; gaṇeviṇu 3.63.

gaṇaṇa S. gaṇanā, calculation 1.1 (-pāḍī), 3.42 (khitta-).
gaṇanā Ts., calculation, mathematics 3.1p (vyavahāra-), 73p, 89p, 6.4–27, 38, 40, 41, 70 (gaṇana), 73 (gaṇana), 80, 81, 83, 84. See gaṇaṇa.
gaṇita Ts., calculated 6.0, 79: calculation 1.14s, 6.49; mathematics 1.93s, 2.17s, 3.104s, 4.64s, 5.33s, 6.94 (-sāra). See gaṇiya.
gaṇima S. gaṇya, calculable 5.21.
gaṇiya S. gaṇita, calculated 3.47 (-phala), 49 (-māṇa), 62 (-phala), 64 (-phala), 66: the sum of a series 3.30; calculation, mathematics 3.32 (-vihi), 55.
gata Ts., gone, being in: subtracted 1.47. See gaya.
gati Ts., going: motion per day 4.13p, 54p. See gaï/gaī.
gabbha S. garbha, the womb: inside 3.74, 75; the middle 3.103.
gama Ts., going: removal 2.3.
gaya S. gata, gone, elapsed, being in 2.16, 17, 4.3, 48, 49, 51: put in 4.63; passed 3.5, 104, 4.11; lost 3.21, 23; subtracted 2.8, 3.104, 4.56.
gayadaṃta S. gaja-danta, elephant's tusk: a figure shaped like the elephant's tusk 3.51, 52.
√gaha S. √grah, to take, seize: gahaṃti 4.52.
gaha^1 S. graha, seizing, a planet: a bhūtasaṃkhyā for nine 4.45.
gaha^2 Pe. gāh, time, place, throne. See kharigaha, vārigaha.
gahiya S. gṛhīta, taken 1.2.
gā = gāthā 3.86s, 6.1–3, 15-27, 32, 35, 38–47, 49–55, 70–73, 75–84.
gāthā Ts., a verse 3.34s, 68s, 104s, 5.33s, 6.0, 28–31, 36, 37, 48, 56, 69, 74, 85, 89, 90, 95, 98. See gā, gāhā.
gāma S. grāma, a village 4.5.
gāra/gārasa S. ekādaśan, eleven 3.93, 4.42. See igārasa.
gāliya S. gālita, melted 3.16.
gāvi S. go, a cow 4.62, 63, 5.14, 16.
gāhā S. gāthā, a verse 3.95s, 100s, 4.37s, 45s, 64s, 6.4–14 (-a). See gā, gāthā.
giri Ts., a mountain 1.1.
giha/geha S. gṛha, a house 4.47 (sura-): a cell of a magic square 4.38, 40–44. Cf. kuṭṭha.
gu S. go. See guvāla.
guṃja S. guñjā, seed of the plant *Abrus precatorius* L.: a weight measure 1.9, 77, 2.13.
guṃmaṭṭa = goṃmaṭa 3.76.
√guṇa S. √guṇ, to multiply: guṇa 3.36; guṇavi 1.29, 90, 2.3, 12, 3.1, 3, 4.11; guṇahu 3.101; guṇi 1.34, 55, 2.6, 3.5, 95, 4.50, 62, 5.22; guṇi-ūṇa(ṃ) 1.39, 72, 3.1, 7; guṇijja 1.27, 35, 54, 2.8, 3.3, 87, 4.22, 52; guṇijjaï 1.18, 89; guṇijjahi 2.8, 4.48; guṇiyaṃti 3.60, 4.50; guṇivi 1.40, 45, 53, 55, 3.31, 100, 4.2; guṇevi 3.15, 21, 28; guṇeviṇu 1.35.
guṇa Ts., merit, property: a multiplier 1.7, 17, 19, 20, 25, 28, 30, 31, 3.11, 26, 30, 34, 39, 43, 46, 48, 55, 63, 77, 81, 96, 4.13, 26–30, 46, 51, 52, 56–59; the product 3.18; a bhūtasaṃkhyā for three 4.42. See uṇa^1.
guṇaka Ts., a multiplier 4.50.
guṇaṇa S. guṇana, multiplication 1.31.
guṇayāra S. guṇa-kāra, a multiplier 3.100, 4.56: multiplication 1.32, 53.
guṇarāsī S. guṇa-rāśi, a multiplying quantity: a multiplier 1.27, 29, 30.
guṇākāra S. guṇa-kāra, multiplication 1.27p, 30p, 53p, 6.5, 13.
guṇita Ts., multiplied 5.25v. See guṇiya.
guṇiya S. guṇita, multiplied 1.7, 16, 22, 29, 39, 41, 47, 53, 54, 63, 81, 86, 2.2, 3, 3.11, 33–35, 43, 45, 46, 55, 70, 74, 76, 78, 81, 83–85, 89, 4.4, 25, 33, 61, 5.3.
guṇiyavva S. guṇitavya, to be multiplied 3.58.
guṇiyarāsi S. guṇita-rāśi, the multiplied quantity: the product 1.30.

gunnarāsi S. guṇya-rāśi, a quantity to be multiplied: a multiplicand 1.27, 30.
gula S. guḍa, jaggery 5.10.
guvāla S. go-pāla, a cowherd 4.63.
gṛha Ts., a house: a cell of a magic square 4.39p. See giha/geha.
go Ts., a cow 4.62p, 64p, 64, 5.15. See gāvi, gu.
goṃmaṭa/goṃmaṭṭa Pe. gumbad, a dome 3.69, 74p, 74.
gomaṭa = goṃmaṭa 6.77.
gorū S. go-rūpa, a cow 2.16. Cf. rū.
golaka Ts., a sphere 5.25v. See gola(ya).
gola(ya) S. gola(ka), a sphere, spherical 3.65, 66, 5.25.
gohuma/gohuva S. godhūma, wheat, *Triticum aestivum* L. 3.97, 5.7.

gh

√ghaṭṭa (used in place of) S. √bhraṃś, to drop, fall: ghaṭṭaï 4.19; ghaṭṭei 3.71.
ghaḍiyā S. ghaṭikā, a small vessel: a time unit 1.11.
ghaṇa S. ghana, solid: cubic (dimension) 1.7, 3.96; the cube (of a number) 1.32, 39–43, 45, 60–62, 2.3, 3.32, 34, 55, 65, 5.25; the cubic series (which consists of the cubes of natural numbers) 3.33.
ghaṇakaṃbiya/-kaṃviya S. ghanakambikā, a cubic kambikā 3.67, 5.26. See kaṃbiya/kamviya/kaṃvī.
ghaṇamūla S. ghana-mūla, the cube root 1.62p, 62.
ghaṇahattha S. ghana-hasta, a cubic hasta 3.60, 61, 71.
ghana Ts., solid, compact: the cube (of a number) 1.39p, 60p, 3.33p, 34p, 5.25v, 6.9, 17, 54, 55. See ghaṇa.
ghanamūla Ts., the cube root 1.43p, 62p, 6.10, 18. See ghaṇamūla.
ghaya S. ghṛta, clarified butter 5.28. See ghiya.
gharaṭṭa Ts., a grindstone 3.64.
ghiya (or ghia) S. ghṛta, clarified butter 3.14, 5.13. See ghaya.

c

ca (in caüṇa 'multiplied by four') S. catur-, four 3.49, 4.23, 31, 52.
caü/caür S. catur-, four 1.6, 8–10, 56, 76, 79, 2.5, 9, 13, 17, 3.2, 6, 8, 13, 20, 56, 58, 73, 90, 98, 100, 104, 4.7, 11, 14, 24, 25, 40–42, 47, 49, 51, 53, 55, 64s (catvāri). See ca, caüka, cattāri, cāra/cāri. Cf. caübhuya, caübbhuva/caübhuva, caüraṃsa, caürasa.
caüka S. catuṣka, group of four, quartet 4.8.
caüṇa. See ca. Cf. paüṇa.
caütīsa S. catustriṃśat, thirty-four 4.39.
caüttha S. caturtha, the fourth 3.65, 5.25, 25v. See cahuṭha, cāuddha.
caüdasa S. caturdaśan, fourteen 3.38, 4.38, 5.8, 29.
caübbhuya S. caturbhuja, one having four arms: a quadrilateral 3.36.
caübbhuva/caübhuva S. caturbhuja, one having four arms: a quadrilateral 3.50, 51.
caür. See caü/caür.
caüraṃsa S. caturasra, that which has four edges or sides: a square figure 3.35, 77, 86, 4.23, 5.23; a rectangular solid 3.79. See caürasa.
caürasa S. caturasra, that which has four edges or sides: a square figure 3.39, 50, 69, 4.31, 6.57 (sama-); a rectangular figure 3.35 (dīha-), 6.58 (dīrgha-). See caüraṃsa.
caüla, chaula (chola) beans, *Vigna catiang* Endl. var. *V. sinensis* Savi. 5.5.
caüvīsa S. caturviṃśati, twenty-four 1.5, 6, 36, 64, 3.67, 75, 102, 4.36, 5.4, 26.
caüsaṭṭha/caüsaṭhi S. catuḥṣaṣṭi, sixty-four 1.28, 4.42.
caühā S. caturdhā, four times 3.36.
caṃda S. candra, the moon 3.51 (paripunna-): a bow-like plane figure 3.51; name of the author's father

1.2, 4.1, 5.33 (caṃdā).
caṃdaṇa S. candana, sandalwood 1.66, 3.10.
caṃdovaya S. candropaka, a canopy 4.33. Cf. caṃdova in AHK.
cakka S. cakra, a wheel 3.51.
caṭṭiya/caṭṭī, an account 4.4, 5, 8, 5.17, 33. Cf. H. ciṭṭhī. Cf. also H. caṭṭī, shortcoming, loss in trade, expense.
√caḍa S. √*caḍh, to rise, raise?: caḍaṃti 1.29.
caṇaya S. caṇaka, chickpea, *Cicer arietinum* L. 5.7.
cattāri S. catvāri, four 4.36.
candra Ts., the moon: name of the author's father 1.93s, 2.17s, 3.104s, 4.64s, 5.33s. See caṃda.
camma S. carman, leather 4.37.
caya Ts., a pile, cover: that which is piled 1.46; the common difference of a series 3.29–31, 4.41.
cayaṇa S. cayana, piling 3.74, 75, 85.
cara Ts., movable? 5.15 (a modifier of kara or 'tax'), 17.
carakha, a wheel? 4.34.
carima Ts., the last 1.38, 2.6, 4.41, 44.
√cala S. √cal, to go, move: calaï 1.70, 71, 3.92, 5.11. See √calla.
cala Ts., moving 5.15 (kanna-).
caliya S. calita, departed 4.14.
√calla S. √cal, to go, move: callae 4.15, 55. See √cala.
cahuṭha S. caturtha, the fourth: a quarter 4.42. See caüṭṭha, cāuddha.
cāuddha S. caturtha, one-fourth 2.16. See caüṭṭha, cahuṭha.
cāja Ṣ. sajja, decorated 4.33.
cāra/cāri S. catvāri, four 1.15, 3.82. See caü/caür.
√cāla S. cālaya, causative stem of √cal, to make move: cāli 1.38; cālivi 1.43. See √cala and √calla.
cālīsa S. catvāriṃśat, forty 1.9, 84, 3.90, 5.6, 32.
ciṃtiya S. cintita, thought 5.59.
cijja, an unidentified substance used for constructing brick walls 5.30.
√ciṇa S. √ci, to collect, gather: ciṇaï 5.29.
ciṇaṇa S. cayana, piling up 6.77.
citi Ts., piling up: a pile of bricks 3.69p, 86s, 6.74.
citta[1] S. citra, a painting, colour 4.33, 35, 36.
citta[2] S. caitra, the Caitra month 4.17.
citta[3] S. cinta, thought 4.58.
cintā Ts., thought, mind 4.58p. See citta[3].
cirāvaṇiya/cirāvaṇī, wages for sawing 3.89, 95. See cīriya. Cf. siyāvaṇiya, H. ciravāī.
cilaṃga, part of an unidentified construction called vāri 4.31.
cīṇaya S. cīnaka, common millet, *Panicum miliaceum* L. 5.5.
cīra Ts., a piece of cloth 1.79. Cf. Habib 2004, item 20.
cīriya S. *cīrita, cut 3.93. Cf. H. √cīranā, to split, cut.
cukkha, pure 5.12.
cunna S. cūrṇa, powder: powdered 5,32; lime (used for brick walls) 5.30, 31.
cuppaḍa, oil 5.1. Cf. H. √cupaṛanā, to oil.
culha, hearth? 5.14. Cf. H. cūlhā, a stove, hearth.
caitra Ts., the Caitra month 5.33s. See citta[2].

ch

cha/chac/chat S. ṣaṣ, six 1.9, 50, 56, 66, 78, 79, 2.4, 5, 3.6, 12, 16, 17, 19, 20, 35, 47, 48, 59, 66, 75, 79, 4.17, 30, 38, 40, 63, 80. See khaḍ, chaha(ya)/chahi, ṣaṭ/ṣaṭa.
chaṃda S. chanda, appearance, shape 3.86
chajjaya, ? 4.34.
chaṭṭha/chaṭṭhama S. ṣaṣṭha, the sixth 1.67, 70, 71, 2.1, 2, 7, 16, 3.45, 5.10, 22.
chatta S. chatra, an umbrella 4.29, 34.
chattīsa S. ṣaṭtriṃśat, thirty-six 3.98.
chappaṃna S. ṣaṭpañcāśat, fifty-six 5.7.
chayāsī S. ṣaḍaśīti, eighty-six 3.9.
chaha(ya) S. ṣaṣ (ṣaṭka), six 3.6, 9, 83, 4.49. See cha/chac/chat.

chāṇaya S. kṣāṇaka, that which is burnt: cow-dung 5.32. Cf. chāṇa in AHK.

chāyā Ts., shadow 3.101p, 101, 102, 103, 104s, 4.33, 6.95, 96 (-sādhanā).

chālīya S. chāgikā, a she-goat 5.14.

chāsaṭṭhi S. ṣaṭṣaṣṭi, sixty-six 1.83.

√chijja S. √kṣi, to decay, decrease: chijjaï 1.21.

cheda Ts., cutting 6.87.

chedita Ts., cut out 3.89p.

cheya(ṇa) S. cheda(na), a divisor, denominator 1.29, 46–49, 50a, 51, 53, 55, 57, 59, 60, 62, 2.1–3, 6, 10, 12, 14, 3.7, 24.

j

ja Ts., born, made from 4.37 (sara-)?. See aṃgaja.

jaï S. yadi, if 1.26, 32, 69–71, 76, 77, 79, 83, 87, 2.5, 16, 3.2, 32, 4.5, 8, 46.

jaṃta[1]/jaṃtara/jaṃtra S. yantra, an instrument, device, diagram, figure: a magic square 4.38p, 38, 39p, 40, 41, 43p, 44, 45, 45s.

jaṃta[2]. See √jā.

jakkha S. yakṣa, a class of demigods 4.47.

jajjāvaya/jajjāvara, a kind of stone 3.68, 5.27.

jaṇa S. jana, people 1.76, 4.3, 48–50, 52, 53, 64, 5.29.

jama S. yama (in the sense of yāmya), south 3.103.

jamāiya S. jāmātṛka, a son-in-law 4.51.

jamātrika S. jāmātṛka, a son-in-law 4.50p.

jaya S. jagat, a world 1.1.

jala Ts., water 2.15, 17, 3.84, 5.29.

jalaṭhāṇa S. jala-sthāna, place for water: watery place, a reservoir 5.31.

java S. yava, barley, *Hordeum*: barleycorn 3.50, 5.7: a linear measure 1.5, 3.94; a weight measure 1.10; a unit for the purity of gold 1.10. Cf. vanna/vannaya/vannī.

jahacchiya S. yathā-iṣṭa, optional 3.87. See jahiccha.

jahiccha S. yathā-iccha, optional 3.87, 4.43. See jahacchiya.

√jā S. √yā, to go: jaṃta 4.47.

jā = jāti 6.33.

jāī S. jāti, class: the 8 classes of reduction of fractions or any one of them 1.15, 2.1a, 2a, 10; denomination (unit) 1.63, 81.

jāīya S. jātīya, of a kind, of a denomination 1.81.

√jāṇa S. √jñā, to know: jāṇa 3.36, 60, 77, 5.2, 6; jāṇaha 1.53, 4.7; jāṇahu 1.57; jāṇijja 4.62; jāṇijjā 5.18; jāṇijjahu 1.92; jāṇeha 1.5, 12, 3.79, 92, 94, 99, 4.10, 5.4, 30. See √najja, √muṇa.

jāṇaṇa S. jñāna, knowing, knowledge 3.28.

jāta Ts., produced 5.25v.

jāti Ts., class: the 8 classes of reduction of fractions or any one of them 2.1p, 2p, 3p, 10p, 14p, 17s, 6.30, 35, 37. See jāī.

√jāya S. √jan, to be born, produced: jāyaï 1.26, 36, 3.37, 47, 56, 67, 82, 99, 4.24, 5.4, 22, 26; jāyae 3.44, 55, 81.

jiṇa S. jina, one of the 24 Jain Arhats or Tirthankaras: a bhūta-saṃkhyā for twenty-four 3.61, 4.40.

jittā/jitti(ya) S. yāvat, as much as 4.6, 12, 5.20, 21.

jivalā = jīvalaï 4.3.

jīra Ts., cumin, *Cuminum cyminum* L. 5.8.

jīva Ts., a living being 1.92p, 92.

jīvalaï, a unit of currency 4.2, 3. See jivalā.

jīvavikraya Ts., selling of living beings 1.92p, 6.27.

jīvā Ts., a chord 3.46–49.

juī S. yuti, union: the sum 1.48, 50, 2.1, 3, 10, 3.3, 7, 23, 31, 33, 34, 51.

juga S. yuga, the four eons or any one of them: a bhūtasaṃkhyā for four 4.38, 45. See juya[2].

juja, a kind of silk 4.18.

jutta S. yukta, yoked: increased by,

added to 1.40, 2.7, 3.104, 4.30, 43, 46, 56; accompanied by 5.31.
juya[1] S. yuta, added 1.47, 4.22: increased by 1.20, 25, 34, 35, 41, 2.7, 3.26, 30, 31, 45, 81, 4.16, 17, 25, 43, 54, 59, 5.22, 25; the sum 1.16, 3.49, 78, 4.22, 26, 29.
juya[2] S. yuga, the four eons or any one of them: a bhūtasaṃkhyā for four 4.38, 42. See juga. Cf. juyala.
juyala S. yugala, a pair: two 3.16, 4.6.
juva S. yuga, a yoke: the sum 3.11, 40, 104; a pair 3.84. Cf. bhāva (for S. bhāga) for the euphonic change. Cf. juga, juya[2].
juvāra, Indian millet, *Sorghum vulgare* L. 5.6.
jūha S. yūtha, a troop, herd 2.17.
jaina Ts., a follower of Jina 1.93s, 2. 17s, 3.104s, 5.33s.
joḍa/joḍaṇa S. yojana, yoking: addition 1.31, 48. Cf. joyaṇa.
√joḍa S. √yuj, to unite, add: joḍijjā 1.39.
joya S. yoga, the sum 1.31, 3.1, 39, 83.
joyaṇa S. yojana, yoking: a linear measure 1.6, 70, 71, 75, 4.15, 55. Cf. joḍa/joḍaṇa.
joyaṇī/joyaṇīya S. yojanin/yojanika, one having yojanas 4.14 (satta-/ caü-).
jja S. jña, a knower 1.1 (deva-).
jñāna Ts., knowledge 4.58p, 60p.

jh

jhallariyā S. jhallarikā, lace or cymbal? 4.32.
jhuṃbukkā D. a cluster, bunch 4.32. Cf. H. jhumakā.

ṭ

ṭaṃka/ṭaṃkaya S. ṭaṅka/ṭaṅkaka, a weight measure 1.9, 10, 83, 3.9?, 4.36: a monetary unit 1.4, 87, 93, 3.9?, 5.16, 18, 19.
ṭippa, ? 4.34.

ṭh

ṭhakkura pheru/pherū, the name of the author 1.93s, 2.17s, 3.104s, 4. 64s, 5.33, 33s.
√ṭhava S. sthāpaya, causative stem of √sthā, to put down, lay: ṭhavaï 4.47; ṭhavi 1.27, 41, 81, 4.38, 64; ṭhavijja 4.62; ṭhavijjaï 1.44, 46; ṭhavijjahi 1.89; ṭhavijjae 1.63. See √thappa.
ṭhaviya/ṭhāviya S. sthāpita, laid down, placed 1.34, 39. See ṭhahiya.
ṭhahiya S. sthāpita, laid down, placed 4.53 (suggested for ṭhāhiya). See ṭhaviya/ṭhāviya.
ṭhāṇa (ṭṭhāṇa after a vowel) S. sthāna, place 3.54, 4.1 (rāya-), 5.31 (jala-): a notational place of the decimal place-value notation 1.34, 37, 39.
ṭhiī S. sthiti, the state 1.49.
ṭhiya S. sthita, remaining, located 4.51.

ḍ

ḍaṃḍa S. daṇḍa, a stick: a linear measure 1.6; a beam 4.23. See daṃḍa.
ḍorī, string 4.36. Cf. H. ḍorī.

ḍh

ḍhilliya, the capital city of Delhi and the region around it 3.67, 4.1, 5.2, 26.

ṇ

√ṇajja. See √najja.
√ṇama. See √nama.
ṇaya S. naya, leading, guiding 4.17 (māī-)? See nayaṇa.

t

taïya S. tṛtīya, the third 1.43, 83, 2.11, 4.32, 5.31. See tṛtīya.
taṃgoṭī, a kind of building or construction? 4.26.
taṃdula S. taṇḍula, rice 1.68, 3.14.
taṃsa[1] S. try-aṃśa, one-third 2.9.
taṃsa[2] S. try-asra, a trilateral 3.39.
tanāva Pe. ṭanāb, a tent rope 4.25, 30. Cf. Abū al-Faḍl [2004, 585: Book III, Āʾīn 9], where the word occurs as a linear measure.
taya S. tata, extended, spread: width 4.25?.

tarakka, a kind of construction? 4.29.

tala Ts., the lower (bottom) part 4.21, 22, 63: below 1.43, 46, 89; the bottom plane of a pit or of a solid 3.54, 78; valley 2.16.

talavaṭṭa, low (but presumably dry) land (as against jalaṭhāṇa, place for water or watery place) 5.31. Cf. talavaṭa, S. upatyakā, in AHK. Cf. also H. talahaṭī.

tāka Ar./Pe. ṭāqa/ṭāq, an arch 3.69, 81p, 81, 82, 84, 6.80.

tākā = tāka? 4.34.

tājikka/tājiya S. tājika, Persian: belonging to Hijrī era 4.17.

√tāḍa S. √taḍ, to beat, multiply: tāḍijjaï 1.89.

tāḍiya S. tāḍita, beaten: multiplied 3.23, 61.

tāṇa S. tāna, a thread, fibre: a kind of fine cotton? 4.18.

ti/tti S. tri-, three 1.1, 10, 39, 40, 41, 44, 45, 50, 52, 54, 56, 58, 61, 63, 66–69, 74, 75, 77, 78, 80, 85, 88, 91, 2.4, 5, 7, 9 (taṃsa¹), 11, 17, 3.6, 11, 13, 14, 17, 20, 31, 33–36, 38 (-koṇa), 39 (-aṃsa² > t-aṃsa), 41 (-kkoṇa), 45, 50, 51, 53, 56, 58, 60 (-kkoṇaya), 62, 72, 73, 79, 80 (-koṇa), 88, 90, 100, 4.7, 12, 25, 40, 42, 46, 51, 53, 58, 5.12, 16–18, 22, 29. See tiuḍū, tiga, tinni, tiya, tīi. Cf. taṃsa¹, taṃsa².

tiuḍū S. trīṇi, three 3.13. See ti. Cf. tiuṇu in AHK.

tikoṇa/tikkoṇa(ya) S. tri-koṇa(ka), that which has three corners: a triangle, triangular 3.38, 41, 60, 63, 80, 5.24.

tiga S. trika, three 3.6 (-saya). Cf. duga.

tiṇa S. tṛṇa, grass 5.17.

tinni S. trīṇi, three 1.42, 79, 2.15, 3.12, 52, 102, 4.46, 57.

tibhuya S. tri-bhuja, one having three arms: a trilateral 3.36. See tibhuva. Cf. taṃsa², tikoṇa/tikkoṇa(ya).

tibhuva S. tri-bhuja, one having three arms: a trilateral 3.50, 51. See tibhuya.

tiya S. trika, three 1.52, 93, 2.7, 9, 13, 3.8, 16, 17, 22, 62, 104, 4.3, 19, 55.

tirāsivihi S. tri-rāśi-vidhi, rule of three quantities 3.11. See tirāsiyaga, trairāsika.

tirāsiyaga S. trairāśika, of three quantities: rule of three or three-quantity operation 1.63. See tirāsivihi, trairāsika. Cf. vitthatiyarāsī, vyastatrairāśika.

tiriya S. tiryañc, transversely 1.5.

tila Ts., sesamum, *Sesamum indicum* L. 3.97, 5.5, 13.

tilla S. taila, oil 5.13, 28.

tihā S. tridhā, three times 1.41.

tihi S. tithi, a lunar day: a bhūtasaṃkhyā for fifteen 4.38, 42.

tīi S. trīṇi, three 1.26.

tīra Ts., shore 2.17.

tīsa S. triṃśat, thirty 1.11, 76, 3.91, 4.3, 8, 40, 5.29.

turiya S. turaga, a horse 3.27.

√tuliya S. √tul, to weigh: tuliyaṃti 1.83.

tuliya¹ S. tulita, weighed 1.83, 3.67, 5.26.

tuliya² S. tulya, equal 4.61, 5.8. See tulla³.

tulla¹ S. taulya, weight 1.83, 3.15, 17, 18, 20, 24, 25.

tulla² S. tola, a weight measure 3.67, 5.26.

tulla³ S. tulya, equal 4.47. See tuliya².

tusaṃnna S. tuṣa-anna, kodrava grain 5.28. See kuddava.

tṛtīya Ts., the third 3.104s. See taïya.

tera/terasa/teraha S. trayodaśan, thirteen 3.38, 40, 61, 91, 4.3, 38, 40, 42.

tevisa S. trayoviṃśati, twenty-three 4.40.

toḍaṇa S. troṭana, plucking 4.49.

tori, Indian rape, *Brassica cumpestris* L. var. *toria* 5.8.

tola(ya) S. tola(ka), a weight measure 1.10, 69, 84, 2.13, 3.16, 19, 20, 25.

tolya S. taulya, weight 6.73.

taulya Ts., weight 3.67p. See tulla[1], tolya.

ttha S. -stha, lying in 2.15.

trikoṇakṣetra Ts., a triangular figure 6.60.

trikoṇavikaṭa Ts., a great or beautiful triangle: a right-angled triangle 6.62.

trairāśika Ts., of three quantities: rule of three, three-quantity operation 6.19.

trairāsika S. trairāśika, of three quantities: rule of three, three-quantity operation 1.63p. See tirāsiyaga. Cf. tirāsivihi.

th

thaṃbha S. stambha, a pillar 2.14, 15, 3.77–79, 101, 102, 4.20, 21, 24–26, 30, 31.

thaṇavāla S. stana-pāla, a breast-holder: a cow 2.16.

√thappa S. sthāpaya, causative stem of √sthā, to lay, place: thappivi 4.50. See √ṭhava.

thala S. sthala, dry gound 2.17.

√thāka, to be, exist: thākaï 4.51. Cf. √thakka in AHK, P. thakkaï (S. *sthakyati).

thiya S. sthita, remained, located 4.49.

thūla S. sthūla, coarse 4.18, 36.

thovaṃkarāsi S. stoka-aṅka-rāśi, the quantity of smaller number of digits: the product of the smaller set of digits 1.72.

d

daüra, a circle 4.25. Cf. Ar-Pe. daur, round, rotation.

daṃḍa S. daṇḍa, a staff 3.101, 102. See ḍaṃḍa.

daṃta S. danta, a tooth: a bhūtasaṃkhyā for thirty-two 4.39. Cf. gaja-/gayadaṃta.

dakkhiṇa S. dakṣiṇa, south 2.16. See dāhiṇa.

damma S. dramma, a monetary unit 1.4, 54, 56, 64–69, 74–76, 78–80, 88, 89, 91, 2.4, 3.2, 9, 10, 12–14, 4.2, 10, 12, 35, 5.9, 18–21, 35.

darśana Ts., showing (a figure) 5.22v, 25p.

dala Ts., a piece, a heap: half 1.17, 25, 2.5, 3.28, 29, 45, 52, 79, 80, 100, 4.27; thickness 3.87, 88.

√dala S. √dal, to halve: dali 4.38.

dalīkaya S. dalī-kṛta, made into two halves, halved 1.22.

davva S. dravya, substance: money, property 3.11, 4.2, 4.

dasa/dasī S. daśan, ten: a numeral for 10 or the name of the second decimal place 1.9, 12–14, 19, 23, 69, 75, 77, 84, 87, 93, 3.6, 9, 10, 16, 19, 22, 25, 44, 57, 64, 72, 85, 92, 93, 4.7, 40, 5.6–8, 14. See dassa, daha.

dasa-avva S. daśa-arbuda, ten arbudas: a numeral for 10^{10} or the name of the eleventh decimal place 1.12.

dasakoḍī S. daśa-koṭi, ten koṭis: a numeral for 10^8 or the name of the ninth decimal place 1.12.

dasakhavva S. daśa-kharva, ten kharvas: a numeral for 10^{12} or the name of the thirteenth decimal place 1.13.

dasanīla S. daśa-nīla, ten nīlas: a numeral for 10^{18} or the name of the nineteenth decimal place 1.13.

dasapaüma S. daśa-padma, ten padmas: a numeral for 10^{16} or the name of the seventeenth decimal place 1.13.

dasama S. daśama, the tenth 3.71, 97, 4.9.

dasalakkha S. daśa-lakṣa, ten lakṣas: a numeral for 10^6 or the name of the seventh decimal place 1.12.

dasasaṃkha S. daśa-śaṅkha, ten śaṅkhas: a numeral for 10^{14} or the name of the fifteenth decimal place 1.13.

dasasahasa S. daśa-sahasra, ten thousand: a numeral for 10^4 or the name of the fifth decimal place 1.14.

dasasahasanīla S. daśa-sahasra-nīla,

ten thousand nīlas: a numeral for 10^{21} or the name of the twenty-second decimal place 1.13.
dasahasa = dasasahasa 1.12.
dassa S. daśan, ten 3.91. See dasa, daha.
daha S. daśan, ten 1.12, 82, 3.8, 43, 46, 53, 73, 4.59, 5.28. See dasa, dassa.
dahalīja Ar./Pe. dihlīza/dihlīz, a vestibule, door 4.20, 34, 35.
√dā Ts., to give: dijjaï 4.2; dijjahi 4.12; dijjā 1.43; diyaï 4.51; dījaï 4.50.
dāra S. dvāra, a door, gateway: the 45 gateways to mathematics 1.15. See duvāra, bāra, vāra², vāri².
dāru Ts., wood, timber 3.87, 88, 89p, 93.
dāhiṇa S. dakṣiṇa, the right-hand side 3.37, 4.39. See dakkhiṇa.
diuḍha S. dvika-ardha, two less by a half: one and a half 1.50. See divaḍḍha. Cf. P. divaddha.
digsādhanā S. dik-sādhana, determination of the cardinal directions by means of a gnomon 6.97.
diṭṭha S. dṛṣṭa, seen: known 3.86.
diṭṭhapatta S. dṛṣṭi-prāpta, obtained through seeing: experience 5.11.
diṇa S. dina, a day 1.11 (-rayaṇi), 17, 71, 76, 2.5, 3.27, 4.11, 13, 14, 16, 17, 54. See divasa.
diṇayara S. dina-kara, the sun: a bhūtasaṃkhyā for twelve 4.38.
dinna S. datta, given 3.8: lent 3.6.
divaḍḍha/divaḍha(ya) S. dvika-ardha-(ka), two less by a half: one and a half 1.74, 3.64, 71, 73, 81, 90, 4.19, 23, 5.11. See aḍḍhāiya, diuḍha, dīvaḍha. Cf. P. divaddha(ya).
divasa Ts., a day 1.70, 76, 4.12, 14, 15, 55: the length of daylight 3.104. Cf. diṇa.
disa S. diś, direction 4.27, 53.
disi S. diś, direction 4.53: a bhūtasaṃkhyā for ten 4.38, 42.
dīrgha Ts., long 6.58 (-caürasa); length 6.86, 92. See dīha.
dīvaḍha S. dvika-ardha(ka), two less by half: one and a half 5.16. See divaḍḍha. Cf. P. divaddha(ya).
dīha S. dīrgha, long 3.35a (-caürasa), 35b (-caüraṃsa), 4.13 (-gaï): length 1.4, 78, 79, 3.52, 54, 55, 57, 61, 62, 66, 70, 72, 73, 77, 78, 81, 82, 84, 87-89, 93, 94, 4.22, 28, 33, 52, 5.3; oblong, rectangular 3.86.
du S. dvi, two 1.42, 77, 80, 85, 91, 2.4, 3.6, 14, 30, 50, 56, 72, 83, 85, 90, 100, 103, 4.7, 8, 15, 19, 21, 30, 31, 40, 42, 46, 5.16. See dui, do. Cf. duṃni, duga, duṇha, dunha, duya, donna/donni.
dui S. dvi, two 1.67, 78, 85, 2.13, 3.17, 26, 4.3, 17. See du.
duṃni S. dvau, two 3.8. See donni. Cf. du.
duga S. dvika, two 1.80, 3.6, 50, 4.26, 35. See duya. Cf. du.
dugudha S. dugdha, milk 4.64p. See duddha.
duṇha S. dvayoḥ, of the two 3.87. See dunha.
dutī = dutīka 6.41.
dutīka S. dvitīya, the second 1.30p, 3.20p, 6.48. See bīya², vīya.
duddha S. dugdha, milk 4.64, 5.13. See dugudha.
dunha S. dvayoḥ, of the two 1.22. See duṇha.
duya S. dvika, two 1.75 (due). See duga. Cf. du.
durehavaṭṭa S. dvi-rekha-vṛtta, a double-lined circle 3.103.
duvāra S. dvāra, a door 4.28, 47, 63. See dāra, bāra, vāra², vāri².
duvālasa S. dvicatvāriṃśat, forty-two 1.88, 2.15, 3.40, 5.24.
duhā S. dvidhā, twice 1.35.
√de S. √dā, to give: de 4.47; dehi 3.12–14; devi 4.58.
deva Ts., a god 1.1.
devaï ? 5.15.
deśa Ts., a place, region 4.17s, 19s, 64s, 5.17s.
deśāṃka S. deśa-aṅka, a digit of the region: a regional method of ac-

countancy 4.9p.
desa S. deśa, a place, region 4.1p, 1, 5, 5.1, 33.
do S. dvau, two 1.42, 72, 90, 3.50, 4.9, 13, 57. See du.
donna/donni S. dvau, two 3.1, 4.28. Cf. du.
dosiya S. *dauṣya, a cloth merchant 1.64, 79.
dvitīya Ts., the second 2.17s. See dutīka.

dh

dhaṇa S. dhana, property, money 1.73, 77, 86, 3.1, 3–6, 4.6, 8, 9: the total sum 1.24, 26, 2.3, 11, 26, 28; the value of a term in a series 3.26; the value of the first term (unusual meaning) 3.30; positive, additive 2.12. Cf. riṇa.
dhaṇiya S. dhānya, coriander, *Coriandrum sativum* L. 5.8.
dhaṇu S. dhanus, a bow: an arc 3.47–49; a segment of a circle 3.47, 48. See dhaṇuha.
dhaṇuha S. dhanuṣa, a bow: an arc 3.49, 50; a segment of a circle 3.46, 50, 51, 6.64. See dhaṇu.
dhanna S. dhānya, grain 3.96, 5.1, 4, 9.
dhammiya S. dhārmika, a pious man 4.47.
√dhara S. √dhṛ, to hold, assume: dharivi 1.44.
dhara Ts., holding: the base? 4.25.
dharā Ts., the earth: the base line of a geometric figure 3.52, 53.
dhānya Ts., grain 5.9s. See dhanna.
dhāvaka Ts., a runner 4.13p.
dhuya S. dhṛta, held? 4.38.
dhura/dhurima S. dhur, a yoke, the foremost place: the first, beginning 1.37, 39, 40; 4.41, 44, 46.
dhuva S. dhruva, constant 3.104.
dhuvaṃ S. dhruvam, certainly 3.43, 59.
dhuvaṇa S. dhāvana, washing 4.19.

n

naṃda S. nanda, the Nandas: a bhūtasaṃkhyā for nine 4.17.
nakha Ts., a nail: a bhūtasaṃkhyā for twenty 4.40.
√najja S. √jñā, to know: najjaï 1.73; nāijjaï 1.86. See √jāṇa, √muṇa.
naṭṭha S. naṣṭa, lost, unknown 2.16, 3.27-29.
√nama S. √nam, to bow to: namiūṇa 1.1.
nayaṇa S. nayana, leading: an eye 4.45 (hara-).
nayara S. nagara, a town 4.5, 63.
narahaḍa, the region around the town Narhar in the modern state of Haryana 5.2.
nava S. navan, nine 1.30, 36, 42, 50, 52, 66, 67, 72, 74, 75, 78–80, 85, 88, 93 2.4, 17, 3.4, 13, 14, 16, 17, 19, 20, 22, 25, 46, 65, 69, 74, 76, 83, 88, 90, 97, 4.3, 5, 8, 11, 17, 40, 53, 57, 60, 61, 63, 5.10, 13, 25, 25v, 30.
navaï S. navati, ninety 1.78.
navama Ts., the ninth 4.14.
navarāśika/navarāsika S. nava-rāśika, of nine quantities: rule of nine or the nine-quantity operation 1.72p, 79p, 6.22.
navasaï S. navaśatī, nine hundred 1.28.
navāra/nāvāra Pe. navār, coarse broad tape with which beds are woven 4.20, 35, 37.
navi, a product from sugarcane juice 5.11.
naṣṭa Ts., lost or unknown (element or factor) 3.21p, 28p, 29p, 30p, 6.47, 50–52. See naṭṭha.
nāijjaï. See √najja.
nāṇavaṭṭa S. nāṇa(ka)-varta(na), exchange of coins 1.65.
nāman Ts., the name 1.14, 6.29.
nāyavva S. jñātavya, to be known 1.3, 17, 49, 3.31, 42, 76, 4.34, 56.
nālaya S. nāli (pl. nālyaḥ), drains 2.5.
nāvāra. See navāra.
nāviya S. nāpita, a shaver or S. nāvika, a sailor or S. nāmita, bent? 5.14. Cf. √nāva for S. √nam in AHK.

nāha S. nātha, a leader, lord 1.1.

niṃva S. nimba, neem tree, *Azadirachta indica* Juss. 3.95.

niggama S. nirgama, going out: an exit 3.84.

nicca S. nityaṃ, always, every day 4.55.

nicchaya S. niścita, determined 1.79, 3.42, 5.18.

√nippa S. niṣ-√pad, to be produced, grow: nippaï 5.2.

nippatti S. niṣpatti, yield, product (of crops) 5.1.

nippanna S. niṣpanna, produced: harvest 3.8, 5.4.

nimma (or nimmi) S. nemi, the rim of a wheel 3.53.

niya S. nija, one's own 3.1, 100, 4.3.

√niyatta S. ni-√vṛt, to return: niyattaī 1.71.

nirutta S. nirukta, declared, told decisively 4.46, 48.

nirega S. nireka, minus one 3.29.

niva S. nṛpa, a king: a bhūtasaṃkhyā for sixteen 4.37.

nivāta S. nipāta, falling: a product from sugarcane juice 5.11.

niviṭṭhiya S. niveṣṭita, made to sit 4.63.

nisāṇa Ar/Pe. nishān, a banner 4.34.

√nisuṇa S. niḥ-√śru, to listen to: nisuṇi 2.16. Cf. √suṇa.

nihāṇa S. nidhāna, treasure: a bhūtasaṃkhyā for nine 4.17. See nihi.

nihi S. nidhi, treasure: a bhūtasaṃkhyā for nine 4.38, 42. See nihāṇa.

nīravasa, S. nīra-vasā, watery marrow: a product from sugarcane juice 5.11.

nīla Ts., blue, blue lotus: blue 4.36, 37; a numeral for 10^{17} or the name of the eighteenth decimal place 1.13, 14.

nīladasalakkha S. nīla-daśa-lakṣa, ten lac of nīlas: a numeral for 10^{23} or the name of the twenty-fourth decimal place 1.14.

nīlasaya S. nīla-śata, hundred nīlas: a numeral for 10^{19} or the name of the twentieth decimal place 1.13.

nīlasahasa S. nīla-sahasra, thousand nīlas: a numeral for 10^{20} or the name of the twenty-first decimal place 1.13.

nīlī Ts., a blue one 5.15.

√nīsara S. niḥ-√sṛ, to go out: nīsariya 4.63.

nyāsa Ts., setting-down: a tabular presentation of numerical data 5. 25v.

P

pa (in paüṇa 'less by a quarter') S. pāda (or pada), a quarter 1.50, 56, 58, 75, 77, 3.13, 17, 95, 99, 100, 4.37, 5.12, 13, 25, 25v , 31. See pā, pāya, vā.

païsa S. pradeśa, a region 4.63. See paesa.

paüṇa. See pa. Cf. caüṇa.

paüma S. padma, a lotus: a numeral for 10^{15} or the name of the sixteenth decimal place 1.13.

paesa S. pradeśa, a region, an area 5.2. See païsa.

paṃgulaya S. paṅgulaka, a lame man 1.70.

paṃka S. paṅka, mud 2.15.

paṃca S. pañcan, five 1.52, 54, 58, 63, 68, 78, 79, 85, 88, 91, 93, 2.13, 3.4, 8, 9, 17, 20, 22, 27, 32, 35, 40, 52, 53, 56, 62, 72, 90, 4.3, 8, 11, 15, 30, 39, 51, 57–59, 63, 5.13. See paṇa.

paṃcakoṇakṣetra S. pañca-, a pentagonal figure 6.61.

paṃcaga S. pañcaka, group of five, quintet 1.73, 3.2, 6, 5.1p, 1, 33, 33s.

paṃcama S. pañcama, the fifth 2.16, 5.10, 19, 23.

paṃcarāśika/paṃcarāsika S. pañcarāśika, of five quantities: rule of five or the five-quantity operation 1. 72p, 73p, 6.20.

paṃcaviṃśati S. pañcaviṃśati, twenty-five 1.93s.

paṃcavīsa S. pañcaviṃśati, twenty-five 1.42. See paṇavīsa.

paṃcasaï S. pañca-śata, five hundred 3.2.

paṃcāsa S. pañcāśat, fifty 1.4a. 3.67,

4.5. See pacāsa, pannāsa.

paṃḍiya S. paṇḍita, a scholar, learned man 1.6, 23, 2.15, 16, 3.44, 47, 59, 75, 4.4, 17, 24, 48, 50, 63, 5.20.

paṃti/paṃtī, S. paṅkti, a sequence 1.29, 38, 44, 45, 4.44.

√pakīra S. pra-√kṛ, to perform: pakīrae 2.12.

pakka S. pakva, cooked: roasted 5.32; refined (gold) 3.18; refinement 3.19. Cf. vipakka.

pakkaviya S. *pakvāpita, refined 3.20. Cf. pakka.

pakkha S. pakṣa, a wing: a side (column) of a two-column table of numbers 1.72.

pakkhaviya S. prakṣepita, made to throw forward: invested 3.8.

pakkheva S. prakṣepa, throwing forward: investment 3.7. See prakṣepaka.

pakva Ts., cooked: refined (gold) 3.18p. See pakka.

pacāsa S. pañcāśat, fifty 5.10. See paṃcāsa, pannāsa.

paccakkha S. pratyakṣa, visible: a visible quantity 2.14, 15.

pacchae S. paścāt, behind 1.44. See pacchā.

pacchā S. paścāt, later 2.1, 3.100, 4.14: back 4.35. See pacchae.

pacchima S. paścima, west 2.16.

paṭṭa Ts., a slab or flat stone 3.64 (gharaṭṭa-); silk, a cloth of silk 3.9, 4.18.

paṭṭaṃsuya S. paṭṭa-aṃśuka, silk 4.19.

paṭṭolaya S. paṭolaka, a kind of cloth: a kind of silk 4.18. Cf. H. paṭolī, a silk-embroidered sari or dhotī. Cf. also Habib 2004, item 73.

paḍa S. paṭa, cloth 4.22, 26, 27: sheet 4.30, 33; plane (dimension), square (dimension) 1.7.

paḍikāiṇi S. pratikākiṇī, a monetary unit 1.3.

paḍirūva S. prati-rūpa, a replica? 4.34.

paḍivissaṃsa S. prativiṃśāṃśa, a monetary unit 1.3.

paḍhama S. prathama, the first 1.34, 3.103, 4.43, 50: first, in the beginning 3.70.

paṇa S. pañcan, five 1.50, 72, 2.1, 2, 3.4, 16, 17, 19, 25, 53, 104, 4.6, 9, 26, 27, 51, 58, 59. See paṃca.

paṇaṭṭha S. praṇaṣṭa, disappeared 2.16.

paṇatīsa S. pañcatriṃśat, thirty-five 4.40.

√paṇama S. pra-√ṇam, to bow to, salute: paṇameviṇu 5.1.

paṇayāla S. pañcacatvāriṃśat, forty-five 2.16 (probably a scribal or typographic error for igayāya). See paṇayālīsa.

paṇayālīsa S. pañcacatvāriṃśat, forty-five 1.15, 76, 5.7. See paṇayāla.

paṇaraha S. pañcadaśan, fifteen 5.5. See panara/panarasa/pannarasa.

paṇavīsa S. pañcaviṃśati, twenty-five 1.15, 65, 3.37, 4.39, 5.28. See paṃcavīsa.

paṇha S. praśna, a question 1.18 (in the sense of paṇhakkhara), 4.64.

paṇhakkhara S. praśna-akṣara, (the number of) letters in question 1.18.

patta[1] S. patra, a leaf 4.32: a bond 3.6.

patta[2] S. pātra, a vessel: a volume measure (for grain) equivalent to one cubic hattha 3.96.

patta[3] S. prāpta, obtained 4.48, 53, 5.11 (diṭṭha-).

patteya(ṃ) S. pratyekam, for each 1.42, 61, 3.7, 49, 66, 4.47.

pattha S. prastha, a volume measure 1.8, 68, 80, 82.

patra Ts., a leaf: a bond 3.5p (in ekapatrīkaraṇa), 6.42 (in egapatrīkaraṇa). See patta[1].

panara/panarasa/pannarasa S. pañcadaśan, fifteen 1.61, 87, 3.38, 40, 47, 4.58. See paṇaraha.

pannāsa S. pañcāśat, fifty 5.26. See paṃcāsa, pacāsa.

√pabhaṇa S. pra-√bhaṇ, to say, speak: pabhaṇaṃti 4.50; pabhaṇei 1.2.

pabhāgajāī S. prabhāga-jāti, the multipart class 2.2a.

pamāṇa S. pramāṇa, standard 1.65 (aggha-), 3.86 (diṭṭha-), 4.34 (ṭippa-).
pamāṇadhaṇa S. pramāṇa-dhana, principal, capital 1.73, 3.1, 3. Cf. mūla.
paya S. pada, a foot, step: the number of terms of a series 1.21, 22, 24, 25, 3.31, 33, 34; a notational place in the decimal place-value system 1.43, 45, 62; the square root 1.20, 24, 3.36, 41, 42, 46, 49; a quarter 2.11.
payaḍa S. prakaṭa, evident 4.37.
payaḍi S. paddhati, a way: a class of writings 4.1 (leha-).
para Ts., other 3.1, 4.17, 38, 49, 58p, 58, 5.8: subsequent 4.50.
parakama S. para-krama, the other order 1.29. Cf. sukama.
parama Ts., highest 1.93s, 2.17s, 3.104s, 5.33s.
parikamma S. parikarman, preparation: the fundamental operations of arithmetic 1.15. See parikarman, parikrama.
parikarman Ts., preparation: the fundamental operations of fractions (8 in number according to Pherū) 1.46p; the fundamental operations of arithmetic (25 in number according to Pherū) 1.93s, 6.2, 28. See parikamma, parikrama.
parikrama, a wrong Sanskritization of parikarman 6.11.
parijñāna Ts., knowledge, judgement 1.30p.
paridhi Ts., circumference 5.22v. See parihi/parihī.
paripunnacaṃda S. paripūrṇa-candra, the perfect full-moon 3.51.
parimāṇa Ts., measure: size 4.24; length 2.14, 15.
parihi/parihī S. paridhi, periphery: the circumference of a circle 3.43–45, 65, 74-76, 85, 96–100, 5.22, 25; the circumference (of a construction) 4.23, 24, 26, 27, 30, 31.
pala Ts., straw, husk: a weight measure 1.9, 66; a unit of time 1.11.
pavaṭṭaṇa S. pravartana, exchange 1.84.
pavāṇa S. pavana, sacred fire, or S. pramāṇa, standard? 5.14.
pavāhiya S. pravāhita, carried: a kind of sand? 5.32 (rakkha-).
pavesa S. praveśa, introduction: entrance 4.23.
pavvaṃgula S. parva-aṅgula, a joint-digit: a linear measure 1.5.
paha S. prabhā, shadow 3.104.
pā S. pāda, a quarter 1.54, 61, 68, 74, 3.74, 4.41, 5.17. See pa, pāya, vā, vāya.
pāī S. pāli or pādi(kā)?, a volume measure 1.8, 80, 5.18.
pāṭī Ts., procedure: procedural mathematics 1.93s, 2.17s, 3.104s, 5.33s, 6.2. See pāḍī.
√pāḍa S. pāṭaya, causative stem of √paṭ, to remove, take off: pāḍaï 4.60; pāḍijja 1.43; pāḍijjā 1.44 (with bhāu), 59 (with bhāu), 62 (with bhāu); pāḍijjaï 1.33, 89; pāḍijjahu 5.19; pāḍevi 1.57 (with bhāeṇa).
pāḍī S. pāṭī, procedure: algorithm 1.1 (gaṇaṇa-).
pāṇī S. pānīya, water 4.63.
pāya S. pāda, a foot: a quarter 2.7, 11, 3.4, 43, 88, 98, 100, 4.41. 5.22, 31. See pa, paya, pā, vā, vāya. Cf. pāyaseva.
pāyara S. prājña, a wise man? 1.49.
pāyaseva S. pāda-seva, service to feet?: a spiral stairway 3.69, 77p, 77, 79, 80, 6.78.
√pāva S. pra-√āp, to reach, obtain: pāvaṃti 1.66, 76, 78, 93, 4.3; pāvijjaï 1.91; pāvei 1.66.
pāsāṇa Ts., stone 3.67p, 6.72, 73, 76: the volume of a solid 3.61p. See pāhaṇa/pāhāṇa. Cf. sela.
pāhaṇa/pāhāṇa S. pāṣāṇa, stone 3.67, 5.26: the volume of a solid 3.61, 63, 64, 66, 72, 5.25, 29, 30.
pāhaṇasaṃkhā S. pāṣāṇa-saṃkhyā, the number of stones: the solid volume 3.71.

piṃḍa S. piṇḍa, round mass: the sum 1.29, 30, 50, 2.14, 3.5, 15, 21, 23, 33, 36, 42, 46, 4.58, 60; the length of an arc 3.47, 48; the thickness of a solid 3.61–64, 73, 81, 82, 94.
piṭṭhī S. pīṭha, foundation? 4.22.
pippali Ts., long pepper 1.67, 91, 3.13.
pihula S. pṛthula, width 3.83.
pukkharaṇī S. puṣkariṇī, a pond 3.56.
puḍa S. puṭa, a cover? 4.32. Cf. karapuḍa.
pula Pe. pul, a bridge 3.69, 84p, 84.
pulabaṃdha S. pulabandha, a bridge structure 3.84p, 84, 6.82 (-na).
puvaddha S. pūrva-ardha, the eastern half 2.16. Cf. puvva.
puvva S. pūrva, being in front of: ancient 1.1; east 4.40. previous 1.34, 38, 41, 45, 55, 92, 3.60, 100, 4.43, 50, 61. Cf. puvaddha.
√pūra S. pūraya, causative stem of √pṝ, to fill: pūraï 4.60; pūrahi 2.5.
pūra Ts., filling: capacity or volume 3.78. See pūraṃtara.
pūraṃtara S. pūra-antara, filling inside: capacity or volume 3.79. See pūra.
prakīrṇaka Ts., miscellaneous: miscellaneous mathematical problems 4.46p.
prakṣepaka Ts., throwing forward: investment (and proportionate distribution of the profit) 6.43. See pakkheva.
prakṣyepaka S. prakṣepaka, throwing forward: investment (and proportionate distribution of the profit) 3.7p. See prakṣepaka.
pratha = prathama 6.40.
prathama Ts., the first 1.93s, 6.39. See paḍhama.
prabandha Ts., plan 6.1, 75.
prabhāgajāti Ts., the multi-part class 2.2p, 6.31.
pravāha. See bhāgapravāha.

ph

phakkiya, grinding or quickly? 3.12.
phala Ts., fruit, result: fruit 4.48; result 1.57, 72, 2.9, 3.49, 101; price 1.92; product 1.28, 54; quotient 1.38, 56, 3.104; yield 5.9s, 12s, 13s, 17s, 21s; yield of grain 5.4; interest 1.73, 3.1–5; quantity to be exchanged in barter 1.90; share 2.3, 4, 3.7, 8; area 3.37–39, 41, 43–47, 52, 60, 63, 4.52p, 5.22, 22v, 25v; volume 3.55–57, 58p, 60, 61p, 62–64, 82, 83, 85, 100, 5.25v, 6.71, 72.
phalaha S. phalaka, a plank 3.87, 88, 93, 94.
phaliha = phalaha 3.87.
√phira S. √*phir, to turn, revolve: phiraṃta 3.78.
√phusa S. √spṛś (in the sense of √mṛj), to wipe off: phusavi 4.7; phusie 4.9; phusijja 4.9; phusivi 4.6.
pheru/pherū, name of the author 1.2, 93s, 2.17s, 3.104s, 4.1, 5.33, 33s. See ṭhakkura pheru/pherū.

b

baïṭṭhiya S. upaveṣṭita, made to sit 4.63.
baṃdha S. bandha, a tie, bond: combined 5.33s, 6.0; construction 3.84p, 84, 6.82 (-na).
battīsa/vattīsa S. dvātriṃśat, thirty-two 1.5, 28, 5.7, 16.
bahu Ts., many, much: many 3.86; much 4.51; greater 4.13.
bahugaï S. bahu-gati, one who has the greater speed: the fast runner or traveller 4.13. Cf. appagaï.
bahuya S. bahula, many, much 5.2.
baheḍā S. vibhītikā, *Terminalia bellerica* Roxb. 3.12.
bāra S. dvāra, a door 3.72. See dāra, duvāra, vāra², vāri.
bārasa/bāraha S. dvādaśan, twelve 1. 10, 11, 69, 5.23, 28. See vāra¹, vārasa/vāraha.
bārasama S. dvādaśa, the twelfth 4.23.
bāraha. See bārasa/bāraha.
bāliṃda S. bāla-indu, a young moon: a plane figure shaped like a crescent moon 3.50, 52.

bāvīsa S. dvāviṃśati, twenty-two 4.24, 5.5.
bāsaṭṭhī S. dvāṣaṣṭi, sixty-two 3.68, 5.27.
bāhira S. bahis, outside 3.74, 75, 77, 99.
bi S. dvi, two 1.6, 18, 20, 24, 25, 3.8, 33, 34, 4.9, 16, 31, 48, 51, 54, 59, 5.30. See bihu, be, vi. Cf. du.
biṃdu S. bindu, a dot: a negative sign (for a numerator) 1.46.
biṃba S. bimba, a disk: an idol, image 4.47.
bīja Ts., seed 6.28. See bīya[1].
bīya[1] S. bīja, seed 3.8.
bīya[2] S. dvitīya, the second 3.38. See dutīka, vīya.
be S. dve, two 4.28. See bi.

bh

bhaṃḍapaḍibhaṃḍakaraṇa S. bhāṇḍa-pratibhāṇḍa-karaṇa, procedure for a commodity and a counter commodity, that is, procedure for barter 1.90.
√bhakkha S. √bhakṣ, to eat: bhakkhahi 4.50, 51.
bhakkhaṇa S. bhakṣaṇa, eating 4.49.
bhakkhiya S. bhakṣita, eaten 4.51.
√bhaṇa S. √bhaṇ, to tell: bhaṇaï 4.1; bhaṇasu 1.23, 3.2, 25, 27, 32, 99; bhaṇaha 1.36; bhaṇahi 3.44, 66, 4.51; bhaṇāmi 5.1; bhaṇi 1.26, 32, 58, 2.15, 3.27, 47, 4.17; bhaṇio 4.21; bhaṇijjahi 4.50; bhaṇijjāsu 1.61; bhaṇiyahi 5.19; bhaṇisu 1.65; bhaṇeha 1.19.
bhaṇiya S. bhaṇita, said, told 1.7, 8, 48, 3.76, 92, 5.11, 17, 33.
bhatta S. bhakta, divided 3.28, 41.
bhaya Ts., fear 2.16.
√bhara S. √bhṛ, to fill: bharaṃti 2.5; bharevinu 4.51.
bhariya S. bharita, maintained: costing 3.10.
bhāṃḍapratibhāṃḍa(ka) S. bhāṇḍa-pratibhāṇḍa(ka), a commodity and a counter commodity: barter 1.90p, 6.26.
bhāga Ts., part 4.6, 12, 64, 5.12: a fraction 2.17s; division 1.63, 3.24, 87; the part class 2.10. See bhāya, bhāva.
bhāgajāī S. bhāga-jāti, the part class 2.1.
bhāgajāti Ts., the part class 2.1p: the eight classes of reductions of fractions (in pl.) 2.17s, 6.29. See bhāgajāī.
bhāgapavāha S. bhāga-apavāha, the part-subtraction class 2.8p. See bhāgapravāha, bhāgāpavāha.
bhāgapravāha, a wrong Sanskritization of bhāgapavāha 6.34. See bhāgapavāha, bhāgāpavāha.
bhāgabhāgajāti Ts., the part-part class 2.3p, 6.32.
bhāgabhāgavihī S. bhāga-bhāga-vidhi, rule of the part-part class 2.3.
bhāgamattīya S. bhāga-mātṛkā, the part-mother class 2.10.
bhāgamātṛjāti Ts., the part-mother class 2.10p, 6.35. See bhāgamattīya.
bhāgahara Ts., division 1.32. Cf. bhāgāhara.
bhāgāṇubaṃdhavihī S. bhāga-anubandha-vidhi, rule for the part-addition class 2.6.
bhāgānubaṃdha S. bhāga-anubandha, the part-addition class 2.6p, 6.33.
bhāgāpavāha Ts., the part-subtraction class 2.8. See bhāgapavāha, bhāgapravāha.
bhāgāhara S. bhāga-hara, division 1.33p, 55p, 6.6, 14. Cf. bhāgahara.
bhāḍaya S. bhāṭaka, wages 1.75.
bhāya S. bhāga, part 1.18, 43, 44, 56, 57, 59, 60, 62, 2.4, 11, 16, 3.48, 65, 5.25, 25v: division 1.33, 38, 55, 72, 86, 90, 2.3, 14, 3.3, 5, 15, 18, 21, 23, 31, 46, 58, 100, 101, 4.4, 11, 13, 16, 54, 5.19; share 4.2; divisor 4.56. See bhāva.
bhāva S. bhāga, part: division 3.11. See bhāya. Cf. juva (for yuga) for the euphonic change.
bhāvyaka Ts., commission of surety

3.3p, 6.41.
bhāsa Ts., light 5.17.
bhiti S. bhitti, a wall 3.74, 75. See bhitti/bhittī.
bhitti/bhittī Ts., a wall 3.69, 70, 72, 73, 77–82, 85, 98, 101, 4.28, 33, 5.29, 30, 6.76, 78. See bhiti.
bhittisuvanna S. bhitti?-suvarṇa, pure gold 1.77.
bhinna Ts., broken, divided: a fraction 1.46p, 46, 49, 50p, 50a, 51, 53p, 53, 55p, 55, 57p, 57, 59p, 59, 60p, 60, 62p, 3.17p, 6.11, 12–18, 46; distributed 3.8 (-phala), 4.3 (-rūva).
bhuṃjiya S. bhukta (*bhuñjita), eaten 4.51.
bhuyā S. bhujā, an arm: a side of a plane figure 3.35, 36.
bhuva S. bhuja (through bhuya and bhua), an arm: a side of a plane figure 3.36–41, 50, 51, 55, 84; the base of a right-angled triangle 3.42. See bhuyā. Cf. caübbhuva, tibhuva, bhūmibhuva, muhabhuva.
bhū Ts., the earth: ground 3.96; the bottom side of a plane figure 3.39, 40, 41; area 5.3 (-saṃkhā). See bhūmi.
bhūmi Ts., the earth 3.103: soil 5.2: the bottom side of a plane figure 3.37, 38. See bhū. Cf. muha.
bhūmibhuva S. bhūmi-bhuja, earth-arm: the base side of a plane figure 3.37. Cf. muhabhuva.
bhūyakaraṇa S. bhūyaḥ-karaṇa, making increase or profit: transactions for profit 4.1.
bheda Ts., splitting: variety, kind 6.5, 25. See bhedi.
bhedi S. bheda, splitting: variety, kind 3.86p.
√bheya S. bhedaya, causative stem of √bhid, to divide: bheijjaï 4.6, 8.

m

maüṭṭha D., moth beans, *Phaseolus aconitifolius* Jacq. 5.4.
maṃḍala S. maṇḍala, circular: a disk 6.63.
maṃḍava S. maṇḍapa, a pavilion 4.27, 28, 29.
magga S. mārga, a way, road: manner, mode 3.78.
majjha S. madhya, middle: middle part (or term) 1.46, 63, 81, 86, 3.40, 41, 50, 53, 54, 74, 80, 85, 103, 4.1, 16, 26, 30, 32, 47, 49, 54, 5.8.
majjhima S. madhyama, middle 3.77, 4.22, 24, 41, 43: middle term 3.26.
maṭṭiya/maṭṭī S. mṛttikā, clay 3.71, 5.28.
maṇa/maṇī Ts., a weight measure 1.9, 75, 83, 87, 91, 3.8, 67, 5.4–8, 10, 13, 19, 20, 26, 28, 30–32.
maṇahara S. manohara, pleasing, attractive 4.38.
maṇiyā S. maṇikā, jewel 4.33.
maṇu S. manu, the father of the human race: a bhūtasaṃkhyā for fourteen 4.40, 42.
mattā S. mātrā, a measure 3.12
mattīya S. mātṛkā, a mother, origin 2.10 (bhāga-).
mamāṇīya/mammāṇī Pe. sang-i marmar, marble 3.68, 5.27.
mayara S. makara, a crocodile, shark: the zodiacal sign Capricorn 3.104.
marddita S. mardita, erased 4.60p. See maliya.
maliya S. mardita, erased 4.60. See marddita.
√mava S. māpaya, causative stem of √mā, to measure: mavi 3.70.
maviya S. māpita, measured 1.82.
maseliya, monthly wages 4.12. Cf. Pe. mahīnā, monthly pay. Cf. mukātaï.
massūra S. masūra, lentil, *Ervum lens* L. 5.7.
mahara Ar/Pe. mahr, dowry? 5.14.
mahākaṇaya S. mahā-kanaka, great gold: pure gold 1.10.
mahisī S. mahiṣī, a she-buffalo 5.14, 16.
mahuva S. madhuka/madhūka, honey tree, *Bassia latifolia* Roxb. 3.95.
māī S. mātṛ, a mother? 4.17 (in māī-

ṇaya).

māṇa(ya) S. māna(ka), measuring: a measure, an amount 3.27, 49 (gaṇiya-), 64, 87, 101, 4.16, 54; a volume measure 1.8, 80, 3.71; area 3.40; measurement 5.1, 5.22p; rate 3.89 (aggha-).

mātṛ Ts., mother 2.10 (bhāga-), 6.35 (bhāga-). See mattīya, māī.

māna Ts., measuring: measurement 5.22p, 24s. See māṇa(ya).

māsa[1] Ts., a month 1.11, 73, 74, 3.2, 4–6, 4.11, 12, 17.

māsa[2]/māsaya S. māṣa/māṣaka, pulse 5.5 (mugga-): a weight measure 1.9, 10, 69, 77, 2.13, 3.17, 22.

√miṇa S. √mā, to measure: miṇavi 3.101.

miriya S. marica, pepper 3.13.

√mila S. √mil, to meet together: milaï 4.15, 55; milaṃtī 4.13; milei 4.14. Cf. √mela.

miśraka Ts., mixture 3.1p, 25s, 6.39, 40. See missa.

miśrita Ts., mixed, added 5.33s.

missa S. miśra, mixture: the sum of the capital and the interest 3.1, 3, 4, 7.

mīra, ? 4.31.

muṃga/mugga/mūṃga S. mudga, kidney beans, mūṃg pulse, *Vigna radiata* (L.) Wilcz. 1.80, 3.14, 97, 5.5 (-māsa).

muṃda S. mudrā, a coin 1.65.

mukataï/mukātaya/mukkātaï Ar/Pe. muqāṭa^{c}a, wages at piece-rate (Ar.), contract for a time (Pe.): annual wages 4.11p, 11. Cf. maseliya.

√muṇa S. √jñā, to know: muṇasi 3.2; muṇaha 4.58, 59; muṇahi 4.38; muṇijjahu 5.19; muṇijjei 3.7, 5.33. See √jāṇa, √najja.

muṇi S. muni, a sage: a bhūtasaṃkhyā for seven 3.104, 4.17, 39, 42, 45.

muṇeyavva S. jñātavya, to be known 3.34, 4.12. Cf. amuṇiyarāsi.

munāraya/munārayā Ar. manār, Pe. mīnār, a minaret 3.69, 80p, 80, 6.79 (munārā).

muruja S. muraja, a drum: a figure shaped like a muraja drum 3.50.

mulla S. maulya, price 1.65, 66, 69, 77-80, 90, 92, 93, 3.9, 10, 11.

muha S. mukha, face, mouth: the top (or face) side of a plane figure 3.37, 39, 40; the two face sides (or the top and the bottom) of a rectangle obtained by transforming a rimlike plane figure 3.53; the (imaginary) top plane of a pit 3.54.

muhabhuva S. mukha-bhuja, the face-/mouth-arm: the face side (top) of a plane figure 3.37. Cf. bhūmibhuva.

muharama Ar/Pe. muḥarram, forbidden, sacred: name of the first month of the Muslim year 4.17.

mūla Ts., the root, original: the lowest edge 3.74; original 1.24, 4.8; the number of terms of a series 1.19, 20; the square root 1.20, 37, 38, 59, 3.30, 43, 48, 4.56, 57, 6.8, 16; the cube root 1.43, 6.10, 18; capital 1.73, 86–88, 3.1–3; basic 6.1, 75. See ghaṇamūla, vaggamūla. Cf. pamāṇadhaṇa.

mūlarāsi S. mūla-rāśi, the root-quantity 1.21.

mṛdaṃga S. mṛdaṅga, a drum 6.67.

metthī D., fenugreek, *Trigonella foenum-graecum* L. 5.9. Cf. H. methī.

√mela S. melaya, causative stem of √mil, to assemble, bring together: meli 1.18, 4.41. Cf. √mila.

y

yala S. tala, ground, floor 4.23.

yuga Ts., an age of the world, four in number: a bhūtasaṃkhyā for four 4.38, 42, 45.

r

√rakkha S. √rakṣ, to protect: rakkhahi 4.63.

rakkha S. rakṣa, a protector or protection: a kind of sand? 5.32 (-pavāhiya). Cf. arakkha.

rayaṇi S. rajani/rajanī, night 1.11.
ravi Ts., the sun: a bhūtasaṃkhyā for twelve 3.103, 4.40, 42.
ravva, thickened juice from sugarcane 5.11.
rasa Ts., taste, essence, juice 5.10 (ikkhu-), 12s (ikṣu-): a bhūtasaṃkhyā for six 4.17, 38, 42, 45.
√raha S. √rah, to leave, abandon: rahahi 4.51.
rahiya S. rahita, abandoned, deprived of: decreased by 2.9; left 2.16.
rāja S. rājan, a king, a master: a mason 5.29. See rāya. Cf. H. rāja.
rāṇaya, ? 4.49.
rāya S. rājan, a king 4.3.
rāyaṭṭhāṇa, Rāja-sthāna, the name of a region 4.1.
rāśi Ts., a heaped mound (of grain) 3.96p, 100s. See rāsi/rāsī.
rāsi/rāsī S. rāśi, a heaped mound (of grain) 3.96–99, 6.90, 91: quantity 1.21, 31, 36, 37, 47, 49, 57, 89, 90, 3.11, 29, 104, 4.3, 9, 10, 16, 54, 56, 57, 60–62, 72. See amuṇiya-, guṇa-, guṇiya-, gunna-, tirāsiyaga, mūla-.
rikkha S. ṛkṣa, a star, a lunar mansion: a bhūtasaṃkhyā for twenty-eight 4.40.
riṇa S. ṛṇa, debt: negative, subtractive 2.12. Cf. dhaṇa, vaya, hīṇaṃsa.
rukkha S. vṛkṣa, a tree 4.63.
rū S. rūpa, colour, form: an integer number 1.46; one (unity) 3.104, 4.38?, 42. See rūva. Cf. gorū.
rūi, a kind of cloth? 4.34. Cf. H. rūī, cotton.
rūva. S. rūpa, colour, form 3.63, 4.34: an integer number 1.47 (-rāsī), 2.6, 8, 4.3; one (unity) 2.3, 14, 4.48, 50. See rū.
reha S. rekhā, a line 3.103. See līha, leha.

l

laṃba(ya) S. lamba(ka), hanging down: a perpendicular 3.39, 40, 41, 53; the upright of a right-angled triangle 3.42; the altitude of a figure shaped like an elephant's tusk 3.52.
lakkha/lakha S. lakṣa, a numeral for 10^5 or the name of the sixth decimal place 1.12, 14, 4.5, 7, 8.
lacchī S. lakṣmī, the name of a goddess 1.1.
laddha S. labdha, obtained 1.43, 45, 53, 56, 57, 60, 2.14, 3.3, 58, 4.48, 49, 52: the quotient 1.18, 44, 3.28, 29, 31, 46, 4.2, 13, 16, 17, 48, 54.
labdha Ts., obtained: an answer 5.25v. See laddha.
√labbha S. √labh, to obtain: labbhaï 1.67, 68, 4.4, 5.21. See √laha. Cf. √le.
lavaṇa Ts., salt 5.28.
√laha S. √labh, to obtain: lahaṃti 1.65. See √labbha. Cf. √le.
lahu S. laghu, small, short 1.90.
lāha S. lābha, profit 1.86–89.
likhita Ts., written down 5.33s.
√liha S. √likh, to write: lihasu 4.7; lihi 1.33, 4.38, 41, 42, 45; lihijja 4.44; lehāṇa 1.1.
līha S. lekhā, a line 3.89, 93, 94. See reha, leha.
lūṇiya S. nava-nīta, newly obtained: butter 5.13. Cf. H. lūnī and lonī.
√le S. √lā, to obtain: lijjahi 1.87, 88; lijjā 3.25. Cf. √labbha, √laha.
leva S. lepa, plastering 5.31.
leha S. lekhā, a line: writing and counting 4.1. Cf. reha, līha.
lehaga S. lekhaka, a scribe 3.4.
lehāṇa. See √liha.
√loḍa, S. √luṭh or √luḍ, to go, move: loḍaï 2.16.
loya S. loka, world, people 1.2.
loyaṇa S. locana, seeing, an eye: a bhūtasaṃkhyā for two 4.45.

v

vaï S. pati, the lord 4.47 (sura-).
vaṃkaï S. vakreṇa, cross-wise 1.89.
√vaṃṭa S. √vaṇṭ, to divide, allocate: vaṃṭivi 4.2.

vaṃṭana S. vaṇṭana, partitioning, dividing: distribution 4.64p.
vaṃsī D., red sandstone 3.68, 5.27.
vakkhara S. vakṣas-kāra, a bag, sack: a container 1.83, 5.20.
vagga S. varga, a class, group: the square (number) 1.32, 34–40, 44, 45, 57, 58, 3.30, 32–34, 41–43, 46, 48, 49, 56–58, 96; the square series (which consists of the squares of natural numbers) 3.33.
vaggamūla S. varga-mūla, the square root 1.20, 32, 59.
vacchara S. vatsara, a year 4.17. Cf. varisa.
vajra Ts., thunderbolt 6.66.
vaṭulā, S. vartula, val pulse, *Dolichos lablab* L.? 5.8.
√vaṭṭa S. √vṛt, to take place: vaṭṭijjā 4.6.
vaṭṭa S. vṛtta, round: circular 3.43, 44, 63, 64, 66, 69, 79p, 79, 80, 86, 94, 98, 4.25; a circle 3.45, 51, 103, 5.22, 23, 24; a rim of a wheel 3.51. See vatta. Cf. vaṭṭaḍa.
vaṭṭakhitta S. vṛtta-kṣetra, a circular plane figure 3.43.
vaṭṭaḍa S. vartula, a circle 5.23. Cf. vaṭṭa, vatta.
vaṭṭha S. pṛṣṭha (AHK), surface, or S. vastra, cloth? 4.22, 24, 26–28. See vattha.
vaḍa S. vaṭa, Indian fig tree, *Ficus benghalensis* L. 3.95.
√vaḍḍha S. √vṛdh, to increase: vaḍḍhaṃta 4.15, 49, 55; vaḍḍhei 1.21.
vaṇa S. vana, a forest 4.49.
vaṇiya S. vaṇij, a merchant 3.14.
vaṇaṃnna S. vana-anna, wild grains 5.28.
vatta S. vṛtta, circular 3.86 (in saṃkhavatta). See vaṭṭa. Cf. vaṭṭaḍa.
vattīsa. See battīsa.
vattha S. vastra, a garment, cloth 4.18, 22, 24, 26, 37, 52, 53. See vaṭṭha.
vatthu S. vastu, an object or a thing 1.88, 89, 3.11, 5.21.
vanna/vannaya/vannī S. varṇa/varṇaka/varṇi(kā), colour 1.79: a unit for the purity of gold 1.10, 69, 77, 84, 3.15–25. Cf. java.
vaya S. vyaya, expenditure 4.3; a negative quantity 1.51. Cf. ādāya, riṇa, hīṇa.
vayalla, a monetary unit (used for tax) ? 5.14.
vara Ts., best, excellent 1.92, 4.53.
varisa S. varṣa, a year: a time unit 1.11, 73, 92, 93, 3.4, 4.11, 17. Cf. vacchara.
varisola S. varṣolaka, a sweetmeat prepared with sugar, rice flour, milk, cardamom, saffron, and camphor 4.50p, 50, 51.
varisolaga D. an unidentified byproduct in sugar manufacture 5.12.
varuṇa Ts., watered, irrigated 5.2.
varga/vargga Ts., a class, group: the square (number) 1.34p, 57p, 3.33p, 34p, 6.7, 15, 54, 55; the square series 3.33p. See vagga.
vargamūla Ts., the square root 1.37p, 59p, 6.8 (vargga-), 16 (vargga-). See vaggamūla.
varṇṇa S. varṇa, colour: purity of gold 3.21p, 6.47.
valaya Ts., a ring 4.28.
valahāra, a necklace? 5.14.
vallahali, ? 4.32.
vallī Ts., a creeper, vine 4.32: a chain of measures 2.13.
vallīsavannaṇavihī S. vallī-savarṇana-vidhi, rule for the reduction of a chain of measures to the same colour, i.e., to the same denominator 2.12.
vallīsavarṇana Ts., reduction of a chain of measures to the same colour, i.e., to the same denominator 2.12p, 6.36.
vavahāra S. vyavahāra, procedure, custom, trade: interest 1.74. See vivahāra.
vasaha S. vṛṣabha, a bull 5.16.
vasu Ts., a class of deities: a bhūta-saṃkhyā for eight 4.38, 42.
vastu Ts., thing 5.33s, 6.0. See vatthu.
vastra Ts., cloth 4.18p, 37s, 52p. See

vattha.
vaha S. vadha, killing: the product (of a multiplication) 3.21, 23.
vahala Ts., or S. bahala, copious, plentiful 4.47.
vahoḍaṇa S. vyāpādana, killing 4.14.
vā (in savā 'with a quarter') S. pāda, a quarter 1.56, 67, 75, 3.17, 95, 102. See pa, pā, pāya, vāya.
vāiyā S. pādikā, a quarter 4.20.
vāīya S. *vājīka, a horse: a bhūtasaṃkhyā for seven 5.5.
vāḍa S. vāṭa, an orchard 4.63.
vāpī Ts., a stepwell 3.69, 86p, 6.84. See vāvi/vāvī.
vāma Ts., the left-hand side 3.37, 4.39.
vāya S. pāda, a quarter 1.61, 66, 74, 84, 3.14, 57, 5.19. See pa, pā, pāya, vā.
vāra[1] S. dvādaśan, twelve 4.21. See vārasa/vāraha.
vāra[2] Ts., a door 4.28, 47. See dāra, duvāra, bāra, vāri[2].
vāra[3] Ts., time 3.70, 4.46, 50, 5.12.
vārasa/vāraha S. dvādaśan, twelve 1.76, 79, 5.29. See bārasa/bāraha, vāra[1].
vāri[1] Ts., water 4.63.
vāri[2] S. vāra or dvār or dvāra, a door 3.81. See dāra, duvāra, bāra, vāra[2].
vāri[3], ? 4.38 (vāritā?).
vārigaha Pe. bārgāh, king's court, palace: an assembly hall 4.29, 31. Cf. Abū al-Faḍl [2004, 89: Book I, Ā'īn 21; Plate XI].
vāluya S. vāluka, sand 2.15.
vāvanna S. dvāpañcāśat, fifty-two 3.37.
vāvi/vāvī S. vāpī, a cistern, stepwell 2.5, 3.69, 86.
vāsaka, a kind of fine cotton? 4.18.
vāsaṇa S. vāsana, an abode, clothes 5.15.
vāhattari S. dvāsaptati, seventy-two 1.83.
vi S. dvi, two 1.34 38, 49, 89, 3.30, 31, 41, 5.14, 18. See du, bi.
vi = vidhi 6.86.
viṃjjha S. vindhya, the name of a mountain 2.17.
vikaṭa Ts., great, beautiful 6.60 (trikoṇa-).
vikkama S. vikrama, the Vikrama era 4.17.
vikkaya S. vikraya, selling 1.92: selling rate 1.86. Cf. kaya.
√vikka S. vi-√kṝ, to sell: vikkijjā 1.88. See √vikkiṇa.
√vikkiṇa S. vi-√kṝ, to sell: vikkiṇijjaṃti 1.87. See √vikka.
vikkhaṃbha S. viṣkambha, the bar of a door: breadth 1.7, 3.10, 35; diameter of a circle 3.43-45, 59, 64, 76, 5.22.
vikraya Ts., selling 1.86p, 92p, 6.25, 27. See vikkaya.
vigama Ts., abandoning 1.44.
viggahaya/vīgaha(ya) S. vigrahaka, stretching out: an area measure commonly known as bīghā 1.4, 5.3, 4, 9.
√viccāra S. vicāraya, causative stem of vi-√car, to consider: viccāri 1.79.
vijja S. vaidya, a physician, doctor 3.12.
viṇā S. vinā, without 4.20, 33, 59, 61, 5.14.
viṇoya S. vinoda, an amusement 4.33.
vittī S. vṛtti, being, action: wages, fee 3.4.
vitthatiyarāsī S. vyasta-trika-rāśi, the inverse three-quantity operation, the inverse rule of three 1.81. See vyastatrairāśika. Cf. tirāsiyaga.
vitthara S. vistara, width 1.4, 78, 85, 3.55, 57, 61, 62, 66, 70, 72, 73, 75, 77, 78, 83–85, 87–90, 103, 4.22, 33, 52, 5.3: diameter of a circle 5.22. See vitthāra.
vitthāra S. vistāra, width 1.79, 3.52, 56, 57, 88, 91, 93: area 1.7; diameter of a circle 3.58. See vitthara.
vidhi Ts., doing, operation, rule 6.3, 68, 70. See vihi/vihī.
vinni S. dvi, two 1.89. See vi. Cf. tinni.
vinneya S. vijñeya, to be known 1.5.
vipakka S. vipakva, cooked: refined

(gold) 3.18. Cf. pakka.
viparīta Ts., inverse 4.56p.
vippa S. vipra, a brāhmaṇa 3.4, 5.15.
√vibhaya S. vi-√bhaj, to divide: vibhaehiṃ 3.61.
vimakaliya S. vyavakalita, separated: subtracted 1.26; the difference of two natural series 1.21; the subtrahend series 1.22, 24, 25; subtraction of natural series 1.26; subtraction of fractions 1.51. See vimalakalita, vyavakalita.
vimakaliyasesa S. vyavakalita-śeṣa, the remainder of subtraction 1.22–24.
vimalakalita, a wrong Sanskritization of vimakaliya 6.4. See vimakaliya, vyavakalita.
√vimucca S. vi-√muc, to open: vimuccahi 2.5.
√viyāṇa S. vi-√jñā, to know: viyāṇāsi 1.32, 58, 3.32; viyāṇāhi 1.7, 26, 60, 5.20; viyāṇehi 3.33. Cf. √jāṇa, √najja, √muṇa.
viracita Ts., written 1.93s, 2.17s, 3.104s, 4.64s, 5.33s.
viloma Ts., against hairs, inverse(ly) 1.27.
vivariya S. viparīta, transposed? 1.72, 90. See vivarīya, viviriya.
vivarīya S. viparīta, reversed 1.55, 90, 92, 4.38. See vivariya, viviriya.
vivahāra S. vyavahāra, conduct, procedure: the 8 types of mathematical procedures 1.15. See vavahāra.
viviriya S. viparīta, inverse? 4.57. See vivariya, vivarīya.
viviha S. vividha, various 3.63.
viśeṣa Ts., difference, particular 5.25p.
viṣama Ts., different, odd, uneven 3.11p: an odd number 4.43p. See visama.
visama S. viṣama, different, odd, uneven 3.11, 38, 54, 6.44: an odd number 1.17, 37, 59, 4.38, 44.
visamatikoṇa S. viṣama-tri-koṇa, an uneven triangle, i.e., a scalene triangle 3.38.
visuva/visova(ga) S. viṃśopa(ka), one-twentieth part 5.23, 24: a monetary unit 1.3, 4, 4.6, 7, 9, 10, 12, 5.20, 21; a weight unit 2.13, 5.13; a linear measure 3.90–93, 4.21; an area measure 5.4. Cf. visuvaṃsaga/vissaṃsa/vīsaṃsa.
visuvaṃsaga/vissaṃsa S. viṃśopaaṃśaka/viṃśa-aṃśa, one-twentieth part of visuva/visova(ga): a monetary unit 1.3, 4.10.
visova(ga). See visuva/visova(ga).
visesa S. viśeṣa, difference 5.23: a peculiar mark, characteristics, quality 3.80, 5.2.
vistara Ts., extension: breadth 6.93.
vissaṃsa. See visuvaṃsaga/vissaṃsa.
viha S. vidha, kinds 3.69, 5.16.
vihatta/vihattha S. vihṛta, taken away: divided 1.50, 56, 81, 3.1, 11 (vihattha), 15, 18, 26, 29, 4.48, 56, 57.
vihi/vihī S. vidhi, doing, rule 1.92, 2.6, 3.11, 24, 42: process 1.55; method 1.58, 3.32; operation 1.38, 2.3, 6, 12.
vihīṇa S. vihīna, deprived of: decreased by 1.25, 3.29, 84.
vīgaha(ya). See viggahaya/vīgaha(ya).
vīya S. dvitīya, the second 1.39, 40. See dutīka, bīya².
vīsa S. viṃśati, twenty 1.3, 4, 28, 68, 77, 82, 83, 85, 3.2, 19, 27, 52, 59, 91, 92, 4.36, 47, 49, 5.3–5.
vīsaṃsa S. viṃśa-aṃśa, one-twentieth part 4.10.
vuḍḍhi S. vṛddhi, increase: the common difference of an arithmetical progression 1.19, 3.30, 4.41, 48.
vuḍḍhiya S. vṛddha + vardhita, old 5.16.
√vuṇa S. √ve, to weave: vuṇijjahi 3.9.
vuṇāvaviya S. vātavya, to be woven 3.9.
vuṇiya S. uta, woven 3.9.
vutta S. ukta, told 1.1.
vṛtta Ts., a circle 5.22v, 6.63. See vaṭṭa, vatta.
vesavāra Ts., spices 5.9.
veha S. vedha, depth 3.55, 58, 60.

vora D., a kind of grain or pulse 3.97. Cf. kora^{1}.

vyavakalita Ts., the difference 1.21p. See vimakaliya, vimalakalita.

vyavahāra Ts., procedure: the 8 kinds of mathematical procedures 3.1p (-gaṇanā, miśraka-), 15p, 25s, 26p, 34s, 35p, 53s, 54p, 68s, 69p, 86s, 87p, 95s, 96p, 100s, 101p, 104s, 6.38, 39, 48, 49, 56, 69, 74, 85, 90, 95, 98. See vavahāra, vivahāra.

vyastatrairāśika Ts., the inverse rule of three, inverse three-quantity operation 1.81p, 6.24. See vitthatiyarāsī. Cf. tirāsiyaga.

ś

śubha Ts., good fortune 6.28.

ṣ

ṣaṭ/ṣaṭa S. ṣaṣ, six 3.39p, 86p. See cha/chac/chat, khaḍ.

s

sa^{1} Ts., accompanied by 5.15: increased by 1.22, 25, 41, 50, 52, 54, 56, 58, 61, 66–69, 74, 75, 77, 84, 2.6, 7, 17, 3.13, 14, 17, 30, 33, 34, 57, 62, 74, 95, 100, 102, 4.20, 43, 5.19, 23, 24. Cf. saḍḍha(ya).

sa^{2} Ts., same. See savaṃnaṇa/savannaṇa, savarṇana, sālaṃba.

sa^{3} S. sva, one's own 3.65. Cf. sukama.

saī S. śatin, an owner of hundred: percentage 4.4, 5, 6, 8.

saṃ = saṃkhyā 6.0, 7, 8, 83, 84.

saṃka S. śaṅku, a peg: a gnomon 3.103.

saṃkalita Ts., added: the sum or the summation of the natural series 1.16p, 3.34p, 6.3, 12, 53, 55; the sum of fractions 1.50p (bhinna-). See saṃkaliya.

saṃkaliti = saṃkalita 3.31p.

saṃkaliya S. saṃkalita, added: the sum or the summation of the natural series 1.16–26, 3.31–34; sum of fractions 1.50 (bhinna-); the sum of an arithmetical progression 4.48.

saṃkha^{1} S. śaṅkha, a conch: a numeral for 10^{13} or the name of the fourteenth decimal place 1.13. Cf. saṃkhavatta.

saṃkha^{2}/saṃkhā S. saṃkhyā, number 1.14, 3.58, 71, 72, 99, 4.13, 20, 43, 44, 48, 64, 5.3.

saṃkhavatta S. śaṅkha-vṛtta, circular like a conch-shell: spiral 3.86.

saṃkhyā Ts., number 4.62p, 6.79. See saṃ, saṃkha^{2}/saṃkhā.

saṃguṇiya S. saṃguṇita, multiplied 1.17.

saṃjāya S. saṃjāta, produced 3.19.

saṃjutta S. saṃyukta, combined: increased by 4.48.

√saṃbhava S. saṃ-√bhū, to be possible: saṃbhave 1.33.

saṃbhūya S. saṃbhūta, produced 3.67, 5.26.

saṃlagga S. saṃlagna, stuck together 3.98.

saṃvat Ts., the year of Vikrama 5.33s.

saṃvatsara Ts., a year 4.17p. See vacchara, varisa.

saṃsaya S. saṃśaya, doubt 1.17, 3.31, 76.

saṃharaṇīya Ts., to be drawn out, to be compressed: the dividend 1.33.

sakkara S. śarkarā, brown sugar 5.10. Cf. khaṃḍa.

sagaḍacakkavaṭṭa S. śakaṭa-cakra-vṛtta, the periphery (or rim) of a wheel of a cart 3.51.

sagavīsa S. saptaviṃśati, twenty-seven 4.40, 5.30. See sattavīsa.

saggavattī S. śāka-patrin, vegetables with leaves, leafy vegetables 5.9. Cf. H. sāgapattī.

saṭṭhi/saṭṭhī S. ṣaṣṭi, sixty 1.4, 11, 28, 70, 73, 80, 3.37, 68, 90, 5.3, 4, 27, 32.

saḍḍha(ya) S. sa-ardha(ka), with a half 1.52, 61, 74, 75, 77, 84, 2.7, 3.57, 62. Cf. aḍḍhāiya.

saṇa S. śaṇa, flax, *Crotalaria juncia* L. 5.6, 31.

sataraha S. saptadaśan, seventeen 1.74, 87, 4.40.

satta S. saptan, seven 1.56, 58, 61, 64, 67, 70–72, 78, 79, 82, 88, 91, 2.4,

17, 3.10, 19, 20, 22, 25, 27, 81, 82, 93, 104, 4.3, 14, 39, 58, 63, 5.13, 24.

sattama S. saptama, the seventh 2.13.

sattavīsa S. saptaviṃśati, twenty-seven 1.28. See sagavīsa.

sattari S. saptati, seventy 4.5, 58.

sattha S. śāstra, a treatise 1.48. Cf. sattha for S. śastra in HGA, p. 445.

satya Ts., true 1.30p.

sadisa S. sadṛśa, same, common, similar 1.47, 48, 50, 51. See sarisa. Cf. sama.

sadṛśa Ts., same, common, similar 4. 61p. See sadisa, sarisa.

saptarāśika/saptarāsika S. saptarāśika, of seven quantities: rule of seven, seven-quantity operation 1. 72p, 78p, 6.21.

sama Ts., equal, even, same 1.20, 24, 30, 31, 2.1, 5, 10, 14, 3.7, 11p, 11-13, 51, 55, 57, 60, 95, 96, 103, 4.20, 31, 64, 5.12, 6.44, 57 (-caürasa): uniformly (circular) 3.44 (-vaṭṭa), 66 (-vaṭṭa); an even number 1.17, 37, 59, 3.11p, 4.38; the arithmetical mean 3.54, 55. Cf. sarisa.

samatta S. samāpta, completed 3.95s. See sammatta.

samaya Ts., time, date, occasion 4.17, 5.33.

samavisama S. sama-viṣama, an evenly-odd number (an integer that can be halved once and only once) 4.38.

samavisamakaya S. sama-viṣama-kraya, purchase of equal or different quantities or purchase in proportion 3.11. Cf. kaya.

samahiya S. samāhita, arranged, assembled 4.38.

samāpta Ts., completed 1.93s, 3.53s, 104s, 5.33s.

samāsa Ts., the sum 3.5.

samāsaṇa S. samāsana, summation 2. 10.

sammatta S. samāpta, completed 3. 34s, 68s, 86s, 100s, 104s, 4.17s, 37s, 45s, 64s, 5.33s. See samatta.

saya S. śata, hundred, i.e., a numeral for 10^2 or the name of the third decimal place 1.12, 13, 26, 28, 42, 68, 69, 73, 74, 79, 85, 91, 93, 2.4, 16, 3.2, 4–6, 8, 90, 4.4, 8, 19, 35, 37, 58, 5.9, 26, 29.

sayama S. śatama, the hundredth 4.9.

sayala S. sakala, all, whole 1.1, 2, 3.39, 63, 80, 4.19, 20, 47, 60.

sara[1] S. śara, an arrow: linear (dimension) 1.7; the height of a circle segment 3.46–49; a bhūtasaṃkhyā for five 4.39, 42.

sara[2], an unidentified substance from which a tape is made 4.37.

sarāi Pe. sarāy, a house, palace, grand edifice, king's court, seraglio, inn: an unidentified part of a building 4.33, 35.

sari S. sarit, a river 4.63.

sarisa S. sadṛśa, similar to, same, equal 3.51, 61, 80. See sadisa. Cf. sama.

sarisama, S. sarṣapa, mustard, *Brassica nigra* Koch 5.7, 13.

sarva Ts., all 5.33s. See savi, savva.

savaṃnaṇa/savannaṇa S. sa-varṇa-na, reduction to the same colour, homogenization: reduction of a composite fraction to a simple fraction 1.47, 51, 2.12, 13.

savarṇana Ts., S. sa-varṇa-na, reduction to the same colour, homogenization: reduction of a composite fraction to a simple fraction 2.12p, 6.36. See savaṃnaṇa/savannaṇa. Cf. kalāsavarṇana, vallīsavaṃnaṇa-vihī, vallīsavarṇana.

savāṇiya S. śrāvaṇika, autumnal (harvest) 5.6 (-sāhā). Cf. H. sāvanī.

savi S. sarva, all 4.47, 51. See savva.

savva S. sarva, all 1.49, 79, 85, 86, 3.26, 28, 29, 58, 67, 4.2, 3, 20–22, 38, 44, 47, 49, 50, 52, 62, 63, 5.8, 9, 26. See savi.

savvattha S. sarvatra, everywhere 3.42, 5.2.

savvadhaṇa S. sarva-dhana, the sum of all terms of a finite series 3.26,

28, 29.
sasi S. śaśin, that which has a rabbit, the moon: a bhūtasaṃkhyā for one 4.42, 45.
sasihara S. śaśa-dhara, that which bears a rabbit, the moon: a bhūtasaṃkhyā for one 4.38.
saha Ts., together with 4.28: increased by 3.26.
sahasa/sahassa S. sahasra, thousand, i.e., a numeral for 10^3 or the name of the fourth decimal place 1.6, 12–14, 23, 4.3, 5, 7, 8, 11.
sahi[1] S. sakhi, a friend 3.62.
sahi[2] D. all, everywhere 4.51, 63.
sahiya S. sahita, accompanied by 3.76, 82, 5.29, 31: added 1.40, 2.7; increased by 1.22, 34, 3.17, 84, 4.59.
sāḍa, a kind of coarse cotton? 4.18.
sādhana Ts., establishment, determination, setting up 3.103p, 6.96 (-ā), 97 (-ā).
sāra Ts., essence (in gaṇitasāra) 1.93s, 2.17s, 3.104s, 4.64s, 5.33s, 6.94 (-sāri): a kind of silk 4.18.
sāricchiya S. ṣādṛśya, likeness, similarity 4.63. See sadisa, sarisa.
sāla S. śāla, sal tree, *Vatica robusta* Steud. 3.95.
sālaṃba S. sa-ālamba, having equal perpendiculars 6.59.
sāsu S. śvaśrū, a mother-in-law 4.51.
sāsuraya S. śvāśuraka, father-in-law's (house) 4.51.
sāhā S. *sādha, gaining: harvest 5.6. Cf. '√sāha[4] grahaṇa karanā' in AHK.
sikkaya, ? 5.8. Cf. seṃkkaya.
siggha S. śīghra, quick 1.36, 3.2, 32.
siṭṭhikara S. sṛṣṭi-kara, one who makes the emission: the creator 5.1.
siya S. sita, stitched or white? 4.21.
siyāvaṇiya, wages for stitching 4.35. Cf. cirāvaṇiya/cirāvaṇī.
sirīsa S. śirīṣa, sirisa tree, *Acacia sirissa* Ham. 3.95.
silā S. śilā, a stone slab 3.62.
sihara S. śikhara, a peak of a mountain 3.98.
sihā S. śikhā, an apex 3.81, 82.
sīmāṇaya, ? 4.25.
sīra, scum from sugarcane juice 5.12.
sīvāṇa S. sīvana, sewing 4.19.
sīsama S. śiṃśapā, Indian rosewood, *Dalbergia* Roxb. 3.95.
sīha S. siṃha, a lion 2.16.
suṃṭhī/soṃṭhi S. śuṇṭhī, dry ginger 1.91, 3.13.
sukama S. su-krama, the correct order (or S. sva-krama, its own order?) 1.27, 29, 39, 3.36. Cf. parakama.
√sujja S. √śvi, to swell, grow: sujji 1.33. Cf. H. sūjanā.
√sujjha S. √śuddh, to become pure: sujjhaṃti 4.33.
√suṇa S. √śru, to hear, listen to: suṇiūṇa 1.2. Cf. √nisuṇa.
sutta S. sūtra, a thread 4.36, 37.
suttahārī S. sūtra-dhārin, one who holds a thread: a carpenter 3.92.
sudi, a day of the white half of a month 5.33s.
suddha S. śuddha, clean, purified: correct 1.30, 3.70.
suddhi S. śuddhi, correction 6.80.
sunna S. śūnya, empty: zero 1.31, 32, 4.59, 60.
sunnuvāri(tā), ? 4.38.
suya S. suta, a son 1.2, 4.1, 59, 5.33.
suyā S. sutā, a daughter 4.59.
sura Ts., a god 4.47 (-vaï): a bhūtasaṃkhyā for thirty-three 4.39.
suradāru Ts. (same as deva-dāru or divine tree), Himalayan cedar, *Pinus deodara* Roxb. 3.95. Cf. dāru.
suragiha S. sura-gṛha, a divine house or a temple 4.47. Cf. giha.
suvaṇṇī S. *suvarṇī, gold 6.45.
suvanna S. suvarṇa, gold 1.77, 3.18, 21.
suvarṇa Ts., gold 6.46, 47.
suvarṇna S. suvarṇa, gold 3.15p, 17p, 18p, 21p. See suvanna.
suhama S. sūkṣma, fine 4.18, 36.
sūtra Ts., a thread: a mathematical rule 1.93s, 3.29p (karaṇa-), 34s, 53s, 61p (karaṇa-), 68s, 86s, 87p (karaṇa-), 100s, 101p (karaṇa-), 104s,

4.17s, 38p, 5.1p, 33s, 6.0, 28, 29, 38, 42, 56, 98. See sutta.
seī/seiya S. seti(kā), a volume measure 1.8, 80, 82, 5.18.
seṃkkaya, ? 5.9. Cf. sikkaya.
seṃvalu S. śālmalī, silk-cotton tree, *Bombax ceiba* L. 3.95.
seḍhiyasaṃkaliya S. śreḍhikā-saṅkalita, the sum of a mathematical series 4.48.
seḍhī S. śreḍhī, line, series: a mathematical series 3.26p, 34s, 6.48, 49.
sera Ts., a weight measure 1.9, 67, 3.12–14, 4.37, 5.19, 20, 31.
sela S. śaila, stone: the volume of a solid 3.65. Cf. pāhaṇa/pāhāṇa.
√seva S. √sev, to serve, attend: seva 4.3.
seva (in pāyaseva) Ts., service 3.69, 77p, 77, 79, 80, 6.78.
sesa/sessa S. śeṣa, remaining 3.104: the remainder 1.20, 22, 24, 30, 52, 92, 2.14, 17, 3.28, 36, 41, 71, 4.10, 12, 47, 56, 57, 59, 60, 5.12, 19.
sopāna Ts., a staircase 3.83p, 6.81. See sovāṇa.
solasa(ga)/solaha S. ṣoḍaśa(ka), sixteen 1.8, 10, 36, 42, 65, 82, 2.16, 3.56, 57, 91, 4.24, 35, 38, 55, 5.5, 6, 28.
solasama S. ṣoḍaśa, the sixteenth 5.10.
sovāṇa S. sopāna, a staircase 3.69, 83.
√soha S. śodhaya, causative stem of √śudh, to purify, subtract: sohi 1.38, 52, 2.14, 3.28; sohijjā 1.45, 52, 3.21, 70; sohivi 1.24, 3.23, 4.44, 60; sohevi 1.23, 37, 4.17.
sohiya S. śodhita, subtracted 1.45.
staṃbhoddesajāti S. stambha-uddeśa-jāti, the pillar-problem class 6.37.
sthaṃbhaṃsakajāti S. staṃbha-aṃśa-ka-jāti, the pillar-part class 2.14p. Cf. thaṃbha.
sthāpanā Ts., setting up, establishing 6.1.
sneha Ts., oil 5.13s.

h

hakka, an unidentified substance used for constructing brick walls 5.30.
√haṇa S. √han, to kill, multiply: haṇijjaï 1.18.
hattha S. hasta, a hand: a linear measure 1.6, 7, 64, 79, 3.38, 40, 53, 56, 59, 62, 64, 82, 89, 4.19, 52; a square measure 1.7, 4.53; a cubic measure 3.60, 61, 71, 96. See attha[3].
haya S. hata, killed: multiplied 1.41, 47, 3.26, 30, 65, 4.22, 25, 5.25.
√hara S. √hṛ, to take, divide: hara 3.7; haraï 1.33; hari 3.5, 30; hariūṇaṃ 1.60; harijja 1.90 (bhāyaṃ -), 3.101; harijjaï 1.18 (bhāu -); harijjae 1.56; harijjahi 4.48; harivi 4.38, 58; harevi 4.4.
hara[1] Ts., taking away: a divisor 1.33, 44, 55; a denominator 2.6, 8.
hara[2] S. dhara, holding, bearing 3.8 (hala-).
haraḍaï/haraḍaī S. harītakī, yellow myrobalan, *Terminalia chebula* Retz. 3.12, 27.
haraṇīya Ts., to be taken off: the dividend 1.55 (suggested for hāraṇīya). See hāraṇīya.
haranayaṇa S. hara-nayana, Hara's (Śiva's) eyes: a bhūtasaṃkhyā for three 4.45.
hariya S. hṛta, divided 3.33, 54.
hariyavva S. hartavya, to be divided 1.53.
hala Ts., a plough 5.14, 16.
halahara S. hala-dhara, a plough-holder: halahari dinna, given/produced by a farmer, harvest 3.8.
hā[1] S. -dhā. See tihā.
hā[2] (in tihā 'one-third' and satihā 'with one-third') S. bhāga, part 1. 52, 58, 66, 69, 3.13, 73, 5.17. See hāya.
hāya S. bhāga, part 1.50, 52, 54, 56, 58, 61, 62, 67, 68, 74, 75, 2.7, 17, 3.88, 100. See hā[2].
hāraṇīya Ts., to be taken off: the dividend 1.33, 55. See haraṇīya.
hārī S. dhārin, holder 3.92 (sutta-).
hālima, garden cress 5.9. Cf. H. hāli-

ma.

hiṭṭha S. adhas, below 3.83, 4.26, 29. See hiṭṭhi/hiṭṭhe.

hiṭṭhi/hiṭṭhe S. adhastāt, to the bottom: below, lower 1.27, 33, 3.10, 83, 4.62. See hiṭṭha.

hiṭṭhima S. adhama, the lowest 1.57, 72, 2.6, 8, 12.

hīṇa S. hīna, deprived of: decreased 1.20, 69, 2.11, 13, 3.30, 36, 49, 4.16, 20, 48, 57; subtracted 4.54, 60; subtraction 1.31, 48; negative, subtracter 1.46, 3.24; smaller 3.24, 5.11, 17. Cf. riṇa, vaya.

hīṇaṃsa S. hīna-aṃśa, defective part: a negative numerator 1.46. Cf. riṇa, vaya.

heu S. hetu, reason 1.2, 4.14.

Bibliography

I. PRIMARY SOURCES

Anuyogadvārasūtra (*Aṇuogaddāra*). [AD]
Ed. with a Hindi translation, notes, etc. by Kanhaiyalal Kamal, et al. Beawar, 1987.

Āryabhaṭa I. *Āryabhaṭīya*. [AB]
1) Ed. with Parameśvara's *Bhaṭadīpikā* by H. Kern. Leiden, 1874; reprinted, Osnabrück: Biblio Verlag, 1973.
2) Ed. with an introduction, English translation, notes, comments, and indexes by K. S. Shukla in collaboration with K. V. Sarma. Āryabhaṭa Critical Edition Series 1. New Delhi, 1976.
3) Ed. with the commentary of Bhāskara I and Someśvara by K. S. Shukla. Āryabhaṭa Critical Edition Series 2. New Delhi, 1976.
For an English translation of Bhāskara I's commentary on the mathematical chapter of the *Āryabhaṭīya* see Keller [2006].

Āryabhaṭa II. *Mahāsiddhānta*. [MS]
Ed. with his own commentary in Sanskrit by Sudhākara Dvivedī. Benares Sanskrit Series 148, 149 and 150, Benares: Braj Bhushan Das & Co., 1910; reprinted, Vrajajīvana Prācya Granthamālā 81, Delhi: Chaukhamba Sanskrit Pratishthan, 1995.

Umāvasāti. *Tattvārthādhigamasūtrabhāṣya*. [UTA]
Ed. by H. R. Kapadia, *Tattvārthādhigamasūtra: A Treatise on the Fundamental Principles of Jainism*, Part I: Chaps. I–V, with the *bhāṣya* and with the commentaries of Devagupta Sūri and of Siddhasena Gaṇi. Sheth Devachand Lalbhai Pustakoddhar Fund Series 67. Bombay: Jivachand Sakerchand Javeri, 1926.

Gaṇeśa II. *Gaṇitamañjarī*. [GM]
A preliminary edition by Takao Hayashi (to be published).

Giridharabhaṭṭa. *Caturacintāmaṇi*. [CCM]
Ed. and tr. by Hayashi [2000]

Jadivasaha. *Tiloyapaṇṇatti*. [TP]
Ed. by Cetanaprakāśa Pāṭanī, with a Hindi commentary of Viśuddhamatī Mātājī. 2 vols. Koṭā: Bhāratavarṣīya Digambara Jaina Mahāsabhā, 1984/86.

Jinapālopādhyāya et al. *Kharataragacchabṛhadgurvāvali*.
Ed. by Jina Vijaya Muni. Singhi Jain Series 42. Bombay, 1956.

Jinabhadra Gaṇi. *Bṛhatkṣetrasamāsa*. [BKS]
Cited by Datta & Singh 1980, 161–62.

Ṭhakkura Pherū. *Ratnaparīkṣādi-saptagranthasaṃgraha*. [SGS]
Ed. by Agaracanda and Bhanwaralāla Nāhaṭā, Rājasthāna Purātana Granthamālā 60. Jodhpur: Rājasthāna Prācyavidyā Pratiṣṭhāna, 1961; reprinted, 1996.

Ṭhakkura Pherū. *Gaṇitasārakaumudī* alias *Gaṇitasāra.* [GSK]
Ed. by Agaracanda and Bhaṃvaralāla Nāhaṭā, SGS, Part 2, pp. 41–74.
Ṭhakkura Pherū. *Jyotiṣasāra.* [JS]
1) Ed. by Agaracanda and Bhaṃvaralāla Nāhaṭā, SGS, Part 2, pp. 1–40.
2) Text with a Hindi commentary by Jyotirvid Śrī Sītārāma Eḍvokeṭa, *Śrī-Ṭhakkura-Pherū-kṛta Prākṛta Gāthā baddha Jyotiṣasāraḥ, 'Sūrya' Hindi ṭīkā-sahita.* Varanasi, 1978.
Ṭhakkura Pherū. *Dravyaparīkṣā.* [DP]
1) Ed. by V. S. Agrawala, 'Ṭhakkura Pherū viracitā Prākṛtabhāṣābaddhā Dravyaparīkṣā,' *Indian Numismatic Chronicle,* 4.1 (1964–65), pp. 75–94.
2) Ed. by Agaracanda and Bhaṃvaralāla Nāhaṭā, SGS, Part 1, pp. 17–38.
3) Ed. with a Hindi translation by Bhaṃvaralāla Nāhaṭā, *Ṭhakkura Pherū viracitā Dravyaparīkṣā aur Dhātūtpatti,* Vaishali: Research Institute of Prakrit, Jainology and Ahimsa, 1976.
English translation of *gāthās* 51–149: Agrawala [1966] and [1969].
Ṭhakkura Pherū. *Dhātūtpatti.*
1) Ed. by V. S. Agrawala with a Hindi translation by Bhanwar Lal Nahata and a Sanskrit *chāyā* by Narottam Das Swami, 'Dhātūtpatti,' *The Journal of the Uttar Pradesh Historical Society,* XXIV-XXV (1951–52), pp. 321–35.
2) Ed. by Agaracanda and Bhaṃvaralāla Nāhaṭā, SGS, Part 1, pp. 39–44.
3) Ed. with a Hindi translation by Bhaṃvaralāla Nāhaṭā, *Ṭhakkura Pherū viracitā Dravyaparīkṣā aur Dhātūtpatti,* Vaishali: Research Institute of Prakrit, Jainology and Ahimsa, 1976.
Ṭhakkura Pherū. *Ratnaparīkṣā.*
1) Ed. by Agaracanda and Bhaṃvaralāla Nāhaṭā, SGS, Part 1, pp. 1–16.
2) Ed. with a Hindi translation by Agarchand Nahata & Bhanwar Lal Nahata, Calcutta, 1963-64; contains also the *Ratnaparīkṣā* by Tattvakumāra Muni, and another text of the same name by Vācaka Ratnaśekhara, both composed in Old Hindi verse.
3) Ed. with an introduction, Sanskrit *chāyā,* English translation and commentary by Sreeramula Rajeswara Sarma, *Ṭhakkura Pherū's Rayaṇaparikkhā, A Medieval Prakrit Text on Gemmology,* Aligarh: Viveka Publications, 1984.
Ṭhakkura Pherū. *Vāstusāra.*
1) Ed. with a Gujarati translation by Pandit Bhagawan Das Jain, *Parama-Jaina Candrāṅgaja Ṭhakkura Pherū viracita Vāstusāra-prakaraṇa,* Jaina Vividha Granthamālā, Puṣpa 4, Jaipur: Atmananda Sabha Bhavan, 1939. With several appendices, including *Sirirayaṇaparikkhā-payaraṇa,* pp. 238–48.
2) Ed. by Agaracanda and Bhaṃvaralāla Nāhaṭā, SGS, Part 2, pp. 75–106.
3) Text with an English translation by R. P. Kulkarni, *Vastusara Prakaranam: A Prakrit Treatise of Vastushastra by Thakkura Pheru,* Pune: Sanskrit-Sanskriti-Samshodhika, 1987.
Nārāyaṇapaṇḍita. *Gaṇitakaumudī.* [GK]
Ed. by Padmākara Dvivedī. Princess of Wales Sarasvati Bhavana Texts 57, Benares, 1936/42. English translation by P. Singh [1998–2002]. New edition and English translation of Chapters 13 and 14 by Kusuba [1993]. In the concordance we use the verse numbers given in Singh [1998–2002].

Pañcaviṃśatikā. [PV]
Ed. and tr. by Hayashi [1991].
Patan Manuscript. [PM]
An anonymous manuscript: No. 8894 in the library of Shree Hemacandracarya Jain Jnan Mandir, Patan, North Gujarat. Ed. by Hayashi [1995a, 464–84].
Prākṛtapaiṅgala.
Ed. with a Hindi translation, commentary and a glossary by Bhola Shankar Vyas, 2 parts, Varanasi, 1959; reprinted in one volume, Ahmadabad: Prakrit Text Society, 2007.
Bakhshālī Manuscript. [BM]
Ed. and tr. by Hayashi [1995a].
Brahmagupta. *Brāhmasphuṭasiddhānta.* [BSS]
1) Ed. with the editor's commentary in Sanskrit by Sudhākara Dvivedī. Benares: Medical Hall Press, 1902.
2) Ed. with Sanskrit and Hindi commentaries by R. S. Sharma. New Delhi: The Indian Institute of Astronomical and Sanskrit Research, 1966. English translation of Adhyāyas 12 and 18 by Colebrooke [2005, 277–328].
Bhāskara I. *Ārybhaṭīyabhāṣya.* [BAB]
See Āryabhaṭa.
Bhāskara II. *Bījagaṇita.* [BG]
1) Ed. with Kṛṣṇa's *Navāṅkura* by Dattātreya Āpaṭe, et al. Ānandāśrama Sanskrit Series 99, Poona: Ānandāśrama Press, 1930.
2) Ed. with Bhāskara II's auto-commentary, Jīvanātha Jhā's *Subodhinī* and his own commentary, *Vimalā*, by Achyutānanda Jhā Śarman. Kashi Sanskrit Series 148, Benares: The Chowkhamba Sanskrit Office, 1949. English translation by Colebrooke [2005, 129–276].
Bhāskara II. *Līlāvatī.* [L]
1) Ed. with Gaṇeśa I's *Buddhivilāsinī* and Mahīdhara's *Līlāvatīvivaraṇa* by Dattātreya Āpaṭe, et al. Ānandāśrama Sanskrit Series 107, 2 vols. Poona, Ānandāśrama Press, 1937.
2) Ed. with Śaṅkara and Nārāyaṇa's *Kriyākramakarī* by K. V. Sarma. Vishveshvaranand Indological Series 66, Hoshiarpur: Vishveshvaranand Vedic Reasearch Institute, 1975. English translation by Colebrooke [2005, 1–127].
Bhuvanadeva. *Aparājitapṛcchā* [AP]
Ed. by P. A. Mankad. Gaekwad's Oriental Series 115. Baroda: Oriental Institute, 1950.
Mahāvīra. *Gaṇitasārasaṅgraha.* [GSS]
1) Ed. with English translation and notes by M. Raṅgācārya. Madras, 1912.
2) Ed. with a Hindi translation by L. C. Jain. Jīvarāja Jaina Granthamālā 12. Sholapur, 1963.
3) Ed. with English transliteration, Kannada translation and notes by Padmavathamma, English translation and notes by M. Rangacharya. Sri Siddhāntakīrthi Granthamālā, Hombuja, Shimoga District, Karnataka: Sri Hombuja Jain Math, 2000.
Muḥarrir. Ḥājjī ᶜAbd al-Ḥamīd Muḥarrir Ghaznāvī, *Dastūr al-Albāb fī ᶜIlm al-Ḥisāb.*

Manuscript: Riyazī Fārsī No. 32. Rampur Raza Library, Rampur.
Śrīdhara. *Gaṇitapañcaviṃśī.* [GP]
Ed. by David Pingree, 'The Gaṇitapañcaviṃśī of Śrīdhara,' *Ludwik Sternbach Felicitation Volume*, 2 parts, Lucknow, 1979, pp. 887–909. For Śrīdhara's authorship see Hayashi [1995b].
Śrīdhara. *Pāṭīgaṇita.* [PG]
Ed. with an anonymous Sanskrit commentary, an English translation, and notes by K. S. Shukla. Lucknow: Lucknow University, 1959.
Śrīdhara. *Triśatikā.* [Tr]
Ed. by Sudhākara Dvivedī. Benares: Chandraprabha Press, 1899.
Śrīpati. *Gaṇitatilaka.* [GT]
Ed. with Siṃhatilaka Sūri's *vṛtti* by H. R. Kāpadīā. Gaekward Oriental Series 78, Baroda: Oriental Institute, 1937.
For reference we use the verse numbers proposed in Hayashi [2005].
Śrīpati. *Siddhāntaśekhara.* [SS]
Ed. with Makkibhaṭṭa's *Gaṇitabhūṣana* and the editor's *vivaraṇa* by Babuāji Miśra. 2 parts. Calcutta: University of Calcutta, 1932/47.
For English translations of the two mathematical chapters see Sinha [1986] and [1988].
Sīrat-i Fīrūzshāhī.
Photo-offset edition of the unique MS, Khuda Bakhsh Library, Patna, 1999.
Someśvara III. *Mānasollāsa.* [MU]
Ed. by G. K. Shrigondekar. Gaekwad's Oriental Series 28, 84 and 138. Baroda: Oriental Institute, 1925/39/61.
Sthānāṅgasūtra (*Ṭhāṇaṃga*). [SA]
Ed. by Sāgarānanda Sūri and Jambūvijayaji, *Sthānāṅga Sūtram and Samavāyāṅga Sūtram*, with the commentary of Abhaydeva Sūri. Lala Sundarlal Jain Āgamagranthamālā 2, Part 1 (pp. 1–411). Delhi, 1985.
Sphujidhvaja. *Yavanajātaka.* [YJ]
Ed. with English translation and commentary by David Pingree. 2 vols. Harvard Oriental Series 48. Cambridge: Harvard University Press, 1978.

II. SECONDARY SOURCES

Abū al-Faḍl.
[2004] *Ā'īn-i Akbarī.* Vol. I, translated from the Original Persian by H. Blochmann, second ed. revised and edited by Lieut. Colonel D. C. Phillot, Calcutta, 1927; Vol. II, translated into English by Colonel H. S. Jarrett, second ed. corrected and further annotated by Jadu Nath Sarkar, Calcutta, 1949; Vol. III, translated into English by Colonel H. S. Jarrett, [second ed.] corrected and further annotated by Jadu Nath Sarkar, Calcutta, 1948. Reprinted in one volume with the title, *The Ā'īn-i Akbarī by Abu 'l-Faẓl 'Allāmī*, Lahore: Sang-e-Meel Publications, 2004.
Agrawala, V. S.
[1943] 'A Note on Medieval Temple Architecture,' *The Journal of the United Provinces Historical Society* 14 (1), pp. 112–17.
[1954] 'The Highest Purity of Gold in India (Solahbānī and Bārahbānī Systems),' *Journal of the Numismatic Society of India* 16 (2), pp. 270–74.

[1966] 'A Unique Treatise on Medieval Indian Coins' (English translation of the *Dravyaparīkṣā*, *gāthā* 51–149), *Ghulam Yazdani Commemoration Volume,* Hyderabad, 1966, pp. 81–101.
[1969] 'Dravyaparīkṣā of Ṭhakkura Pherū' (English translation of the *Dravyaparīkṣā*, *gāthā* 51–149), *Indian Numismatic Chronicle,* vol. VII (1969), pp. 100–14.
Āpaṭe, Varman Shivaram.
[Apte] *The Practical Sanskrit-English Dictionary.* Poona, 1890; revised and enlarged by P. K. Gode, C. G. Karve, et al. Poona, 1957; reprinted, Kyoto, 1978.
Colebrooke, Henry Thomas.
[2005] *Classics of Indian Mathematics: Algebra, with Arithmetic and Mensuration, from the Sanskrit of Brahmagupta and Bhāskara*, with a Foreword by S. R. Sarma. Delhi: Sharada Publishing House, 2005. Originally published with the title, *Algebra with Arithmetic and Mensuration from the Sanscrit of Brahmegupta and Bhácara.* London, 1817; reprinted, Wiesbaden, 1973.
Crooke, William.
[1968] *The Popular Religion and Folklore of Northern India.* New edition, revised and illustrated. London: Archibald Constable, 1896; reprinted, Delhi: Munshiram Manoharlal, 1968.
Datta, Bibhutibhusan and Singh, Avadesh Narayan.
[1980] 'Hindu Geometry,' *Indian Journal of History of Science* 15 (2), pp. 121–88.
[2001] *History of Hindu Mathematics, A Source Book.* Lahore, 1935/38; reprinted, Delhi, 2001.
Deyell, John S.
[1990] *Living Without Silver: The Monetary History of Early Medieval India.* New Delhi: Oxford University Press.
Dube, R. K.
[2006a] 'Copper Production Process as described in an early fourteenth Century Prakrit Text composed by Ṭhakkura Pherū,' *Indian Journal of History of Science* 41, pp. 297–312.
[2006b] 'The Extraction of Lead from its Ores by the Iron-Reduction Process: A Historical Perspective,' *The Journal of the Minerals, Metals & Materials Society* 58 (10), pp. 18–23.
Dundes, Alan and Ved Prakash Vatuk.
[1974] 'Some Characteristic Metres of Hindi Riddle Poetry,' *Asian Folklore Studies* 33 (2), pp. 85–153.
Führer, A.
[1984] 'Prabhosâ Inscriptions,' *Epigraphia Indica* 2, pp. 240–44.
Gosvāmī, Śāradā.
[2000] *Itihāsa Puruṣa Śrī Agaracanda Nāhaṭā: Jīvana aura Sāhitya.* Calcutta.
Gupta, Parmeshwari Lal.
[1957] 'The Coinage of the Khalji Sultans of Delhi,' *Journal of the Numismatic Society of India* 19, pp. 35–46.
Gupta, Radha Charan.
[1979] 'Jaina Formulas for the Arc of a Circular Segment,' *Jain Journal* 13

(3), pp. 89–94; reprinted, *Ancient Jain Mathematics*, ed. by Bhuvanendra Kumar, Jain Humanities Press (Mississauga, Canada), 2004, pp. 63–69.
[1991] 'On the Volume of a Sphere in Ancient India,' *Historia Scientiarum* 42, pp. 33–44.
[2001a] 'A new Formula from Babylonian Mathematics,' *HPM Newsletter* 46 (March 2001), pp. 2–3.
[2001b] 'World's Longest Lists of Decuple Terms,' *Gaṇita Bhāratī* 23, pp. 83–90.
[2002] 'Mensuration of a Circle Segment in Babylonian Mathematics,' *Gaṇita Bhāratī* 24, pp. 12–17.

Habib, Irfan.
[1982] 'Agrarian Economy,' in: Raychaudhury and Habib [1982], Part 1, Chap. III.
[2004] 'Textile Terms in Medieval Indian Persian Texts: A Glossary,' *Proceedings of the Indian History Congress: 64th Session,* Mysore 2003, Indian History Congress, Patna, pp. 525–47.
[2008] 'Indian Textile Technology: Thirteenth-Fifteenth Centuries,' paper presented at the Indian History Congress, 68th Session, Delhi, 28–30 December 2007 (to be published).

Hassan, Zafar.
[1915-20] *Delhi Province: List of Muhammadan and Hindu Monuments.* 3 vols. Calcutta.

Hayashi, Takao.
[1987] 'Varāhamihira's Pan-diagonal Magic Square of the Order Four,' *Historia Mathematica* 14, pp. 159–66.
[1991] 'The *Pañcaviṃśatikā* in Its Two Recensions: A Study in the Reformation of a Medieval Sanskrit Mathematical Textbook,' *Indian Journal of History of Science* 26, pp. 399–448.
[1992] 'Mahāvīra's Formulas for a Conch-like Plane Figure,' *Gaṇita Bhāratī* 14, pp. 1–10.
[1995a] *The Bakhshālī Manuscript: An ancient Indian mathematical treatise.* Groningen Oriental Studies 11, Groningen: Egbert Forsten.
[1995b] 'Śrīdhara's Authorship of the Mathematical Treatise *Gaṇitapañcaviṃśī*,' *Historia Scientiarum* 4 (3), pp. 233–50.
[1997] 'Calculations of the Surface of a Sphere in India,' *The Science and Engineering Review of Doshisha University* 37 (4), pp. 194–238.
[2000] 'The *Caturacintāmaṇi* of Giridharabhaṭṭa: A Sixteenth-Century Sanskrit Mathematical Treatise,' *SCIAMVS* 1, pp. 133–208.
[2005] 'A Note on Authenticity of Verses in Śrīpati's *Gaṇitatilaka*,' *The Science and Engineering Review of Doshisha University* 46 (2), pp. 39–46.
[2006] 'A Sanskrit Mathematical Anthology,' *SCIAMVS* 7, pp. 175–211.
[2008] 'Magic Squares in Indian Mathematics,' *Encyclopaedia of the History of Science, Technology, and Medicine in Non-Western Cultures*, ed. by Helaine Selin, 2nd edition, Berlin etc.: Springer, pp. 1252–59.

Hayashi, T., Kusuba, K., and Yano, M.
[1997] *Indo Suugaku Kenkyuu* (*Studies in Indian Mathematics: Series, Pi and Trigonometry*, in Japanese). Tokyo: Kouseisha Kouseikaku.

Keller, Agathe.
[2006] *Expounding the Mathematical Seed: A Translation of Bhāskara I on*

the Mathematical chapter of the Āryabhaṭīya. 2 vols. Basel: Birkhäuser.

Kumar, Naresh.

[AHK] *Apabhraṃśa-Hindī Kośa.* New Delhi: D. K. Print World, 1987; revised, 1999.

Kusuba, Takanori.

[1993] *Combinatorics and Magic Squares in India, A Study of Nārāyaṇa Paṇḍita's Gaṇitakaumudī, Chapters 13–14.* Dissertation, Brown University.

McGregor, R. S.

[OHED] *The Oxford Hindi-English Dictionary.* Oxford/Delhi: Oxford University Press, 1993.

McHugh, James Andrew.

[2008] *Sandalwood and Carrion: Smell in South Asian Culture and Religion.* Dissertation, Harvard University.

Mehta, B. N. and Mehta, B. B.

[MGED] *The Modern Gujarati-English Dictionary.* Baroda: M. C. Kothari, 1925.

Michaels, Axel.

[1978] *Beweisverfahren in der Vedischen Sakralgeometrie.* Alt- und Neu-Indische Studien, Universität Hamburg, No. 20. Wiesbaden: Franz Steiner, 1978.

Monier-Williams, Monier.

[MMW] *A Sanskrit-English Dictionary.* Oxford, 1899; reprinted, Delhi, 1963.

Motīcandra.

[1962] *Kāśī kā Itihāsa.* Bambaī: Hindi Grantha Ratnakara.

Nahata, Agar Chand and Nahata, Bhanwar Lal.

[1976] *Nāhaṭā-Bandhu Abhinandana Grantha.* Varanasi.

[1986] *Śrī Bhaṃwaralālā Nāhaṭā Abhinandana Grantha.* Calcutta.

Pillai, L.D. Swamikannu.

[1982] *An Indian Ephemeris: A.D. 700 to A.D. 1799, showing the daily solar and lunar reckoning according to the principal systems current in India, with their English equivalents, also the ending moments of tithis and nakshatras and the years in different eras, A.D., Hijra, Saka, Vikrama, Kaliyuga, Kollam etc., with a perpetual planetary almanac and other auxiliary tables.* Madras: Government Press, 1922; reprinted, Delhi: Agam, 1982.

Pingree, David.

[CESS] *Census of the Exact Sciences in Sanskrit, Series A.* 5 vols. Memoir of the American Philosophical Society 81, 86, 111, 146, and 213, Philadelphia: American Philosophical Society, 1970, 1971, 1976, 1981, and 1994.

Pischel, R.

[Pischel] *A Grammar of the Prākrit Languages,* translated from German by Subhadra Jhā, 2nd revised ed. Delhi: Motilal, 1981.

Pope, Arthur Upham (ed.), with Ackerman, Phyllis (assist. ed.).

[1977] *A Survey of Persian Art: From Prehistoric Times to the Present.* 7 vols. London: Oxford University Press, 1938–58; 3rd ed., 16 vols., Tehran: Soroush Press, 1977.

Prinsep, James.

[1995] *Essays on Indian Antiquities, Historic, Numismatic and Palaeographic, of the Late James Prinsep, F.R.S., Secretary to the Asiatic Society of Bengal: To Which Are Added His Useful Tables, Illustrative of Indian History, Chronology, Modern Coinages, Weights, Measures, etc.*, ed. with notes and additional matter by E. Thomas. 2 vols. London, 1858; reprinted, 1971, 1995.

Ray, Joges Chandra.

[1918] 'Sugar Industry in Ancient India,' *The Journal of the Bihar and Orissa Research Society*, 4 (1918), pp. 435–54.

Raychaudhury, Tapan and Habib, Irfan (eds.).

[1982] *The Cambridge Economic History of India*, Vol. I: *c. 1200–c. 1700.* Cambridge: Cambridge University Press.

Sarma, Sreeramula Rajeswara.

[1983] '*Varṇamālikā* System of Determining the Fineness of Gold in Ancient and Medieval India,' *Aruṇa-Bhāratī: Professor A. N. Jani Felicitation Volume*, Baroda, pp. 369–89.

[1984] *Ṭhakkura Pherū's Rayaṇaparikkhā: A Medieval Prakrit Text on Gemmology.* Aligarh.

[1986] 'Thakkura Pheru and the Popularisation of Science in India in the Fourteenth Century,' in: *Śrī Bhaṃwaralālā Nāhaṭā Abhinandana Grantha*, Calcutta, Part 4, pp. 63–72; reprinted in: *Jain Journal* 21.3 (January 1987) 86-95; and again in: *Jain Anthology*, Calcutta, 1991, pp. 146–56.

[1990] 'Islamic Calendar and Indian Eras,' *History of Science and Technology in India*, ed. by K. Kuppuram and K. Kumudamani, Delhi: Sundeep Prakashan, vol. 2, pp. 433–41.

[1991] 'Thakkura Pheru: The Versatile Jaina Scientist,' *Proceedings of the International Seminar on Jaina Mathematics and Cosmology*, Hastinapur, pp. 179–86.

[1996] 'Sanskrit Manuals for Learning Persian,' *Adab Shenasi*, ed. by Azarmi Dukht Safavi, Aligarh, pp. 1–12.

[2002a] 'Rule of Three and Its Variations in India,' *From China to Paris: 2000 Years Transmission of Mathematical Ideas*, ed. by Y. Dold-Samplonius, J.W. Dauben, M. Folkerts, B. van Dalen, Stuttgart: Franz Steiner, pp. 133–56.

[2002b] 'From Yāvanī to Saṃskṛtam: Sanskrit Writings Inspired by Persian Works,' *Studies in the History of Indian Thought*, Kyoto, 14, pp. 71–88.

[2006] 'Mathematics and Iconography in *Līlāvatī* 263,' *Journal of the Asiatic Society of Mumbai* 80, pp. 115–26.

Schram, Robert.

[1908] *Kalendariographische und Chronologische Tafeln.* Leipzig: J. C. Hinrichs.

Sewell, Robert and Dikshit, Sankara Balkrishna.

[1896] *The Indian Calendar: with tables for the conversion of Hindu and Muhammadan into A. D. dates, and vice versa*, London: S. Sonnenschein.

Sharma, Dasharatha.

[1959] *Early Chauhān Dynasties: a Study of Chauhān political history, Chauhān political institutions, and life in the Chauhān dominions, from c. 800 to 1300 A. D.; with a foreword by K. M. Panikkar.* Delhi: S. Chand, 1959; 2nd revised edition, Delhi: Motilal, 1975.

Sheth, Haragovind Das T.

[PSM] *Pāiya Sadda Mahaṇṇavo: A Comprehensive Prakrit-Hindi Dictionary.* 4 vols. Calcutta, 1923–28; 2nd edition in one volume, Varanasi: Prakrit Text Society, 1963; reprinted, Delhi: Motilal, 1986.

Singh, Paramanand.

[1998–2002] 'The Gaṇita Kaumudī of Nārāyaṇa Paṇḍita (Translation with Notes),' *Gaṇita Bhāratī* 20, 1998, pp. 25–82 (Chaps. I–III); 21, 1999, pp. 10–73 (Chap. IV); 22, 2000, pp. 19–85 (Chaps. V–XII); 23, 2001, pp. 18–82 (Chap. XIII); and 24, 2002, pp. 35–98 (Chap. XIV).

Sinha, Kripa Nath.

[1982] 'Śrīpati's *Gaṇitatilaka*: English Translation with Introduction,' *Gaṇita Bhāratī* 4, pp. 112–33.

[1986] 'Algebra of Śrīpati: An Eleventh Century Indian Mathematician (Translation of avyaktagaṇitādhyāya of *Siddhāntaśekhara*),' *Gaṇita Bhāratī* 8, pp. 27–34.

[1988] 'Vyaktagaṇitādhyāya of *Siddhāntaśekhara* (English Translation with Introduction),' *Gaṇita Bhāratī* 10, pp. 40–50.

Srinivasan, Saradha.

[1979] *Mensuration in Ancient India.* Delhi: Ajanta Publications.

Steingass, Francis Joseph.

[1996] *A Comprehensive Persian-English Dictionary, including the Arabic Words and Phrases to be met with in Persian Literature.* London, 1982; reprinted, Delhi, 1996.

Strauch, Ingo.

[2002] *Die Lekhapaddhati-Lekhapañcaśikā: Briefe und Urkunden im mittelalterlichen Gujarat*, Text, Übersetzung, Kommentar, Glossar (Sanskrit-Deutsch-Englisch). Monographien zur Indischen Archäologie, Kunst und Philologie, Band 16. Berlin: Dietrich Reimer.

Subbarayappa, B. V. and Sarma, K. V.

[1985] *Indian Astronomy: A Source-Book, based primarily on Sanskrit Texts.* Bombay: Nehru Centre.

Tagare, G. V.

[HGA] *Historical Grammar of Apabhraṃśa.* Deccan College Dissertation Series 5. Poona: Deccan College, 1948; reprinted, Delhi: Motilal, 1987.

Velankar, H. D.

[1946-47] 'Prākṛta and Apabhraṃśa Metres (Classified Lists and Alphabetical Index),' *Journal of the Bombay Branch of the Asiatic Society* 22–23, pp. 5–32.

Wright, H. Nelson,

[1974] *The Coinage and Metrology of the Sultans of Delhi.* New Delhi.

Yule, Henry and Burnell, A. C.

[Hobson-Jobson] *Hobson-Jobson: A Glossary of Colloquial Anglo-Indian Words and Phrases, and of Kindred Terms, Etymological, Historical, Geographical and Discursive.* London: Murray, 1886; new edition by William Crooke, London: Murray, 1903; reprinted many times.

Index of Mathematical Terms

(2n + 1)-quantity operation, 117, 118

add, xxxii, 47–51, 58, 59, 65, 67, 70, 73, 77, 78, 80–82, 89, 104, 105, 109, 113, 130, 166, 167, 173, 175, 179, 180, 236, 247, 252, 254
addition, xxiv–xxvi, xxxii, xxxviii, xl–xliii, xlvi, 47, 48, 50, 57, 58, 92, 129, 130, 180, 236, 245
additive, xxxii, xliii, 48, 64, 109, 179, 226, 231, 240
algorithm, xxi, xxii, xxvi, xxviii, xxxii, 45, 97, 106–10, 115, 118, 122, 144, 208, 210, 212, 243
altitude, 67, 248
angle, xxxv, 231
annulus, xxxvi, 146
arc, xxii, xxviii, xxxiv, xxxv, 67, 146, 147, 240, 244
area, xxi, xxii, xxvi–xxx, xxxiv, xxxviii, xl, 46, 54, 65–69, 78, 82, 85, 87, 97, 100, 141–47, 149, 150, 152, 154–56, 167, 184, 190, 191, 231, 244, 246, 247, 250
area measure, xxxvii, xxxix, 45, 98, 250, 251
area unit, 97, 98
arithmetical progression, xxi, xxxiv, xxxviii, xliii, 140, 141, 166, 169, 228, 251, 252
arrow (height of a circle segment), xxii, xxviii, xxxv, 66, 67, 147, 253
arrows (5), 79, 80, 253

barleycorn (figure), xxxvi, 67, 146, 152, 235
barleycorn (linear measure), 46
base (side, line, etc.), xxvi, xxvii, xxxv, 66, 67, 78, 142, 143, 146, 148, 226, 240, 246
beautiful triangle, 238
bottom (plane, side, etc.), xxxv, 62, 67, 70, 77, 78, 136, 149, 150, 155, 237, 246, 247
bow (arc), xxviii, 67, 240
bow (circle segment), 66, 67, 146, 240
bow field, 145
bow-like figure, xxxv, 67, 93, 233
breadth, 45, 46, 54, 65, 67–71, 77, 85, 98, 145, 150, 154, 156, 250

calculation, xxii, xxviii, xxix, xliv, 65, 67, 73, 88, 93, 105, 109, 116, 129, 144, 145, 151, 152, 156, 159, 167, 185, 190, 191, 227, 232
capacity, xxxiv, xxxviii, 58, 67, 68, 93, 100, 148, 149, 154, 155, 231, 244
cart-wheel-circle (figure), xxxvi, 146
chord, xxii, xxviii, xxxv, 66, 67, 145–47, 235
circle, xviii, xxi, xxii, xxvii, xxviii, xxx, xxxv, xxxviii, xl, 66–68, 70, 73, 78, 87, 93, 143–47, 151, 152, 154, 190, 191, 238–40, 243, 249–51, 253
circular (figure), xvii, xviii, xxviii, xxxviii, 66, 68–72, 78, 93, 153, 155, 156, 246, 249, 252, 253
circumference, xviii, xxvii, xxxv, xl, 66, 68, 69, 71, 72, 78, 87, 88, 143–45, 151, 154, 156, 158, 159, 190, 243
class of fractions, xix–xxii, xxiv–xxvi, xxxii, xxxvii, xxxviii, xli–xlvi, 47, 57–59, 92, 112, 125, 126, 129–131, 208, 235, 242, 244, 245, 255
common difference, xxxiv, xliii, 47, 64, 65, 93, 140, 171, 203, 228, 234, 251
common ratio, xliii
component square, 80, 170, 171
composite fraction, xxiv, xxv, xxxii, 112, 230, 253
composite number, xxiv

computation, xvii, xxxii, 45, 47, 66, 67, 69, 71, 81, 91–94, 97, 118, 131, 134, 136, 145, 175, 180, 231
computational rule, 133, 134, 137
corner, xxxv, 72, 159, 230, 231, 237
crescent moon (figure), 67
crescent-like figure, xxxv
cube (figure), xxxvi, 153
cube (number), xxxiii, xxxiv, xxxvii, xl–xliii, 49, 50, 52, 65, 67, 68, 88, 91–93, 109, 115, 141, 233
cube (operation), 49, 108
cube place, 50, 52, 110
cube root, xxxiii, xxxvii, xl–xliii, 50, 52, 91, 92, 110, 115, 233, 247
cubic (dimension), 46, 100, 233
cubic (measure), 230, 255
cubic (solid), xxxiv, 148
cubic series, xxxiv, xxxviii, 65, 93, 141, 233

decrease, xxvii, xxxii, xxxiii, xliii, 47, 48, 73, 82, 86, 88, 105, 131, 161, 180, 228, 235, 248, 251, 256
denominator, xxiv, xxxii, 50–52, 57–59, 62, 64, 111–13, 118, 122, 125–28, 130, 135, 225, 235, 255
depth, xxxiv, 67, 68, 149, 227, 251
diagonal, xxxv, 79, 168, 169, 229
diagram, 168, 170, 235
diameter, xxvii, xxxiv, xxxv, 66, 68, 69, 87, 88, 143–45, 149, 151, 154, 190, 191, 228, 250
difference, xxi, xxiii–xxv, xxxiii, xxxiv, xxxvii, xli–xliii, 47–49, 51, 54, 55, 66, 70, 76, 87, 91, 104, 105, 112, 113, 122, 140, 166, 167, 190, 208, 217, 221, 225, 251, 252
digit, xxii, xxv, xxx–xxxii, xxxix, 46–50, 53, 55, 63–65, 75, 76, 83, 106, 111, 118, 166, 176, 180, 181, 225, 238, 239
directions (10), 79, 80, 239
disk, 93, 245, 246
divide, xxxiii, xxxv, xliii, 47, 49–55, 57–59, 61–68, 71, 72, 75–77, 81, 82, 104, 108, 110, 113, 116, 118, 120, 122–24, 126–28, 145, 159–61, 164, 166, 168, 169, 174, 175, 178, 179, 181, 212, 245, 246, 248, 251, 255
dividend, xxxiii, 49, 51, 252, 255
division, xxi, xxiii, xxix, xxxiii, xxxvii, xxxix–xliv, 49, 51, 58, 73, 76, 91, 106, 107, 114, 115, 164, 178, 181, 245
divisor, xxvi, xxxiii, 49–51, 82, 110, 111, 120, 150, 179, 235, 245, 255
door-junction (multiplication method), 48, 230
double-lined, 73, 239
drum (figure), xxxvi, 67, 93, 146, 247

elephant's tusk (figure), xxxvi, 67, 93, 146, 231, 232, 248
eleven-quantity operation, xx, xxiv, xxxvii, 117, 229
elongated circle, xxviii, 146, 147, 152
equatorial line, 152
equi-bilateral (quadrilateral, trilateral), xxvii
equi-perpendicular (quadrilateral), xxvii, 143
equi-perpendicular figure, 93, 142
equi-trilateral (quadrilateral), xxvii
equilateral (quadrilateral, triangle, trilateral), xxvii, xxx, 191
even number, xxx, xxxii, 47, 79, 103, 169, 253
even order, xxxix, 168, 169
even place, 49, 52, 108
evenly-odd number, xxx, xxxii, 79, 169, 253
evenly-odd order, xxxix, 168, 169
excavation, xxxvi, xlii, 93, 148, 231
eyes (2), 80, 248

face (side of a figure), xxvii, 66, 67, 247
figure, xix, xxix, xxxi, xxxii, xxxiv–xxxvi, xxxviii, 37, 65–67, 87, 88, 93, 143, 146, 148, 152, 156, 168, 174, 190, 191, 225, 230, 232, 233, 235, 238, 240, 241, 247, 248
fires (3), 79, 225
first term of a series, xxxiv, xliii, 47, 64, 65, 81, 93, 140, 141, 171, 175, 226, 227, 240
five-quantity operation, xxxvii, xlii, 117, 241
fraction, xxiv, xxv, xxxii, xxxvii, xl–xliii, xlv, xlvi, 7, 50–52, 57–59, 63, 64, 81, 91, 92, 111–15, 118, 125, 127, 135, 211, 225, 230, 243, 245, 246, 251, 252

fruit, 48, 55, 123, 124, 126, 127, 133, 148, 244
fruit (digit), 53, 55
fruit (field-), 65, 145
fruit (length), 67
fruit (number), 117
full moon (figure), xxxvi, 67, 146
fundamental operation, xix–xxi, xxiii, xxiv, xxxvii, xl, xlii, xliv, 45, 47, 50, 55, 91, 92, 97, 103, 104, 106–08, 110, 111, 113–15, 117, 120, 121, 123, 124, 243

geometric (computation), 66, 231
geometric (figure), 65, 191, 225, 240
geometrical progression, xxi, xliii
geometry, xvii, xxxiv
gods (33), 79, 254

hands (2), 79, 80, 82, 230
Hara's eyes (3), 80, 255
height, xxviii, xxxiv, xxxv, xxxix, 46, 62, 67–73, 77, 78, 81, 88, 94, 98, 100, 132, 136, 145, 146, 148, 149, 151, 153–56, 158–60, 191, 227, 228, 253
hemisphere, 152
hexagonal, 71, 78, 231
homogenization (of fraction), xxiv, xxvi, xxxii, xliii, 57, 92, 112, 230, 253
horses (7), 85, 250
hypotenuse, xxxv, xxxviii, 66, 143, 229

improper fraction, 112, 113
increase, xxxii, xliii, xliv, 47, 48, 50, 52, 53, 58, 59, 63–69 , 73, 76–78, 80–82, 86–88, 103, 105, 144, 161, 166, 173–77, 179, 180, 225, 226, 228, 235, 236, 249, 251, 252, 254
inequi-perpendicular (quadrilateral), xxvii, 142
inequilateral (quadrilateral), xxvii
integer, xxi, xxiii, xxiv, xxxii, xxxvii, xl, xlii, xliii, xlv, 47, 50, 91, 103, 104, 106, 111, 112, 115, 181, 225, 226, 248, 253
integer part, 104, 105, 144, 166, 167
integer root, 104, 105
integer solution, 132
inverse five-quantity operation, 121
inverse operation, xxx, xli, xlii, xlv, 82, 179
inverse rule of three, xxxiii, 54, 92, 250, 252
inverse three-quantity operation, xxxvii, xlii, xliv, 100, 101, 120, 250, 252
investment, xxxiii, 57, 62, 75, 92, 126, 135, 163, 225, 242, 244
investment procedure, xxii
irregular (breadth), 67
irregular (figure), xxxv, xxxviii, 146
irregular (magic square), xxix
irregular (side), 148
Īśvara (11), 79, 227

Jinas (24), 68, 79, 235

kings (16), 79, 241

last term of a series, xxxiv, 64, 225
lateral side (of a plane figure), 66
length, xxii, xxvii–xxix, xxxiv, xxxviii, 45, 46, 53, 65, 67–73, 77–79, 82, 85, 94, 97, 98, 100, 116, 119, 120, 141, 145, 146, 148–50, 152–62, 178, 184, 227, 239, 243, 244
less by, xxxiii, 51–53, 58, 62, 64, 72, 78, 79, 86, 87, 89, 239, 241
line, xxix, xxxv, 71–73, 100, 136, 157, 158, 248
linear (dimension), xxii, xxiii, xxxiv, 46, 100, 253
linear (measure), xxii, xxxvii, 46, 77, 98, 99, 121, 132, 184, 225, 229–31, 235, 236, 243, 251, 255
linear equation, xxvi, 208
linear expression, 105
longitudinal line, 152
lunar days (15), 79, 80, 237
lunar mansions (28), 79, 248

magic square, xvii, xx, xxix, xxx, xxxix, 79, 80, 168–72, 230, 232, 233, 235
Manus (14), 79, 80, 246
mathematics, vii, xi, xii, xvi–xix, xxi, xxix, xxxvii, xlii, xliv, 47, 55, 59, 73, 83, 89, 91, 94, 103, 183, 226, 232, 259
mean, xxxviii, 67, 148, 253
mixed amount, 61, 62
mixed fraction, xxiv, xxvi, 112, 132
mixed number, xxiv, 51, 111–13
mixed quantity, 227

moon (1), 79, 80, 254
moon (figure), xxxv, 67, 146
moon's creascent shape, 67
moon's digits (16), 80, 230
multiplicand, xxxiii, 48, 106, 114, 233
multiplication, xxi–xxiv, xxxiii, xxxvii, xl–xlv, 48, 49, 51, 91, 106, 107, 114, 123, 125, 127, 144, 177, 181, 232, 250
multiplication method, xxxiii, 106, 230
multiplier, xxxiii, 48, 72, 81, 82, 106, 114, 120, 159, 175–77 , 179, 228, 232
multiplier operation, xlv
multiply, xxii, xxvii, xxxiii, xliii, 46–55, 57, 58, 61–72, 75–79, 81–83, 85, 87, 88, 103, 104, 116, 120, 122, 126–28, 144, 145, 154, 155, 158, 160, 161, 165, 175–82, 232, 233, 237, 252, 255

nails (20), 79, 240
Nandas (9), 77, 240
natural number, xxi, xxiii, xxxiv, 80, 227, 233, 249
natural series, xxiii, xxiv, xxxiv, xxxviii, xlii, 47, 48, 65, 103, 105, 140, 141, 221, 251, 252
negative fraction, 111
negative number, xxxi, 51, 113, 248, 256
negative numerator, xxxii, 50, 256
negative part, 208
negative quantity, xxxi, 249
negative sign, 112, 214, 245
negative term, 111, 153
nine-quantity operation, xxxvii, xlii, 117, 240
notational place, xxiv, xxxi, xxxii, 46, 48, 106–110, 236, 243
number, xvii, xx, xxiii, xxiv, xxix, xxxi, xxxvi, xxxviii, xxxix, xliii, xliv, 30, 31, 46–51, 53, 55, 67, 68, 71, 72, 75–77, 79–83, 87, 88, 91, 94, 100, 102, 103, 108, 110, 111, 116–21, 124, 132, 139, 148, 149, 153, 154, 157, 164, 165, 168–82, 188, 215, 216, 225, 233, 238, 242, 243, 247, 248, 252
number (think of), xxx
number of terms, xxxiv, xliii, 47, 48, 64, 65, 93, 103–05, 141, 171, 231, 243, 247
numerator, xxxii, 50–52, 57–59, 111–13, 122, 127, 128, 131, 225, 245
numerical figure, xxxii

oblong, xxvii, xxxv, xxxviii, 141–43, 146, 152, 239
octagonal, 71, 78, 225
odd number, xvi, xxx, xxxii, 47, 79, 80, 103, 169, 251
odd order, xxxix, 80, 168, 169, 171
odd place, 49, 52, 108
optional (number, quantity), xxiii, 139
optional quantity operation, xlv, xlvi, 208, 209

pan-diagonal, 168, 262
partnership, xxii
pentagonal, xix, 93, 241
percentage, xxix, xxxix, 75, 163, 164, 187, 252
perpendicular, xxvi, xxvii, xxxiv, xxxv, xxxviii, 66, 142, 143, 146, 148, 248, 254
pit, xxxv, xxxvi, xxxviii, 67, 148, 149, 227, 231, 237, 247
place-value notation, xxx, 225, 236
place-value system, xxxii, 243
plane, xxxv
plane (dimension), xxii, xxiii, xxxiv, 242
plane figure, xxxv, 226, 231, 233, 244, 246, 247, 249
planets (9), 80, 232
positive integer, xxii, 144, 164
positive integer (think of), 179, 180
positive number, xxxi, 51, 113, 240
positive part, 208
positive quantity, xxxi, 227
procedural mathematics, xxxii, 55, 59, 61, 73, 89, 243
procedure, xviii, xx, xlii, xliii, 106, 166, 167, 176, 179, 230, 243, 245, 249, 251
procedure for barter, 55, 245
procedure for bricks, 69
procedure for excavations, xxviii, xxxviii, 67, 69, 93, 148
procedure for gold, xxii, xxxviii, 62, 92, 137
procedure for mixture, xxxviii, xlii, xliv, xlvi, 61, 64, 92, 133, 202
procedure for mounds of grain, xxxviii, 72, 94, 158
procedure for piling, xxviii, xxix, xxxviii, 69, 71, 93, 153

procedure for plane figures, xix, xxxviii, 65, 67, 93, 141
procedure for sawing, xxxviii, 71, 72, 94, 157
procedure for series, xxxviii, 64, 65, 93, 140
procedure for shadows, xxix, xxxix, 72, 73, 94, 159
procedures (8 types of), xix–xxi, xxxviii, xlii, xliv, 47, 61, 73, 92, 94, 133, 251, 252
procedures (9 types of), xlii
procedures (fundamental), 91
product, xxii, xxvii, xxxiii, xli, xliii, 48, 51, 53–55, 57, 61, 63, 65, 67, 70, 76, 88, 104, 116–18, 120, 122–24 , 126, 160, 175, 180, 181, 213, 232, 238, 244, 250
proper fraction, 113
properties (3), 80
proportion, xviii, xxii, xxxiii, xxxvii, xli, xliv, xlv, 52, 62, 115, 136, 253
proportional, xxix, 116, 120, 124, 136, 157
proportionate distribution, xxii, xxv, xxix, xxxiii, xxxviii, xxxix, 62, 64, 92, 126, 133, 135, 136, 139, 163, 244
Pythagorean Theorem, xxxviii

quadratic equation, xxvi, 208
quadrilateral, xxvi, xxvii, xxxv, xxxviii, 66, 67, 142, 143, 146, 233
quantity, xxiii, xxxi, xxxix, xl, xlv, xlvi, 47, 50, 52–55, 59, 62, 67, 73, 76, 82, 98, 105, 113, 116–18, 120–23 , 127, 130, 136, 137, 157, 160, 166, 175, 187, 188, 192, 193, 229, 232, 233, 237, 238, 240, 241, 244, 248, 253
quarter square, 80, 168, 170
quotient, xxxiii, 49–51, 59, 64–66, 73, 75–77, 81, 108, 110, 160, 161, 244, 248

rectangle, 65–67, 229, 247
rectangular, xxviii, xxxvi, xxxviii, 68, 70, 71, 78, 93, 120, 148–50, 153, 155, 233, 239
reduction of fraction, xix–xxii, xxiv–xxvi, xxxvii, xlii, xlv, 47, 50, 51, 57, 92, 112, 125, 230, 235, 245, 253
reduction to the same colour, xxiv
reduction to the same denominator, xxiv, 131, 249
regular (figure), 146
remainder, xxxiii, xli, xliii, xliv, 47, 48, 51, 59, 64, 66, 67, 69, 76, 77, 80–82, 87, 104–06, 108, 110, 131, 161, 179, 180, 209, 210, 226, 251, 255
right-angled triangle, xxxiv, xxxv, xxxviii, 66, 93, 143, 229, 238, 246, 248
rim-like figure, 146
rule of eleven, xxxiii, 53, 54, 92, 120, 229
rule of five, xxxiii, 53, 55, 92, 241
rule of nine, xxxiii, 53, 54, 92, 240
rule of seven, xxxiii, 53, 92, 253
rule of three, xxxiii, xlii, 52, 54, 62, 92, 237, 238
rule of three quantities, 237

sages (7), 73, 77, 79, 80, 247
scalene triangle, xxxv, 66, 251
scalene trilateral, xxvii
seas (7), 79, 228
seed (of rule), 92, 245
segment (of a circle), xxi, xxii, xxvii, xxviii, xxxv, xxxviii, 66, 145–47, 240, 253
segment (of a sphere), 151, 152
segment (of the base), xxxv, xxxviii, 66, 143, 226
semicircle, xviii
sense organs (5), 79, 80, 227
series, xxi, xxiii, xxxiv, xlii, 47, 48, 64, 65, 81, 175, 225–27, 231, 232, 234, 240, 243, 247, 253, 255
setting-down, xxx, 88, 111, 112, 127, 128, 165, 166, 174, 175, 177, 178, 182, 241
seven-quantity operation, xxxvii, xlii, 117, 253
show (a figure), xxxv, 87, 238
side (of a figure), xxvii, xxxiv, xxxv, 66, 67, 70, 142, 143, 148, 152, 155, 178, 225, 226, 233, 246, 247
sign, xxxi, xxxii, 87, 111, 112
simple fraction, xxiv, xxv, 112, 253
size, xxxiv, 71, 78, 81, 121, 150, 154, 178, 192, 243
sky (0), 48, 82, 231
solid (dimension), xxii, xxiii, xxxiv, 100, 145
solid (figure), xxviii, xxxiv, xxxvi, 70, 148, 155, 233, 237, 243, 244,

255
solid (volume), 68, 69, 243
solid geometry, xvii
sphere, xxii, xxviii, xxx, xxxvi, xl, 68, 88, 144, 151, 152, 191, 228, 233
square (dimension, power), 46, 82, 100, 145, 157, 178, 242
square (figure), xvi, xvii, xxvii, xxviii, xxx, xxxv, xxxviii, xl, 65, 66, 69–71, 78, 82, 87, 93, 141–43, 153, 155, 178, 190, 191, 229, 233
square (measure), 230, 255
square (number, quantity), xxxiii, xxxiv, xxxvii, xl–xliii, 49–52, 65–68, 72, 91–93, 107, 108, 114, 141, 145, 179, 214, 249
square (operation), 82, 107
square root, xxvii, xxxiii, xxxvii, xl–xliii, 47–49, 52, 65–67, 82, 91, 92, 104, 108, 115, 144, 179, 243, 247, 249
square series, xxxiv, xxxviii, 65, 93, 141, 249
subtract, xxxii, 48–51, 58, 59, 64, 66, 67, 69, 70, 73, 77, 79, 80, 104, 105, 112, 172, 180, 232, 251, 255, 256
subtracter, 256
subtraction, xxiv–xxvi, xxxviii, xlii–xliv, 48, 50, 51, 57, 58, 92, 105, 112, 129–31, 179, 245, 251, 256
subtractive, 179, 248
subtrahend, xxxii, 64, 82, 111, 228
subtrahend series, 47, 48, 251
sum, xxi, xxiii–xxv, xxvii, xxxii, xxxiv, xxxvii, xxxviii, xli–xliv, 47–49, 51, 57, 61–67, 73, 77, 78, 81, 82, 91, 93, 103–06, 112, 113, 126–28, 134, 135, 139–41 , 152, 161, 166, 168, 172, 175, 176, 179, 180, 210, 214, 225, 227–29, 232, 235, 236, 240, 244, 247, 252, 253, 255
summation, xxxii, xliii, 47, 48, 58, 104, 112, 131, 252, 253
sun (12), 73, 79, 80, 239, 248
surface (area), xxviii, xxxviii, xl, 68, 78, 88, 136, 151, 152, 154, 191

tastes (6), 77, 79, 80, 248
teeth (32), 79, 238
thickness, xxix, xxxiv, 68, 70–72, 100, 136, 149, 150, 154–58 , 238, 244
three-quantity operation, xxiv, xxxvii, xlii, xliv, xlv, 115, 118, 120, 136, 237, 238
thunderbolt (figure), xxxvi, 67, 93, 146, 148, 230, 249
top (plane, side, etc.), xxvi, xxvii, xxxv, 62, 67, 70, 77, 78, 136, 142, 143, 148, 149, 155, 247
trapezium, xxvi, xxvii, xxxvi, 136, 142, 143, 230
treasures (9), 77, 79, 80, 241
triangle, xviii, xxxv, xl, 66, 68, 70, 87, 191, 226, 237
triangular (figure), 70, 93, 238
trilateral, xxvii, xxxv, xxxviii, 66, 67, 142, 143, 146, 236, 237
type problem, xxvi, xl–xliii, xlv, 207, 208

unknown number, 82
unknown quantity, 82, 131, 176, 226
upright (of a right triangle), xxxv, 66, 143, 248

Vasus (8), xxxvi, 79, 80, 249
verification, xxiii, 176
verification (of multiplication), xxiii, xxxvii, 106
verification by nine, xxiii, 106
visible quantity, xliv, 210, 242
volume, xvii, xxii, xxviii, xxxiv, xxxvi, xxxviii, xl, 68–72, 88, 89, 93, 94, 117, 120, 136, 144, 145, 148–56, 158, 159, 187, 191, 231, 243, 244, 255
volume measure, xxxvii, 46, 100, 230, 242, 243, 247, 255

well-born trilateral, xxxiv, 93
wheel of a cart (figure), 252
whole number, 58
width, xxxiv, 46, 62, 67–71, 78, 79, 82, 85, 94, 100, 119, 120, 136, 141, 146, 148, 149, 151–58, 178, 184, 236, 244, 250

young moon (figure), xxxv, 146, 244
Yugas (4), 79, 80, 235, 236, 247

zero, xxi, xxiii, xxxi, xxxvii, xli–xlvi, 48, 49, 83, 107, 112, 164, 180, 231, 254

Index of Things Mentioned in the Text

account, 75, 86, 89, 163, 189, 193, 234
accountancy, xxix, 76, 184, 240
alloy, xv, 54, 63, 64, 120, 121, 137–39, 229
apex, 70, 155, 254
arch, xvii, xxviii, xxxviii, 69, 70, 79, 94, 153, 155, 156, 237
army, 81, 174, 175, 229
assembly hall, 78, 167, 250

banner, 79, 241
barber?, 86, 189
barley, 85, 185–87, 235
beam, 78, 236
black stone, 69, 88, 153, 230
blanket, 53, 54, 119, 121, 229
bond, 61, 62, 92, 134, 135, 228, 229, 242
brāhmaṇa, 61, 86, 251
brick, xvii, xxviii, xxix, 69, 88, 93, 153, 154, 193, 227, 234
brick (wall), xxviii, xl, 193, 234, 255
bridge, xvii, xxviii, 69, 153, 156, 244
bridge structure, xxxviii, 70, 94, 156, 244
brown sugar, 86, 188, 252
bull, 86, 189, 249
butter, 86, 188, 248
butter (clarified), 62, 137, 233
buying rate, xxv, 55, 121, 122, 136, 229

calculator, xlii, xliv, 61, 133
camel, xxv, 55, 76, 82, 124, 166, 230
candied sugar, 86, 185, 231
canopy, 79, 234
capital, xxii, xxv, xxxviii, 54, 55, 61, 75, 118, 121–23, 133–35, 243, 247
cardinal direction, 73, 94, 239
carpenter, xxix, 71, 98, 153, 254
cattle, xxxix, 185, 188
chaula beans, 85, 186, 233
chickpea, 85, 186, 234
chir-pine tree, 158, 230
cistern, 58, 128, 129, 250
clay, xxiii, xl, 69, 88, 153, 192, 246
cloth, xx, xxix, xxx, xxxix, 52, 54, 77–79, 82, 86, 100, 116, 135, 167, 178, 229, 234, 240, 242, 248–50
coin, xi, xii, xv, 52, 116, 240, 247
commission, 61
commission of surety, 61, 92, 133, 246
coriander, 85, 186, 240
cotton, 77, 85, 158, 167, 185–87, 229, 230, 237, 248, 250, 254, 255
country, 75
cow, xxxix, 46, 59, 83, 86, 132, 181, 182, 188, 189, 231–33, 238
cow-dung, 89, 193, 235
cowherd, 83, 182, 233
creator, 85, 183, 254
cumin, 85, 186, 235
cutch tree, 158, 231
cutting, xxix, 71, 77, 94, 157, 158, 167, 229, 231, 235
cymbal, 78, 236

daughter (digits), 82, 83, 180, 254
day, xxix, xxxix, 47, 53, 58, 64, 76, 77, 82, 89, 101, 103, 117, 119, 128, 129, 140, 160–62, 165, 166, 178, 179, 218, 220, 221, 231, 232, 239, 241, 254
day and night, 101
daylight, 73, 161, 162, 239
distance, 53, 116, 117, 119, 166, 178
divine tree, 254
dome, xvii, xxviii, xxxvi, xxxviii, 69, 93, 153–55, 233
door, 48, 69, 70, 77–79, 81, 153, 231, 239, 244, 250
door (-junction), 48, 230
dowry?, 86, 189, 246
dry ginger, 55, 62, 137, 254
dry land, 89
dry place, 193

elephant, xvi, 59, 120, 132
emblic myrobalan, 137
empty space, 69, 70, 148, 154–56, 231

entrance, 78, 173, 174, 243
exemption (from tax), xxxix, 189

father-in-law, 81, 254
fee, 61, 133, 250
fenugreek, 86, 247
flax, 85, 89, 186, 187, 193, 252
flower, xxx, xxxix, 80, 81, 172–74, 176, 230
forest, 81, 174, 182, 249
fruit, xxx, 57, 62, 81, 174, 175, 183–85, 225, 227, 244

garden cress, 86, 256
gate, 83, 182
gateway (to mathematics), xix, xxxvii, 47, 103, 183, 239
ghee, xviii, xl, 86, 88, 188, 192
ginger, 123
gnomon, xxix, xxxix, 73, 160–62, 239, 252
god, 45, 81, 174, 239, 254
goddess, 248
gold, xii, xv, xvi, xxii, xxiii, xxxvii, xxxviii, 46, 53, 54, 62–64, 92, 93, 101, 117, 119–21, 137–40, 227, 229, 235, 242, 246, 249, 251, 254
grain, xviii, xxiii, xxx, xxxviii–xl, 46, 54, 72, 85, 87, 94, 117, 121, 158, 159, 183, 185, 187, 226, 229, 231, 240, 242, 244, 248, 252
granite, 69, 88
grass, 78, 229, 237
grindstone, 68, 151, 233
ground, xxxviii, 59, 72, 73, 78, 158, 246, 247

harvest, 62, 85, 86, 135, 183, 241, 254, 255
harvest (autumnal), 85, 186, 253
harvest (spring), 85, 186, 227
hearth, 86, 189, 230, 234
honey tree, 158, 246
horse, 64, 140, 226, 237, 250
horse beans, 72
horse grain, 85
horse gram, 159, 185–87, 230
horse move, 80, 172, 226
house, 69, 81, 86, 168, 177, 227, 232, 233, 253, 254
household goods, xxxix, 188

image, xi, 81, 173, 245
Indian fig tree, 158, 249
intercalary month, 77, 166, 167
interest, xxii, xxxviii, 53, 61, 118, 133–35, 244, 247, 249
intermediate space, 70, 225
investment, xxii, xxxiii, 57, 62, 75, 92, 126, 135, 163, 225, 242, 244
irrigated (region), xxxix, 85, 183, 249

jaggery, 86, 188, 233

kidney beans, 85, 159, 186, 247
kodrava grain, 85, 88, 192, 237
kora grain, 86, 231

lace, 78, 236
lame man, 53, 117, 241
leaf, 78, 252
learned, 68
learned man, 46, 48, 59, 66, 67, 69, 75, 77, 78, 81, 83, 87, 242
leather, 79, 234
lentil, 85, 186, 187, 246
lime, 88, 89, 193, 234
limestone, 89, 193, 229
linseed, 85, 186, 226
lion, xvi, 59, 254
living being, 55, 124, 235
living thing, xli, xlv, 124
log, 71, 72, 158
long pepper, 53, 55, 62, 117, 123, 137, 244
low land, 237
lunar day, 77, 167, 237

mango, xxxix, 81, 124, 174, 175, 185, 225, 227
marble, 69, 88, 153, 246
mason, xvii, 88, 193, 248
masonry, xix, xxx, xl, 88, 98, 183, 192
meeting (of two runners), xxix, xxxix, 165, 166, 178
merchant, xi, xiii, xiv, xvii, xxii, 52, 54, 62, 121, 240, 249
milk, xviii, 81, 86, 188, 239, 249
milk (buffalo's), 188
milk (cow's), xxxix, 83, 182, 188
millet (common), 85, 186, 234
millet (Indian), 85, 186, 236
millet (Italian), 85, 185–87, 229
minaret, xvii, xviii, xxviii, xxxviii, 69, 70, 94, 153, 155, 247
mind reading, xxx, xxxix, 179, 180

money, xii, xv, xxxix, 53, 58, 62, 75, 76, 136, 140, 163, 164, 238, 240
month, xix, 7, 46, 53, 61, 62, 76, 77, 89, 101, 117, 118, 135, 165–67, 234, 247, 254
mortar, 69, 88, 89, 153, 193, 231
moth beans, 186, 246
mother-in-law, 81, 175, 177, 254
mound (of grain), xxxviii, xlii, 72, 94, 158, 159, 248
mud, 59, 241
mustard, 85, 185–88, 253
myrobalan, 62, 137, 227

necklace, 121
necklace (pearl), 132
necklace (royal), 181
necklace?, 86, 189, 249
neem tree, 158, 241
noon, 160
noon shadow, xxix, xxxix, 73, 160–62

oil, xxx, xxxix, xl, 85, 86, 88, 183, 187, 188, 192, 234, 237, 255
orchard, 83, 250

painting, 79, 167, 234
pavilion, 77, 78, 246
peak, 72, 254
pearl, 132
pepper, 62, 117, 123, 137, 247
physician, 62, 250
pillar, xix, xxvi, xxxviii, xxxix, xlii, 59, 70, 72, 73, 92, 98, 131, 132, 159, 160, 208, 238, 255
plank, xxxviii, 71, 72, 136, 157, 158, 244
plastering, xl, 89, 193, 248
plough, 86, 189, 255
plough-holder, 62, 255
pole, 77, 78
pond, 68, 148, 149, 244
price, xiv, xxv, xxx, xxxix, xl, 52–55, 62, 85, 87, 116, 117, 119–21, 123, 124, 136, 137, 157, 183, 188–90, 225, 226, 244, 247
price rate, 71, 225
price standard, 52, 116, 225
principal, 53, 55, 243
profit, xxii, xxv, xxxiii, 55, 75, 121, 122, 228, 244, 246, 248
property, 75, 238, 240
pulse, xviii, 54, 62, 72, 85, 120, 137, 186, 247, 249, 252

purchase, 62, 92, 136, 229, 253
purchase and sale, xxi, xxv, xxxvii, xli, xliv, 54, 92, 121, 122
purchase in proportion, xxii, xxxiii, 62, 136, 253
purity (of gold), xii, xxiii, xxxvii, 63, 64, 93, 101, 117, 119–21, 137–39, 235, 249

rape (Indian), 85, 186, 237
refinement (of gold), 63, 92, 138, 139, 242
rice, 53, 62, 85, 117, 137, 185–87, 230, 236
rising (of the sun), 73, 228
river, 83, 253
rosewood, 158, 254
runner, xxix, xxxix, 76, 116, 165, 166, 178, 226, 240, 244

safflower, 85, 86, 185–88, 230
sal tree, 158, 254
sale of human beings, xxv, 124
sale of living beings, 55, 92
sale of living things, xxv, xlii
salt, xl, 88, 192, 248
sand, 59, 89, 193, 226, 243, 247, 250
sandalwood, xvi, 52, 62, 116, 135, 136, 234
sandstone, 150, 153, 249
saw, 71, 72, 94, 157, 230, 231
sawing, xxix, xxxviii, xlii, 71, 72, 94, 98, 99, 157, 158, 234
scribe, xix, 3, 39, 61, 133, 248
seed, 46, 62, 85, 135, 185, 229, 232, 245
selling of living beings, 124, 235
selling of living things, xxxvii
selling rate, xxv, 121, 122, 250
sesamum, xxxix, 72, 85, 86, 159, 186–188, 192, 237
setting (of the sun), 73, 226, 228
sewing, xxix, 77, 167, 254
shadow, xxix, xxxix, xlii, xliv, 72, 73, 79, 94, 159–62, 235, 243
share, 57, 61, 62, 64, 75, 127, 128, 135, 163, 178, 244, 245
she-buffalo, 86, 189, 246
she-goat, 86, 189, 235
sheet, 78, 79, 242
silk, 77, 135, 167, 226, 235, 242, 254
silk-cotton tree, 158, 255
sirisa tree, 158, 254
soil, 85, 183, 246
solar day, 77, 166

soldier, 81, 174
son (digits), 82, 83, 180, 254
son-in-law, 81, 82, 175–77, 235
specific gravity, xxiii, xxxviii, xl, 152, 187, 192
spice, 86, 251
staff, 72, 73, 159, 160, 238
staircase, xxviii, xxxviii, 69, 70, 94, 155, 156, 255
stairway, xvii, xviii, 69, 70, 93, 153, 155, 243
stepwell, xxviii, 69, 71, 94, 153, 156, 250
stick, 46, 79, 227, 236
stitching, 79, 254
stone, xvii, xxiii, xxxiv, xxxviii, xl, 68, 69, 88, 93, 98, 120, 148–53, 191, 192, 228–30, 235, 242, 243, 255
stone slab, 68, 150, 151, 254
sugar candy, 86
sugarcane, 85, 86, 186, 187, 227, 248
sugarcane juice, xviii, xxx, xxxix, 86, 183, 187, 227, 240, 241, 254

tape, 77, 79, 240, 253
tax, xviii, xxx, xxxix, 85–87, 89, 183, 187–89, 193, 230, 234, 249
temple, xiv, xv, xxx, xxxix, 81, 172–74, 254
tent, xviii, xxix, 77, 78, 167, 231
tent pole, 77, 167
tent rope, 78, 236
thread, 79, 237, 254
timber, xxix, xxxviii, 71, 94, 98, 154, 157, 158, 239
timber sawing, xxix
time, xxix, xxxvii–xxxix, 46, 53, 58, 61, 73, 101, 116–19, 128, 133–35, 160, 161, 165, 166, 178, 230, 233, 243, 249
tooth, 86, 226
town, xi, 75, 83, 182, 240
tree, xxxviii, 83, 182, 248

umbrella, 78, 79, 227, 234

val pulse, 85
valley, 59, 237
vegetable, 86, 252
vibhītikā, 62, 137, 244
village, 75, 232
vine, 78, 249

wage table, xxix
wages, xxix, xxxviii, xxxix, 53, 71, 72, 76, 79, 98, 119, 157, 158, 165, 234, 245–47, 250, 254
wall, xviii, xxviii, xxxvi, xxxviii, xl, 69–72, 78, 79, 88, 89, 93, 148, 153–56, 159, 160, 193, 234, 246, 255
washing, 77, 167, 240
water, 59, 70, 83, 128, 182, 187, 235, 237, 243, 250
water pot, 88, 193
watering spot, 83
watery marrow, 86, 188, 241
watery place, 89, 193, 235, 237
weight, xii, xv, xix, xxii, xxiii, xxxvii, xliii, xlv, 45, 46, 54, 63, 64, 69, 85, 86, 93, 97, 101, 116, 117, 119–21 , 130, 137–40, 152, 153, 185, 187, 188, 192, 224, 232, 235–38, 243, 246, 247, 251, 255
well, xxviii, xxxviii, 68–71, 93, 94, 149, 153, 155, 156, 161, 230
wheat, 72, 85, 159, 185–87, 233
wild grain, 88, 192, 249
wood, 69, 71, 72, 153, 157, 158, 229, 230, 239
wood-apple, 124
wool, 54
worker, 53, 119, 193, 229
worm, 53, 117, 230

year, 53, 55, 61, 76, 77, 101, 117, 118, 124, 160, 165–67, 218, 220, 221, 247, 249, 252, 263
yellow myrobalan, 62, 64, 137, 140, 255
yield, xviii, xxix, xxx, xxxix, 61, 85–87, 89, 183–89, 193, 241, 244

Index of Sanskrit/Prakrit Authors and Titles

Authors

Abhayadeva, vi, 103
Āryabhaṭa I, vi, vii
Āryabhaṭa II, vi, xxiii, xliv, 122, 126, 146
Umāsvāti, vi, 146
Gaṇeśa I, 111, 147
Gaṇeśa II, vi, 124
Giridharabhaṭṭa, vi
Jadivasaha, vi
Jinabhadra Gaṇi, vi
Ṭhakkura Pherū, vi–viii, xi–xxvi, xxviii–xxx, xxxii, xxxiv, xxxvi, xlv, xlvi, 3, 6, 7, 9, 15, 18, 26, 27, 33, 38, 45–47, 49, 50, 52, 55, 57, 59, 73, 75, 83, 87, 89, 92, 97, 98, 100–02, 104, 109, 111–18, 122–24, 126, 127, 130, 134, 135, 142, 144, 146, 151–54, 157, 161, 163, 166–69, 171, 172, 183, 184, 187, 188, 193, 196, 208, 229, 236, 243, 244
Nārāyaṇa, 176
Nārāyaṇa Paṇḍita, vi, xxiii, xxx, xlv, 124, 146, 208
Brahmagupta, vi, xlii, 118
Bhāskara I, vi, 40
Bhāskara II, vi, xx, xlv, 111, 124, 145, 208
Bhuvanadeva, vi, 144
Mahāvīra, vi, xvii, xxi, xxviii, xliii, 124, 139, 146, 151, 191
Rāghavabhaṭṭa, 145
Śrīdhara, vi, xvii, xix–xxi, xxiii, xxv–xxviii, xlii, xliii, xlvi, 92, 118, 120, 124, 126, 127, 143–45, 150, 208
Śrīpati, vi, xliv, 124, 132, 146, 150, 208
Siṃhatilaka Sūri, vi, xlv, 111
Someśvara III, vi
Sphujidhvaja, vi

Titles

Aṇuogaddāra (see also *Anuyogadvārasūtra*), xxiii, 100
Anuyogadvārasūtra (see also AD and *Aṇuogaddāra*), vi
Aparājitapṛcchā (see also AP), vi, 144, 150
Āryabhaṭīya (see also AB), vi, xxx
Kharataragaccha-bṛhadgurvāvalī, xiii
Kharataragaccha-yugapradhāna-catuḥpadikā, xi, xii, xiv, 6, 7
Kharataragacchālaṃkāra-yugapradhānācārya-gurvāvalī, xiii
Gaṇitakaumudī (see also GK), vi, xxiii, xxx, xlv
Gaṇitatilaka (see also GT), vi, xliv, xlv
Gaṇitapañcaviṃśī (see also GP), vi
Gaṇitamañjarī (see also GM), vi
Gaṇitasāra (see also GSK), vi, vii, xvi, 83
Gaṇitasārakaumudī (see also GSK), vi, vii, xi, xvi–xix, xlv, 3, 4, 6, 7, 55, 59, 73, 89, 195, 196
Gaṇitasārasaṃgraha (see also GSS), vi, xvii, xx, xxi, xliii
Caturacintāmaṇi (see also CCM), vi
Jyotiṣasāra (see also JS), vi, xi, xiv, xix, xxix, 6, 87
Ṭhāṇaṃga (see *Sthānāṅgasūtra*), 103
Tattvārthādhigamasūtra, vi
Tiloyapaṇṇatti (see also TP), vi, xxvii, 145
Triśatikā (see also Tr), vi, xvii, xlii
Dravyaparīkṣā (see also DP), vi, xi–xiii, xv, xvi, 6, 7, 46, 87
Dhātūtpatti, xi, xv, xvi, 6, 7
Pañcaviṃśatikā (see also PV), vi, xxi, xxix, 144
Patan Manuscript (see also PM), vi, xlvi, 6
Pāṭīgaṇita (see also PG), vi, xvii, xx, xlii

Bakhshālī Manuscript (see also BM), vi, vii, 122
Bījagaṇita (see also BG), vi
Bṛhatkṣetrasamāsa (see also BKS), vi
Brāhmasphuṭasiddhānta (see also BSS), vi, xlii
Mahāsiddhānta (see also MS), vi, xxiii, xliv
Mānasollāsa (see also MU), vi
Yavanajātaka (see also YJ), vi
Ratnaparīkṣā, xiv, 6
Ratnaparīkṣādi-saptagranthasaṃgraha (see also SGS), vi, xi, 6
Līlāvatī (see also L), vi, xx, xlv, 147
Vāstusāra, xi, xiv, 6, 7, 87
Siddhāntaśekhara (see also SS), vi, xliv
Sthānāṅgasūtra (see also SA and *Ṭhāṇaṃga*), vi

Abbreviated Titles (cf. p. vi)

AB, vi, 197–206
AD, vi, 100, 102
AP, vi, 144, 150
ASA, vi, 103
BAB, vi, 97, 100, 147
BG, vi, 111
BKS, vi, 147
BM, vi, xxv, 122, 196–206
BSS, vi, xl–xlii, 118, 142, 158, 161, 197–206
CCM, vi, 147
DP, vi, 52, 116, 117
GK, vi, xxiii, xl, xli, xlv, 107, 124, 129, 136, 142, 147, 156, 196–206, 208–14
GM, vi, 124
GP, vi, 147
GSK, vi, xvii–xxxiv, xxxvi, xl, xli, xlv, xlvi, 4, 18, 45, 46, 52, 54, 57, 59, 69, 72, 75, 80–82, 85, 88, 91, 93, 97–126, 128–60, 163–69, 171–76, 178–84, 186–93, 197–206, 208, 224
GSS, vi, xx–xxiii, xxv, xxviii, xl, xli, xliii, xlvi, 47, 102, 122, 124, 129, 132, 139, 140, 142, 147, 151–53, 156, 158, 160–62, 181, 191, 192, 197–206, 208–14
GT, vi, xl, xli, xliv, 97, 112, 124, 132, 134, 196–206, 208–14
JS, vi, xxix, 161
L, vi, xx, xl, xli, xlv, xlvi, 62, 72, 97–99, 101, 107, 111, 118, 120, 121, 124, 129, 132, 142, 151, 158, 187, 197–206, 208, 209, 211, 212
MS, vi, xxiii, xxv, xl, xli, xliv, 97, 107, 116, 122, 123, 126, 129, 132, 136, 142, 147, 150, 151, 158, 161, 184, 197–206, 209, 210
MU, vi, 81, 111
PG, vi, xx, xxi, xxiv, xxv, xxvii, xl–xliii, xlv, xlvi, 47, 50, 62, 103–10, 112–21, 123–27, 129–44, 146, 148, 196–206, 208–12
PM, vi, xlvi, 6, 17, 26, 28, 31–33, 73, 81, 83, 104, 126–29, 159, 160, 165, 166, 173–78, 180–82
PV, vi, xxi, xxix, 99, 144, 147, 161, 162, 196–206
SA, vi, 103
SGS, vi, xi, xiv–xvi, 6, 161
SGT, vi, xl, xli, xlv, 111
SS, vi, 132, 142, 147, 150, 151, 161, 197–206
TP, vi, 145, 147
Tr, vi, xx–xxii, xxiv–xxvii, xl–xliii, xlvi, 62, 72, 97–99, 101, 103–10, 112–21, 123–27, 129–38, 140–52, 154, 157–59, 161, 197–206, 208
UTA, vi, 147
YJ, vi, 161, 162

The ***Gaṇitasārakaumudī*** was composed in the early fourteenth century at Delhi by the Jain polymath Ṭhakkura Pherū who held a high position at the treasury of ᶜAlā' al-Dīn Khaljī, and contributed to the popularization of science by producing six treatises in Apabhraṃśa verse on diverse scientific subjects.

The *Gaṇitasārakaumudī* extends the range of mathematics far beyond the traditional framework. The first three chapters are structured like the earlier mathematical texts in Sanskrit and treat traditional topics like fundamental operations, fractions, series, proportion, plane and solid geometry and so on. The remaining two chapters contain supplementary material derived from diverse aspects of contemporary life where numbers play a role such as mathematical riddles, conversion of dates from Vikrama era to Hijrī era, magic squares, and, most remarkably, average yield per *bīghā* of several kinds of grains and pulses — topics that were not touched upon in any mathematical text before.

The present volume offers, besides an introduction, a critically emended text, an English translation, and a detailed mathematical commentary where efforts are made to interpret Pherū's formulas and algorithms in modern notation and to invite attention to parallel procedures laid down in other mathematical treatises. There are several appendices, including a comprehensive glossary-index.

Sreeramula Rajeswara Sarma, formerly Professor of Sanskrit at Aligarh Muslim University, published Ṭhakkura Pherū's *Ratnaparīkṣā* on gemmology with an English translation and commentary (Aligarh 1984). Recent publications include *Astronomical Instruments in the Rampur Raza Library* (Rampur 2003) and *The Archaic and the Exotic: Studies in the History of Indian Astronomical Instruments* (New Delhi 2008).

Takanori Kusuba, Professor of History and Philosophy of Science at Osaka University of Economics, earned his PhD from Brown University in 1993 for his thesis *Combinatorics and Magic Squares in India: A Study of Nārāyaṇa Paṇḍita's Gaṇitakaumudī, Chapters 13-14*. Jointly with Professor David Pingree, he published *Arabic Astronomy in Sanskrit: Al-Birjandī on Tadhkira II, Chapter 11 and its Sanskrit Translation* (Leiden 2002).

Takao Hayashi, Professor of History of Science at Doshisha University, Kyoto, wrote extensively on the history of Indian mathematics and translated Bhāskara's *Līlāvatī* and Narayaṇa's *Gaṇitakaumudī* (14th chapter on magic squares) into Japanese. He was awarded the Salomon Reinach Foundation Prize by the Institut de France, Paris, in 2001 for *The Bakhshālī Manuscript: An ancient Indian mathematical treatise* (Groningen 1995).

Michio Yano, Professor and Dean of the Faculty of Cultural Studies at the Kyoto Sangyo University, and the chief editor of *SCIAMVS*, published numerous studies on Indology and history of science (esp. astronomy and astrology in India, Islam, and China), including Japanese translations of Āryabhaṭa's *Āryabhaṭīya*, Varāhamihira's *Bṛhatsaṃhitā* and the *Carakasaṃhitā* (Sūtrasthāna).

The *Studies in Indian Mathematics: Series, Pi and Trigonometry* (in Japanese, Tokyo 1997) by Takao Hayashi, Takanori Kusuba and Michio Yano was awarded the **Publication Prize from the Mathematical Society of Japan** in 2005.